# Mastering PLC Programming

The software engineering survival guide to automation programming

**M.T. White**

&lt;packt&gt;

BIRMINGHAM—MUMBAI

# Mastering PLC Programming

Copyright © 2023 Packt Publishing

*All rights reserved.* No part of this book may be reproduced, stored in a retrieval system, or transmitted in any form or by any means, without the prior written permission of the publisher, except in the case of brief quotations embedded in critical articles or reviews.

Every effort has been made in the preparation of this book to ensure the accuracy of the information presented. However, the information contained in this book is sold without warranty, either express or implied. Neither the author, nor Packt Publishing or its dealers and distributors, will be held liable for any damages caused or alleged to have been caused directly or indirectly by this book.

Packt Publishing has endeavored to provide trademark information about all of the companies and products mentioned in this book by the appropriate use of capitals. However, Packt Publishing cannot guarantee the accuracy of this information.

**Group Product Manager**: Mohd Riyan Khan
**Publishing Product Manager**: Suwarna Patil
**Senior Editor**: Tanya D'cruz
**Technical Editor**: Arjun Varma
**Copy Editor**: Safis Editing
**Project Coordinator**: Prajakta Naik
**Proofreader**: Safis Editing
**Indexer**: Pratik Shirodkar
**Production Designer**: Shyam Sundar Korumilli
**Senior Marketing Coordinator**: Nimisha Dua
**Marketing Coordinator**: Agnes D'souza

First published: March 2023

Production reference: 1220223

Published by Packt Publishing Ltd.
Livery Place
35 Livery Street
Birmingham
B3 2PB, UK.

ISBN 978-1-80461-288-0

www.packtpub.com

*For Jennie. I like to think you helped me write this and how cool it would have been to have two writers in the family.*

*Jennie Branch Bolton*

*1981-2020*

# Contributors

## About the author

**M.T. White** has been programming since the age of 12. His fascination with robotics flourished when he was a child programming microcontrollers such as Arduinos. M.T. currently holds an undergraduate degree in mathematics, a master's degree in software engineering, and is currently working on an MBA in IT project management. M.T. is currently working as a software developer for a major US defense contractor and is an adjunct CIS instructor at ECPI University. His background mostly stems from the automation industry where he programmed PLCs and HMIs for many different types of applications. M.T. has programmed many different brands of PLCs over the years and has developed HMIs using many different tools.

## About the reviewers

**Oleg Osovitskiy** is a senior firmware engineer with more than 23 years of experience in industrial automation. He is a certified IEC-61508 functional safety engineer (#11605/15), and a certified IEC-62443 CySec specialist (#658/22). He worked as a control engineer for the gas industry, implementing technological and emergency algorithms for various factories and plants. He has extensive hands-on experience with PLCs, I/O drivers, and communication drivers for various industrial protocols, including Modbus, HART, CANopen, EtherNet/IP, EtherCAT, and others. He currently lives and works in Canada, Quebec, and is responsible for developing firmware for several mission-critical, safety PLCs.

*I'd like to thank my wife and two lovely daughters, who support me and understand the time and commitment it takes to learn new skills and obtain new knowledge in our constantly changing and demanding environment. They are the purpose and joy of my life.*

**Keith Lyding** is an electrical engineer for a manufacturing company in Columbus, OH. He has over 15 years of experience in the electrical field, as well as more than 9 years of experience in automation. He graduated from Thomas Edison State University in 2019. He served in the US Navy for six years, and has worked for Nucor Steel, and currently, for Sonoco Products Company where he works primarily with Allen Bradley PLCs, Inductive Automation's Ignition platform, EXOR and Panelview HMIs, and many other platforms. He enjoys troubleshooting, as well as automating complex operations. In his spare time, he loves to serve in his church, coach his son's baseball team, and spend time with his family.

*I am thankful for Paul Cassidy and Brian Babin, who discipled me as a young Christian. I would also like to thank Kyle Ahrendt and Will Carleton, former coworkers and experts in their fields. My competitive nature drove me to relentlessly follow your example. Finally, I'd like to thank my wife Katie, my amazing wife of 12 years. She is so gracious with me, especially when I forget to tell her I'm working late.*

**Tony LeRoy** has worked in the automation field since 2013, starting as a machine operator, transitioning to industrial maintenance, and then to controls engineering and design. Specializing in PLC programming, HMI design, and SCADA development, Tony has developed a passion for making the physical world and the digital world come together. Tony holds three associate degrees in mechatronics, industrial electronics, and general engineering technology, all from Tri-County Technical College. Currently working for a system integrator, focusing on the research and development of control solutions for emerging technologies, he also does consulting and freelance work, hoping to own a business one day.

*I would like to thank my family and my friends for their understanding about the time, dedication, and passion that I devote to my work, and for still sticking by my side. I also would like to thank my professors at Tri-County Tech for giving me a love of learning and paving the way for my success today. Thank you for your hard work and for giving students brighter futures!*

# Table of Contents

Preface xix

# Part 1 – An Introduction to Advanced PLC Programming

## 1

## Software Engineering for PLCs 3

| | | | |
|---|---|---|---|
| Technical requirements | 4 | Introducing CODESYS | 9 |
| Software engineering for PLCs | 4 | Testing CODESYS | 10 |
| Understanding the IEC 61131-3 standard | 5 | Creating the program | 11 |
| What does the IEC 61131-3 standardize? | 7 | Summary | 15 |
| Programming a PLC – The five IEC languages | 7 | Questions | 15 |

## 2

## Advanced Structured Text — Programming a PLC in Easy-to-Read English 17

| | | | |
|---|---|---|---|
| Technical requirements | 18 | Understanding pointers | 25 |
| Understanding error handling | 18 | Representing PLC memory | 26 |
| Variables | 19 | General syntax for pointers | 26 |
| The main program | 19 | The ADR operator | 27 |
| The division by 0 error | 19 | Dereferencing pointers | 28 |
| Checking for 0 code | 21 | Handling invalid pointers | 29 |
| TRY-CATCH blocks | 21 | Understanding references | 31 |
| FINALLY statements | 23 | Declaring a reference variable | 31 |
| Identifying and handling errors | 23 | | |

| | | | | |
|---|---|---|---|---|
| Example program | 32 | Understanding state machines | 37 |
| Checking for invalid references | 32 | Variables for the state machine | 38 |
| **Understanding documentation** | **33** | Exploring state machine logic | 38 |
| Self-documenting code | 33 | **Summary** | **40** |
| Code to variables | 34 | **Questions** | **41** |
| Code commenting | 35 | **Further reading** | **41** |

# 3

# Debugging — Making Your Code Work                                         43

| | | | |
|---|---|---|---|
| **Technical requirements** | **43** | The CODESYS debugger tool | 52 |
| **What is debugging?** | **44** | Forcing variables | 57 |
| Types of bugs | 44 | **Troubleshooting – a practical example** | **59** |
| Testing versus debugging | 45 | Case 4 – a while loop | 64 |
| Breaking down the debugging process | 45 | **Summary** | **65** |
| **Understanding debugging tools and techniques** | **47** | **Questions** | **65** |
| Print debugging | 47 | **Further reading** | **66** |

# 4

# Complex Variable Declaration — Using Variables to Their Fullest           67

| | | | |
|---|---|---|---|
| **Technical requirements** | **68** | Declaring a struct | 78 |
| **Auto declaring variables** | **68** | **Getting to know enums** | **81** |
| **Understanding constants** | **69** | **Exploring persistent variables** | **82** |
| **Investigating arrays** | **71** | Persistent variable list | 83 |
| Initialized arrays | 73 | **Final project – motor control program** | **83** |
| Multidimensional arrays | 75 | **Summary** | **86** |
| **Exploring global variable lists** | **76** | **Questions** | **86** |
| Creating a GVL | 76 | **Further reading** | **86** |
| **Understanding structs** | **78** | | |

# Part 2 – Modularity and Objects

## 5
### Functions — Making Code Modular and Maintainable 89

| | | | |
|---|---|---|---|
| Technical requirements | 90 | Understanding arguments | 99 |
| What is modular code? | 90 | Named parameters | 100 |
| Why use modular code? | 90 | Default arguments | 102 |
| Exploring functions | 91 | Final project – temperature unit converter | 104 |
| What goes into a function? | 91 | Summary | 106 |
| Creating a function | 92 | Questions | 106 |
| The PLC_PRG file | 96 | Further reading | 106 |
| Examining return types | 96 | | |
| The RETURN statement | 97 | | |

## 6
### Object-Oriented Programming — Reducing, Reusing, and Recycling Code 107

| | | | |
|---|---|---|---|
| Technical requirements | 108 | getter and setter | 120 |
| What is OOP? | 108 | Getter method | 120 |
| Why use OOP? | 109 | Setter method | 121 |
| The four pillars – A preview | 110 | Understanding recursion and the THIS keyword | 122 |
| Understanding function blocks | 110 | THIS keyword | 123 |
| Getting to know objects | 113 | Recursion in action | 123 |
| Getting to know methods | 114 | Final project – creating a unit converter | 125 |
| Adding a method | 115 | Summary | 127 |
| Getting to know properties | 118 | Questions | 128 |
| Adding a property | 119 | Further reading | 128 |
| Understanding the purpose of a | | | |

# 7

## OOP — The Power of Objects — 129

| | | | |
|---|---|---|---|
| Technical requirements | 129 | Composition in practice | 140 |
| Understanding access specifiers | 130 | Examining interfaces | 143 |
| Calculation program | 130 | Getting to know design patterns | 147 |
| Exploring the pillars of OOP | 133 | Final project – creating a simulated assembly line | 148 |
| Encapsulation versus abstraction | 133 | | |
| Inheritance | 134 | Summary | 150 |
| Polymorphism | 138 | Questions | 150 |
| Inheritance versus composition | 139 | Further reading | 150 |
| When to use composition | 139 | | |

# Part 3 – Software Engineering for PLCs

# 8

## Libraries — Write Once, Use Anywhere — 153

| | | | |
|---|---|---|---|
| Technical requirements | 153 | Rule 4 – Documentation | 160 |
| Investigating libraries | 154 | Building custom libraries | 164 |
| Why do we need libraries? | 154 | Requirements | 164 |
| Libraries versus frameworks | 154 | Implementation | 165 |
| Distribution | 155 | Final project – part computation library | 169 |
| Third-party libraries | 155 | | |
| Installing a library | 156 | Requirements | 169 |
| Guiding principles for library development | 158 | Implementation | 169 |
| | | Summary | 171 |
| Rule 1 – Keep it simple, stupid (KISS) | 158 | Questions | 172 |
| Rule 2 – Abstraction and encapsulation | 159 | Further reading | 172 |
| Rule 3 – Patterns make for perfection | 160 | | |

# 9

## The SDLC — Navigating the SDLC to Create Great Code  173

| | | | |
|---|---|---|---|
| Technical requirements | 173 | Final project – creating a simple library | 189 |
| Understanding the SDLC | 174 | Gathering requirements for the library | 189 |
| Why care about the SDLC? | 174 | Designing the library | 190 |
| How is the SDLC implemented? | 175 | Building the library | 191 |
| Investigating the general steps of the SDLC | 176 | Testing the library | 192 |
| Requirements/planning | 177 | Deploying the library | 194 |
| Design | 178 | Maintaining the library | 194 |
| Build | 182 | Summary | 194 |
| Test | 183 | Questions | 195 |
| Deployment | 188 | Further reading | 195 |
| Maintenance | 189 | | |

# 10

## Advanced Coding — Using SOLID to Make Solid Code  197

| | | | |
|---|---|---|---|
| Technical requirements | 198 | The Liskov substitution principle | 208 |
| Introducing SOLID programming | 198 | The interface segregation principle | 213 |
| Benefits of SOLID programming | 198 | The Dependency inversion principle | 215 |
| The governing principles of SOLID programming | 199 | Final project – a painting machine | 219 |
| The single-responsibility principle | 199 | Summary | 221 |
| The open-closed principle | 203 | Questions | 221 |
| | | Further reading | 221 |

## Part 4 – HMIs and Alarms

# 11

## HMIs — UIs for PLCs  225

| | | | |
|---|---|---|---|
| Technical requirements | 226 | Understanding HMIs | 226 |

| | | | |
|---|---|---|---|
| Why create and use an HMI? | 226 | Exploring wireframing | 232 |
| How are HMIs created? | 228 | Final project – creating an HMI | 233 |
| Programming languages to develop HMIs | 229 | Summary | 236 |
| What should an HMI do? | 230 | Questions | 236 |
| HMIs versus SCADA | 230 | Further reading | 236 |
| How the SDLC applies to HMIs | 231 | | |

# 12

# Industrial Controls — User Inputs and Outputs 237

| | | | |
|---|---|---|---|
| **Technical requirements** | 238 | Text field | 246 |
| **Exploring common HMI controls** | 238 | Control properties | 247 |
| Flip switches | 238 | **Final project – creating a simple HMI** | 249 |
| Push switches | 239 | Requirements for the HMI | 249 |
| Buttons | 239 | Design of the HMI | 249 |
| LEDs | 240 | Building the HMI | 250 |
| Potentiometers | 241 | | |
| Sliders | 242 | **Summary** | 255 |
| Spinners | 243 | **Questions** | 255 |
| Measurement controls | 243 | **Further reading** | 255 |
| Histogram | 245 | | |

# 13

# Layouts — Making HMIs User-Friendly 257

| | | | |
|---|---|---|---|
| **Technical requirements** | 258 | **Organizing the screen into multiple layouts** | 268 |
| **The importance of colors** | 258 | Creating visualizations screens | 269 |
| Backgrounds | 258 | Changing the default screen | 271 |
| Red, yellow, and green | 260 | Navigating between screens | 273 |
| Control colors | 260 | **Final project – creating a user-friendly HMI** | 275 |
| Labeling colors | 261 | | |
| **Understanding grouping/position** | 261 | **Summary** | 279 |
| **Best practices for blinking** | 263 | **Questions** | 280 |
| Blinking a component | 264 | **Further reading** | 280 |
| Animation | 268 | | |

## 14

### Alarms — Avoiding Catastrophic Issues with Alarms — 281

| | | | |
|---|---|---|---|
| Technical requirements | 282 | Setting up an alarm table | 290 |
| What are alarms? | 282 | PLC alarm logic | 292 |
| When should you use an alarm? | 282 | Alarm acknowledgment | 299 |
| What should an alarm say? | 283 | Final project – motor alarm system | 301 |
| Alarm configuration – I, Warning, and Error setup | 283 | Requirements | 302 |
| | | Design/implementation of the HMI | 302 |
| Alarm groups | 286 | Summary | 304 |
| Alarm HMI components | 288 | Questions | 304 |
| Setting up an alarm banner | 289 | Further reading | 304 |

# Part 5 – Final Project and Thoughts

## 15

### Putting It All Together — The Final Project — 307

| | | | |
|---|---|---|---|
| Technical requirements | 308 | Implementing the PLC code | 318 |
| Project overview | 308 | PLC_PRG file | 318 |
| Getting the requirements | 309 | Alarms function block | 319 |
| HMI design | 309 | Door function block | 320 |
| HMI implementation | 310 | Oven function block | 321 |
| LED variables | 311 | Testing the application | 322 |
| Acknowledgment variable | 311 | Testing the door lock | 323 |
| Spinner variables/setup | 312 | Testing the gauge | 324 |
| Gauge variable/setup | 312 | Summary | 327 |
| Alarm table variables/configuration | 314 | Questions | 327 |
| PLC code design | 316 | | |

# 16

## Distributed Control Systems, PLCs, and Networking — 329

| | | | |
|---|---|---|---|
| Technical requirements | 330 | EtherCAT | 338 |
| What are computer networks? | 330 | DeviceNet | 339 |
| Network topology | 330 | Protocol conversion | 342 |
| | | Other communication topics to explore | 342 |
| Common IT protocols | 331 | | |
| TCP/IP | 331 | Understanding distributed control systems | 343 |
| UDP | 332 | | |
| PLC/automation device communication | 334 | The differences between DCSs and PLCs | 344 |
| Modbus | 334 | Summary | 345 |
| Profibus | 335 | Questions | 345 |
| Profinet | 336 | Further reading | 346 |

## Assessments

| | | | |
|---|---|---|---|
| Chapter 1: Software Engineering for PLCs | 347 | Chapter 8: Libraries — Write Once, Use Anywhere | 349 |
| Chapter 2: Advanced Structured Text — Programming a PLC in Easy-to-Read English | 347 | Chapter 9: The SDLC — Navigating the SDLC to Create Great Code | 349 |
| Chapter 3: Debugging — Making Your Code Work | 347 | Chapter 10: Advanced Coding — Using SOLID to Make Solid Code | 350 |
| Chapter 4: Complex Variable Declaration — Using Variables to Their Fullest | 348 | Chapter 11: HMIs — UIs for PLCs | 350 |
| | | Chapter 12: Industrial Controls — User Inputs and Outputs | 350 |
| Chapter 5: Functions — Making Code Modular and Maintainable | 348 | Chapter 13: Layouts — Making HMIs User Friendly | 351 |
| Chapter 6: OOP — Reducing, Reusing, and Recycling Code | 348 | Chapter 14: Alarms — Avoiding Catastrophic Issues with Alarms | 351 |
| Chapter 7: OOP — The Power of Objects | 349 | Chapter 15: Putting It All Together — The Final Project | 351 |
| | | Chapter 16: Distributed Control Systems, PLCs, and Networking | 351 |

## Index — 347

## Other Books You May Enjoy — 360

# Preface

Object-oriented programming and the principles that govern the concept rule the modern IT world and automation programming is no different. Though modern technology is progressing rapidly in the automation realm, software development practices are not. As such, this book is meant to be a bridge between automation programmers and modern software engineer practices.

## Who this book is for

This book is for automaton programmers with a background in software engineering topics such as object-oriented programming and general software engineering knowledge. Automation engineers, software engineers, electrical engineers, PLC technicians, hobbyists, and upper-level university students with an interest in automation or robotics will also find this book useful and interesting.

To get the most out of this book, you should have a basic knowledge of PLCs, PLC programming, and modern structured text. Though not totally necessary, a rough idea about object-oriented programming would also be beneficial.

## What this book covers

*Chapter 1*, *Software Engineering for PLCs*, establishes the basics of software engineering and why it is important for PLC programmers. The chapter also walks you through installing CODESYS and creating a sample project to ensure the setup is working.

*Chapter 2*, *Advanced Structured Text — Programming a PLC in Easy-to-Read English*, explores some of the lesser-used concepts of structured text, such as error handling and pointers. This chapter also covers the basics of state machines and proper code documentation.

*Chapter 3*, *Debugging — Making Your Code Work*, introduces troubleshooting PLC code. The chapter covers concepts such as print debugging, using built-in debugging tools, and more.

*Chapter 4*, *Complex Variable Declaration — Using Variables to Their Fullest*, is about complex variables. Topics covered include variable lists, auto-declaring variables, structs, and much more.

*Chapter 5*, *Functions — Making Code Modular and Maintainable*, introduces code modularity. To do this, the concept of functions is covered, along with arguments, return types, and more.

*Chapter 6*, *OOP — Reducing, Reusing, and Recycling Code*, introduces the power of objects and how they can be used. The chapter explores basic **object-oriented programming** (**OOP**) principles such as function blocks, methods, and getter and setter methods.

*Chapter 7*, *OOP — The Power of Objects*, is a continuation of *Chapter 6* and covers more complex object-oriented principles such as the pillars of OOP, composition, access specifiers, interfaces, and more.

*Chapter 8*, *Libraries — Write Once, Use Anywhere*, explores the whole process of creating a library from scratch to consuming the library. This chapter essentially is applied OOP.

*Chapter 9*, *The SDLC — Navigating the SDLC to Create Great Code*, introduces the full software development life cycle (SDLC). The goal of this chapter is to teach you how to navigate the full SDLC process to properly build and implement PLC code.

*Chapter 10*, *Advanced Coding — Using SOLID to Make Solid Code*, shows you how to create SOLID PLC code. The goal of this chapter is to teach you how to create well-engineered code that can be adapted and will age well. In short, this chapter explains how to properly implement OOP.

*Chapter 11*, *HMIs — UIs for PLCs*, introduces the concept of **Human Machine Interface** (**HMIs**). The goal of this chapter is to introduce the core idea behind HMIs, wireframing, setting up a basic HMI project, and why HMIs are used.

*Chapter 12*, *Industrial Controls — User Inputs and Outputs*, covers some of the commonly used CODESYS HMI widgets. The goal of the chapter is to introduce the widgets, what they do, and how they work.

*Chapter 13*, *Layouts — Making HMIs User-Friendly*, explores how to make functional HMIs. In other words, the goal of this chapter is to lay down principles that can be used to create high-functioning and user-friendly HMIs in CODESYS.

*Chapter 14*, *Alarms — Avoiding Catastrophic Issues with Alarms*, covers one of the most important aspects of automation programming – alarms. This chapter introduces the concept of alarms and how to set up an alarm, its layout, and even how to trigger them.

*Chapter 15*, *Putting It All Together — The Final Project*, is the last hands-on chapter. This chapter cherry-picks concepts from the whole book and incorporates them into a final project.

*Chapter 16*, *Distributed Control System, PLCs, and Networking*, is theoretical in nature, unlike all the previous chapters. This chapter covers the basics of networking, as well as introducing the basics of common networking protocols for automation.

## To get the most out of this book

This book covers some advanced PLC programming topics. As such, it is recommended that you read the book from cover to cover. It is also recommended that you have some knowledge of PLC programming and at least a basic grasp of structured text.

| Software/hardware covered in the book | Operating system requirements |
| --- | --- |
| CODESYS | Windows |

If you are using the digital version of this book, we advise you to type the code yourself or access the code from the book's GitHub repository (a link is available in the next section). Doing so will help you avoid any potential errors related to the copying and pasting of code.

## Download the example code files

You can download the example code files for this book from GitHub at `https://github.com/PacktPublishing/Mastering-PLC-programming`. If there's an update to the code, it will be updated in the GitHub repository.

We also have other code bundles from our rich catalog of books and videos available at `https://github.com/PacktPublishing/`. Check them out!

## Download the color images

We also provide a PDF file that has color images of the screenshots and diagrams used in this book. You can download it here: `https://packt.link/bqJiM`.

## Conventions used

There are a number of text conventions used throughout this book.

`Code in text`: Indicates code words in text, database table names, folder names, filenames, file extensions, pathnames, dummy URLs, user input, and Twitter handles. Here is an example: "As can be seen in the code, the keyword `EXTENDS Felion` is added to the function block code."

A block of code is set as follows:

```
//turn on motor
IF turnOnMotor = FALSE THEN
    turnOnMotor := TRUE;
END_IF
```

When we wish to draw your attention to a particular part of a code block, the relevant lines or items are set in bold:

```
//turn on motor
IF turnOnMotor = FALSE THEN
    turnOnMotor := TRUE;
END_IF
```

Any command-line input or output is written as follows:

```
$ mkdir css
$ cd css
```

**Bold**: Indicates a new term, an important word, or words that you see onscreen. For instance, words in menus or dialog boxes appear in **bold**. Here is an example: "Select **System info** from the **Administration** panel."

> **Tips or important notes**
> Appear like this.

# Get in touch

Feedback from our readers is always welcome.

**General feedback**: If you have questions about any aspect of this book, email us at `customercare@packtpub.com` and mention the book title in the subject of your message.

**Errata**: Although we have taken every care to ensure the accuracy of our content, mistakes do happen. If you have found a mistake in this book, we would be grateful if you would report this to us. Please visit `www.packtpub.com/support/errata` and fill in the form.

**Piracy**: If you come across any illegal copies of our works in any form on the internet, we would be grateful if you would provide us with the location address or website name. Please contact us at `copyright@packt.com` with a link to the material.

**If you are interested in becoming an author**: If there is a topic that you have expertise in and you are interested in either writing or contributing to a book, please visit `authors.packtpub.com`.

# Share Your Thoughts

Once you've read *Mastering PLC Programming*, we'd love to hear your thoughts! Scan the QR code below to go straight to the Amazon review page for this book and share your feedback.

`https://packt.link/r/180461288X`

Your review is important to us and the tech community and will help us make sure we're delivering excellent quality content.

# Download a free PDF copy of this book

Thanks for purchasing this book!

Do you like to read on the go but are unable to carry your print books everywhere? Is your eBook purchase not compatible with the device of your choice?

Don't worry, now with every Packt book you get a DRM-free PDF version of that book at no cost.

Read anywhere, any place, on any device. Search, copy, and paste code from your favorite technical books directly into your application.

The perks don't stop there, you can get exclusive access to discounts, newsletters, and great free content in your inbox daily

Follow these simple steps to get the benefits:

1. Scan the QR code or visit the link below

https://packt.link/free-ebook/9781804612880

2. Submit your proof of purchase
3. That's it! We'll send your free PDF and other benefits to your email directly

# Part 1 – An Introduction to Advanced PLC Programming

This section of the book is designed to give you an in-depth look at advanced Structured Text programming and general software development principles. The goal of this section is to teach you how to write PLC programs using advanced programming techniques. The section explores advanced elements of Structured Text, complex variables such as variable lists and arrays, and finally, debugging so you can learn how to properly debug PLC code. In short, this section will lay the framework for the more advanced concepts that are explored throughout this book.

This part includes the following chapters:

- *Chapter 1, Software Engineering for PLCs*
- *Chapter 2, Advanced Structured Text — Programming a PLC in Easy-to-Read English*
- *Chapter 3, Debugging — Making Your Code Work*
- *Chapter 4, Complex Variable Declaration — Using Variables to Their Fullest*

# 1
# Software Engineering for PLCs

Software engineering is a pivotal, yet often overlooked aspect of **Programmable Logic Controller** (**PLC**) programming. There is a core problem with automation engineering that stems from most PLC projects usually being viewed as hardware first. Many books, workshops, and so on are focused on PLC projects as hardware-first systems. Usually, programming is secondary to the overall hardware design of the system. In other words, the software is there to operate the hardware.

Many PLC programmers are not formally trained software developers and have backgrounds ranging from electricians to electrical and mechanical engineers. Though there is nothing wrong with a PLC developer not being a formally trained programmer, there are techniques that are usually taught in programming classes that are often lost when a non-formally trained programmer tries to program a PLC. This book aims to teach and apply software engineering practices to PLC programming. By learning these techniques, PLC developers can utilize the full gamut of the **IEC 61131-3** standard and create advanced software faster and cleaner.

The hot topic in today's fast-paced industrial world is **Artificial Intelligence** (**AI**) and automation. In short, machines are getting smart, and a major component of that is the software that controls the systems. The first PLC was introduced around the late 1960s and early 1970s; as such, PLCs (and by extension, automation) are nothing new. However, what has changed is the complexity of the systems that PLCs control. With the lower costs and rising computing power of PLCs, the applications that PLCs control are now becoming more complex seemingly by the day. The days of PLC programmers getting through the day with basic programming techniques and ladder logic are quickly becoming a thing of the past. To survive and be competitive in today's market, a new way of thinking about PLC code is needed. Today's world now needs PLC programmers that can function as software engineers.

In this chapter, we're going to cover the following topics:

- Software engineering for PLCs
- The IEC 61131-3 standard
- Ways of programming a PLC
- CODESYS
- A ladder logic `Hello, World!` program to test the installation of CODESYS

## Technical requirements

This book is designed to have a very low bar to get started. The only items that are needed to get started on your journey to mastering advanced PLC programming are a Windows computer and a free program called **CODESYS**. CODESYS is an all-in-one PLC development environment that contains a built-in simulator that can run PLC code without the need for physical hardware. CODESYS can be downloaded for free here: `https://www.codesys.com/download.htmlhttps://us.store.codesys.com/`.

To get CODESYS up and running, it is recommended to have the following specs:

- Windows 8 or later (32/64 bit)
- 12 GB free hard drive space
- 8 GB of RAM

Installation of CODESYS is quite simple. All you have to do is follow the link, create an account, and follow the installation wizard. We'll explore CODESYS a bit more later, but for now, all you need to worry about is downloading and installing the software.

All code examples for this book will be housed on GitHub. Although you don't need a GitHub account to get the code down, it is recommended that you do create an account and download the GitHub desktop tool. As you're working on examples throughout this book, you will be encouraged to put your spin on them. As such, GitHub will allow you to commit the code without fear of losing past iterations of it. The source code for this project can be found here: `https://github.com/PacktPublishing/Mastering-PLC-programming/tree/master/Chapter%201`.

## Software engineering for PLCs

Software engineering is more than just writing programs. Software engineering is the art of effectively solving problems. A major problem with the current industrial programming mindset is that software is often treated almost as a second-class citizen to the hardware. In other words, PLC software is treated as a complement to hardware. More often than not, the software is treated as a throwaway component. It is not uncommon for software practices to be thrown to the wind in automation programming. As such, code\bases that can be easily modified and last for years will often have to be discarded long before they should. Many books and training courses treat PLC software in this way, which, in turn, continues a cycle of treating PLC software as a complement to PLC hardware. Overall, this is a flawed philosophy. Software is every bit as important as (and to certain extents, more important than) hardware. In all, when properly written programs are implemented, machines will be more easily modifiable and correctable. Software can then be transferred to other machines, which will minimize coding defects and yield successful manufacturing operations.

For many non-traditional software developers, a very bad philosophy has taken root. Many of these developers feel that a working solution is a good solution. However, imagine that you're a car mechanic.

Would it be wise to weld the hood shut so that every time you needed an oil change, the hood would have to be cut off with a plasma cutter? Technically, the hood would function the same way as it would on your vehicle now. It would still protect the elements of your engine but at the cost of needing to cut it open for routine maintenance. Though welding your hood onto your vehicle would work, it would not be a wise engineering choice. Software development should be approached in the same way. Just because a solution works does not make it a good solution.

How should a software engineer approach a problem? The answer to that question is the same way any other type of engineer would approach a problem: by first understanding the issue and then developing an effective solution for it. When software engineers approach a problem, they need to try to implement a solution that solves the problem in a way that is simple, efficient, and as easy to maintain as possible. In much the same way as an electrical or mechanical engineer would design their product, a software developer will need to do the same. A software engineer will have to learn to develop solutions that fulfill the requirements of the original problem as well as concoct a solution that can be easily modified in the future. A software engineer must have the following in mind when developing software:

- Does the solution solve the problem?
- Is the solution overcomplicated?
- Can the solution be easily modified if changes are needed?
- Can the solution be verified to ensure it works (can it be easily tested)?

Often, this mentality is lost on PLC programmers. Many PLC programmers do not see themselves as software engineers; however, it must be understood that the moment a keyboard or mouse is touched with the intent of programming something to solve a problem, the programmer becomes a software engineer. When code is developed with this mentality, the same mentality that electrical engineers would use to implement their design, a codebase is created that is clean, easy to maintain, and easy to upgrade, and it will pass the test of time and allow for adaptation for the future.

A key feature of modern software and a key feature of a quality software developer is reusability. Quality code can be used for many different projects without rewriting it. In the automation realm, this can be a bit challenging, as every PLC producer has their own take on PLC development software. However, many PLCs follow what is known as the IEC 61131-3 standard, which provides some uniformity across PLC platforms.

# Understanding the IEC 61131-3 standard

PLCs generally are not cross-compatible. Most PLC programming environments are vendor-specific, meaning that a program written for one device, and even from the same manufacturer, will not compile and run on a device produced by another manufacturer. This means that without standardization, this could lead to utter chaos in the field. Each PLC could easily have not only its unique programming environment but also its own set of rules that govern that environment. A developer migrating from one PLC brand to another may have to take extra time to learn the new programming system. However, the purpose of the 61131-3 is to provide a standard so a developer can easily switch from a PLC of one brand to a PLC of another without having to learn a whole new programming system. In short, the IEC 61131-3 standard makes migrating from one compliant PLC to another as simple as writing the code in a new environment.

This is where the IEC 61131-3 standard comes into play. The IEC 61131-3 is a vendor-neutral and hardware-independent PLC programming standard. The goal of the IEC 61131-3 is designed to provide uniformity across all compliant PLCs that follow the standard. The IEC 61131-3 standard is to PLCs what ECMAScript is to JavaScript. In other words, the best way to think of the standard is as a set of rules that govern the programming interfaces for PLCs from different vendors. As such, by learning the rules on one device, a developer can easily port their knowledge over to another compliant device with relative ease. As such, the overall cost and time it takes to develop a PLC program will drastically decrease, as the developer will not have to learn a new programming syntax.

It is important to understand that just because a PLC follows the IEC 61131-3 standard does not mean that the code is cross-compatible. As stated before, PLC code is generally not cross-compatible. A program written for an RSLogix device will not run on a Beckhoff device. This is mainly due to the hardware architecture, the compilation process, and so on. However, considering that the device is compliant, the code can be ported over by creating a new project, copying the code into the new file, and tweaking the code to meet the requirements of the new device.

The IEC 61131-3 standard is not a language, as inexperienced PLC developers sometimes confuse it with. The IEC 61131-3 is simply a set of rules that compliant PLCs use for developing software. Not every PLC is 61131-3 compliant nor does every 61131-3 compliant PLC utilize every feature of the standard. Common IEC 61131-3 compliant PLCs are as follows:

- Beckhoff
- Wago
- Allen-Bradley
- Omron
- Siemens

This list is by no means an exhaustive list and the available features will vary from brand to brand. There are many more PLCs that are compliant. For the most part, all the major PLC manufacturers are 61131-3 compliant, especially for their newer devices. However, if you need to ensure that the device is compliant, all you have to do is simply check with the manufacturer. Usually, compliance is posted on the manufacturer's website.

Adopting the standard is not a badge of quality, and non-compliant PLCs should not be viewed as inferior to PLCs that are compliant. There are many PLCs that do not follow the standards that are excellent and reliable devices to work with. Many non-compliant devices also share similarities with the standard at the basic level. However, due to the interoperability of IEC 61131-3 programming practices, using compliant devices will ultimately cut down the overhead cost of education. As such, compliant devices are usually favored for industrial automation projects. However, it should be noted that compliant PLCs will often cost more than non-compliant PLCs.

## What does the IEC 61131-3 standardize?

Now that a little background on the IEC 61131-3 standard has been established, it is important to look at what is governed. The biggest aspect of PLC programming that IEC 61131-3 standardizes is language syntax, data types, and supported programming interfaces (programming languages). If you've ever programmed an RSLogix PLC in ladder logic, Structured Text, or another interface, and then programmed an Omron, Beckhoff, or other compliant PLC, you may have noticed that the general syntax, data types, and so on are very similar. Usually, the only programming components that vary are things like function blocks, as many function blocks are just canned functions that were built and included in the programming environment by the manufacturer. In other words, the gross similarities are the standard at work.

Recently, the IEC 61131-3 standard introduced what is known as object-oriented programming. It can be argued that the introduction of this concept is quite revolutionary as it means that the advanced techniques that are used to develop traditional programs can now be applied to the realm of automation. If you are familiar with a language such as C++, Java, C#, Python, or any of the modern traditional programming languages, you are most likely familiar with object-oriented programming. As such, understanding object-oriented programming for PLCs will be as easy as learning the syntax since the same rules apply to PLC programming. However, if your background does not include object-oriented programming, the principles that govern the paradigm will be explored in detail later, starting in *Chapter 6*.

## Programming a PLC – The five IEC languages

The IEC 61131-3 standard includes several different types of language interfaces to program a PLC. In short, you can choose from multiple interfaces to program a PLC. These interfaces are akin to different languages, and each of the interfaces has its strengths. Some of the interfaces are graphically similar to what you would find in a system such as LabView, while others are text-based and akin to what you would find in a programming system such as C++ or BASIC. In the way the 61131-3 standard is set up, all the systems are compatible with each other, meaning that whatever can be done in one interface can also be done in another, and modules such as functions written in one interface can be used in another. The five IEC languages are described in the following sections. Let's take a look.

### *Ladder logic*

If you are reading this book, chances are you know ladder logic and you know it well. Ladder logic is the unspoken standard for programming PLCs. Ladder logic was the programming interface that was developed to allow programmers to program in complex relay logic circuits without the need for bulky hardware or miles of wire. Of all the ways to program a PLC, ladder logic is probably the most common. To be a PLC programmer, a basic understanding of ladder logic is required.

Ladder logic is an excellent and very important PLC programming interface. However, ladder logic does have some drawbacks. Those of you that have had to program complex systems, such as systems for motion control, complex state machines, or the like, know that Ladder diagrams can easily become an unmaintainable nightmare. Ladder logic is an excellent tool for relatively simple applications or for beginners who are just starting their journey. However, as software becomes more complex and new features such as machine learning become more integrated into everyday automated systems, ladder logic is going to become an increasingly difficult tool to work with.

## *Sequential Flow Charts*

Similar to ladder logic **Sequential Flow Charts** (**SFCs**) are another graphical tool for programming PLCs. However, instead of SFCs simulating relay logic, they allow programmers to essentially program a PLC using flow charts. SFC is best used to program processes that can be broken down into steps. SFC allows complex programs to be broken down into smaller modules and govern the flow between the modules. The big advantage of an SFC is that it graphically shows the flow of a program. This is a great advantage for developers who are working on process-driven projects.

## *Function Block Diagrams*

The **Function Block Diagrams** (**FBDs**) interface is the final form of graphical programming language supported by the IEC 61131-3 standard. Much like SFCs and ladder logic programs, FBDs are a widely used language for programming PLCs. The core benefit of FBDs is that they can be used to simplify the programming of closed-feedback loops as they mostly work off of inputs and outputs and can provide feedback to themselves. For most IEC systems, the blocks are interconnected with lines that represent the flow of data from one block to another.

The FBD language is an excellent language choice for developers who are working on high-level projects. For example, suppose you're working on a PLC program for a water treatment plant. You may have a process called water intake, water purifier, and collection process. As the developer, you may already have the functionality for these processes and as such, it is your job to string them together. For applications like these, it is very easy to employ FBDs to diagram out the process as a means of programming the PLC.

## *Instruction List*

**Instruction List**, or as it is more commonly known, **IL**, is a text-based language that is governed by the IEC 61131-3 standard. IL is an offbeat language that is not used much in PLC programming. Users have to turn this on as a feature in CODESYS. The language itself is similar to the old Assembly language.

IL is arguably the most unpopular language in the IEC 61131-3 standard. It is complex to use and requires acute attention to detail. It is very easy to create an infinite loop, computational errors, and so on. It is also extremely difficult to debug. However, programs written in the IL language are generally considered quicker and require less memory. The language has all but fallen out of favor and should only be used if necessary.

### Structured Text

**Structured Text** is arguably the second most popular programming language in the IEC standard. Structured Text is the closest to a traditional, text-based programming language that can be used to program a PLC. The syntax draws heavily from languages such as PASCAL and Ada. Many of the PLC programmers that I have encountered in the past have always seen Structured Text with a bit of fear. However, Structured Text is nothing to be afraid of. In fact, Structured Text can actually make things easier. Anyone that has ever had to sift through hundreds of rungs of ladder logic code will know that it is often difficult to figure out which rung does what and get a grasp on the overall flow of the program, especially when the code is poorly documented and there are many jumps used in the program.

In short, Structured Text will be the way of the future. As PLC technology progresses and applications become more advanced, Structured Text will gradually become the new standard in PLC programming. In other words, the days of simply turning machinery on and off at certain intervals are quickly coming to an end. The modern world is edging into complex machine learning and motion control, which means that it will be difficult, if not impossible, to fully implement these new, complex systems solely in ladder logic. Though it is possible to implement new concepts, such as object-oriented programming, in ladder logic, SFC, FDB, and so on, it can be awkward. Overall, due to the rising complexity of new automation systems, it is well worth the time to learn Structured Text and the advanced functionality that it provides.

Structured Text is the language that is going to be the focus of this book. To get the most out of this book, you should have a basic understanding of Structured Text. However, Structured Text is pretty easy to follow, as it is a simple, human-readable format. The examples in this book will be advanced Structured Text concepts but they will be easy enough to follow. If you feel that you do not have a great grasp of Structured Text, I recommend reviewing some basics such as loops, `if` statements, `switch` statements, and basic data types to get rolling. You will only need to have a loose grasp of these concepts to begin with.

As can be seen, there are many different ways to program a PLC. Now that a background in the different PLC programming languages has been established, we can begin experimenting with some basic code. To do this we will need a development environment. The development environment that we will use is called CODESYS.

## Introducing CODESYS

The most common tool for learning the full gamut of the IEC 61131-3 standard is CODESYS. CODESYS is a free-to-download and free-to-use PLC programming environment that is developed by the German company CODESYS. The programming system has a built-in editor, syntax-checking tools, and a built-in simulator that will allow you to compile and run your code virtually. Not only that, but CODESYS also has a built-in HMI development tool that we'll use in later sections of this book that can be used to develop fully working HMIs. As such, you can learn the full breadth of the IEC 61131-3 standard without having to spend a dime on expensive hardware or software and still be able to develop and watch your code in action.

CODESYS is much more than just a virtual development tool. Currently, it is set up to program a wide variety of PLCs and is the basis for other development environments. CODESYS can best be thought of as a true **Integrated Development Environment** (**IDE**) for PLCs. CODESYS comes with many advanced tools such as debuggers, library management tools, and so on that are used to speed up the development process. Those of you who are familiar with IDEs such as Visual Studio will already be somewhat familiar with the overall gist of CODESYS. Above all else, CODESYS supports the full spectrum of the IEC 61131-3 protocol, including object-oriented programming.

Systems such as Beckhoff's TwinCat and Wago's e!COCKPIT are all built on top of CODESYS. In short, CODESYS is a prime tool for learning PLC software development as well as creating production code for supported PLCs. So, upon completion of this book, you should not only have a pretty decent grasp of the IEC 61131-3 standard but should also have a good idea of how to use multiple other PLC development environments.

If you have not already installed CODESYS, it is important to install it now. The remainder of the book will require the software to be installed. The link for installation can be found in the *Technical requirements* section of this chapter. Installation is pretty straightforward. All you have to do is follow the provided link and follow the wizard. Since CODESYS is a German company, the download website will be in German. I suggest using Chrome to translate the text. At the time of writing this book, you will need to provide some information such as your email to create an account so that you can download the software. Outside of that, CODESYS is a pretty heavy software package, so downloading it may take a little while.

## Testing CODESYS

Usually, the first program a person writes in a new language is called `Hello, World!`. It is a simple program that will display the words `Hello` and `World` on the screen. The PLC equivalent of this is turning a coil off and on. To get familiar with and test our CODESYS installation, we're going to create that simple ladder logic program:

1. Once CODESYS is installed, launch the program, and you should see a page on which you can create a new project. This page is called the **Start** page and it will have a **New Project** link.

2. Click **New Project** and you should see a **New Project** window. Here, click **Standard project**, name the project `Chapter1`, and then click **OK**.

3. Now, you should see a standard project box. This step is the step where you select the programming interface for the project. By default, it will be set to **FBD**. This will need to be changed to **Ladder Logic Diagram**. To do this, click the **PLC_PRG** drop-down box, select **Ladder Logic Diagram (LD)**, and press **OK**.
4. After the project is created, a file tree will appear in the device tab to the left of the screen. Double-click on **PLC_PRG** and you will see a ladder logic development screen.

## Creating the program

The aforementioned steps will create a ladder logic project. The project that's generated will have all the necessary files and dependencies you need to implement your code. As such, all you will need to focus on is implementing the program's logic. The file that we are going to implement our logic in is labeled PLC_PRG.

### The PLC_PRG file

This is the PLC_PRG file that serves as the main entry point for the PLC program:

```
PROGRAM PLC_PRG
VAR
END_VAR
```

Figure 1.1 – PLC_PRG ladder logic development

This is the first file that will be called when a PLC program is run. This is the file in which we will develop our Hello, World! ladder logic program.

To break this area down, the bottom of *Figure 1.1* is a rung. This is where the actual Ladder commands will go. Above that, in the text area, is where variables are declared. The ladder logic tools can be found to the right of the screen, as shown in *Figure 1.2*.

## ToolBox

**ToolBox** is where all the ladder logic commands can be found for use in the rungs:

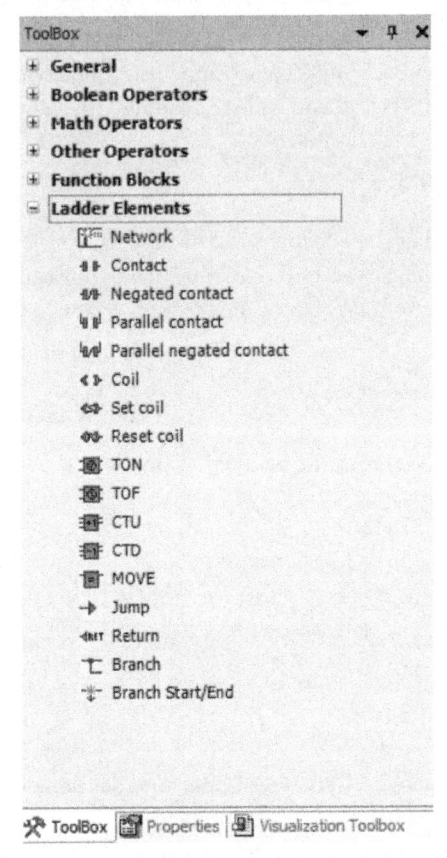

Figure 1.2 – Ladder logic ToolBox

As can be seen in *Figure 1.2*, there are many drop-down menus. The menus contain many different ladder logic instructions. For our purposes, click **Ladder Elements**. Once you expand that menu, drag over both a contact to the **Start here** box and a coil to the **Add output or jump here** box and insert the instructions in the rung area. Also, add two Boolean variables to the variable area (see the following format).

## Variable code

This is the full code that is needed to declare all the variables needed for the program:

```
PROGRAM PLC_PRG
VAR
    input : BOOL;
```

```
    output: BOOL:
END_VAR
```

This code creates two Boolean variables called `input` and `output`. Assign the `input` variable to the contact and the `output` variable to the coil by clicking on **???**, then click on the three dots and select the appropriate variable. The name of the variable can also be typed in directly in place of **???**. The `input` variable will be used to change the state of the `output` variable. In short, the purpose of our `Hello, World!` program will be for the `output` variable to mirror the state of the `input` variable.

## *Completed Hello, World! project*

When you are finished setting up your project, it should reflect what is in *Figure 1.3*:

Figure 1.3 – Completed PLC Hello, World! program

*Figure 1.3* is the code needed to run a `Hello, World!` program. Essentially, this code will turn the `output` variable on when the `input` variable is on, and off when the `input` variable is off.

To test the simulator to see the program work, click **Online** on the ribbon at the top of the screen and select **Simulation**. This will tell CODESYS that there is no physical hardware, and that you want to run the program virtually. Click the button that is shown in *Figure 1.4*.

## *Login button*

This button is the **Login** button that will log you into the virtual hardware. When the button has been pressed, the icon next to it will enable:

Figure 1.4 – The login button

Login will *activate* the program; however, it may not always run the program. To run the program, you must press the *Play* button next to the grayed-out icon in *Figure 1.4*.

You should now have a development screen that resembles *Figure 1.5*.

## A running ladder logic program

*Figure 1.5* is the running PLC program with all of the variables in a FALSE or off state:

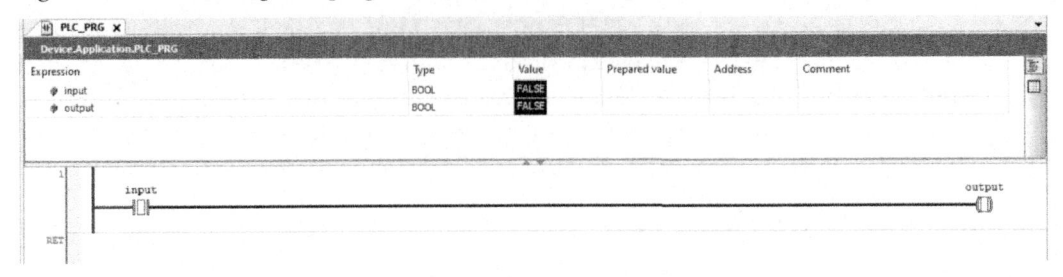

Figure 1.5 – Hello, World!

To turn the output variable on, you will need to change the false variable of the input variable to a true value. To do this, double-click the **Prepared value** field in the input row until it says TRUE. Once you have a blue box that says **TRUE** in the cell, right-click the cell and press **Write All Values Of 'Device. Application'**. Once you do this, your program should resemble *Figure 1.6*.

## Toggling input to true

This is the output when the input variable is set to TRUE:

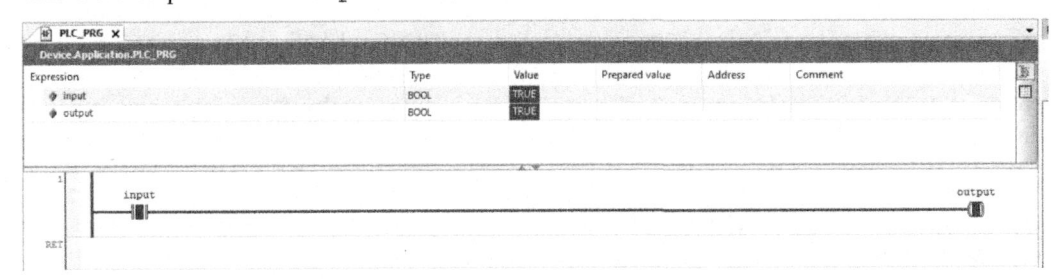

Figure 1.6 – Hello, World! with a TRUE input

When the input variable is set to TRUE, the whole line turns blue and the inner square in the output contact is also turned blue. This means that the rung is activated and is on. Essentially, when you see blue, that means that the rung is active and is doing whatever logic you have programmed in.

The input can also be toggled back to FALSE. The steps are the same for toggling the input variable to a FALSE state with the only exception being that you will set the prepared value to FALSE instead of TRUE.

## Toggling input to false

This is the output when the input variable is set to FALSE:

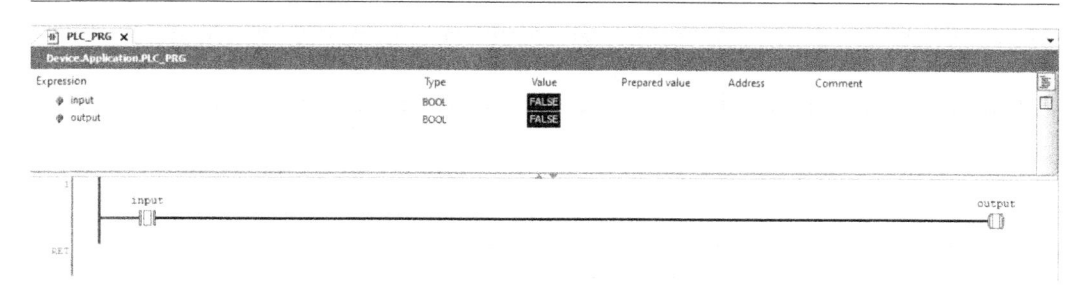

Figure 1.7 – Hello, World! with a FALSE input

As can be seen, setting the `input` variable to `FALSE` changes all the blue back to black. Blue meant the rung was running, and black means the rung is off.

## Summary

As a PLC programmer, it is of absolute importance that you understand the IEC 61131-3 standard. It is also of absolute importance that PLC software is not treated as a throwaway component. The heart and soul of any PLC-based system is the software. Great diligence must be given to the software when it is first being developed. As we have seen, many languages can be used to develop a PLC program. However, for complex software, such as the software that will be explored in this book, Structured Text will be the primary language used. As such, the following chapter will be dedicated to the more advanced concept of Structured Text language.

At this point, you should have CODESYS installed and working. If everything went according to plan, you should be able to run the `Hello, World!` program that was explored previously. The program that was presented is by no means a significant program and its only real purpose is to test the CODESYS installation and get you familiar with logging into and running a program in the CODESYS environment. In all, the main takeaway for this chapter should be that a well-engineered PLC program will ultimately save time and money in the long run, as it will be flexible and stable enough to support any changes that may arise in the project.

## Questions

Answer the following questions based on what you've learned in this chapter. Cross-check your answers with those provided at the end of the book, under *Assessments*.

1. What PLC language is used as a replacement for relay logic?

    A. Structured Text

    B. Function Block Diagrams

    C. Ladder Logic

    D. Instruction List

    E. Sequential Flow Charts

2. Which are IEC 61131-3 programming languages?

    A. Structured Text
    B. Ladder Logic
    C. Java
    D. C++
    E. Instruction List

3. What is IEC61131-3?

    A. A PLC programming language
    B. A vendor-independent standard
    C. A vendor-specific standard
    D. An Allen-Bradley programming environment
    E. A CODESYS standard

4. What PLC language is most like Assembly?

    A. Ladder Logic
    B. Ada
    C. Structured Text
    D. Instruction List
    E. C++

5. What PLC language is most like a traditional language such as BASIC?

    A. Structured Text
    B. Ladder Logic
    C. Instruction List
    D. Function Block Diagram
    E. Sequential Flow Charts

6. What is the most popular PLC programming language?

    A. Structured Text
    B. Instruction List
    C. Ladder Logic
    D. Function Block Diagram
    E. Java

# 2
# Advanced Structured Text — Programming a PLC in Easy-to-Read English

This chapter is dedicated to exploring concepts that many PLC programmers may find exotic. However, if you have a background in a traditional language, these concepts will seem very familiar.

Structured Text has many attributes that are like traditional programming languages such as C/C++, C#, Java, and the like. The IEC 61131-3 standard has adopted many of the features that are standard in most modern programming languages. However, due to the limited computer programming background of many PLC programmers, coupled with many PLC programmers relying solely on ladder logic, these concepts are often not known or fully understood.

The goal of this chapter is for you to learn how to write software that can fail gracefully without killing your PLC, how to access data directly from memory, how to properly document code, and more.

The topics covered in this chapter are not necessarily complex but are often misunderstood due to the few resources that openly present these topics. However, the concepts are powerful and, when implemented, can drastically improve the quality of your code. These topics are as follows:

- Understanding error handling
- Understanding pointers
- Understanding documentation
- Understanding state machines

## Technical requirements

To follow along with this chapter, you will need to create a new Structured Text project and name it `Chapter2`. The project is created in the same way the `Hello, World!` project was created in the previous chapter. When creating the project, ensure you select Structured Text instead of FBD or Ladder Diagrams. The code for this chapter is also available on GitHub. It is recommended that you follow along with the examples in this chapter and then pull down the code from the cloud so you can modify it. The source code for this chapter can be pulled from the following link:

https://github.com/PacktPublishing/Mastering-PLC-programming/tree/master/Chapter%202

## Understanding error handling

Errors can kill the execution of a program, which, in turn, can lead to injury or death. A fatal error will halt the execution of a program. These fatal errors are typically called exceptions. Essentially, an exception occurs when the PLC encounters a problem that it cannot handle at runtime. The ultimate fate of the PLC when an unhandled exception occurs is the program locking up and the PLC needing a reboot. On top of all that, if the condition that caused the error originally occurs again, the program will crash again, and the system will need to be rebooted. In essence, the only safe way to handle the condition is to modify the code to ensure that the condition does not happen again.

Exception errors will not show up during the compilation process. Instead, exceptions occur when the program is running. Due to their nature, it is often difficult or impossible to fully predict when an exception will occur. To make matters worse, some exceptions can take very specific conditions to trigger, and, as a result, there might be long intervals between exceptions. As such, great diligence must be given to possible errors when developing the software.

Many different things can cause an exception and crash a program. A common exception that can often occur is a *division by 0* error. In short, this error occurs when a divisor is accidentally set to 0 or gets extremely close to a decimal point where the PLC will treat it as 0. Other common errors that can throw an exception are null pointers or *array out of index* errors. However, depending on what you're working on, there could be others as well. Generally, it is good to use some form of error handling when working with any of the following:

- Division
- Pointers
- Arrays

To explore what an error looks like in CODESYS, let's create a simple program that will attempt to divide by 0.

# Understanding error handling

## Variables

These are the variables that will be needed for the division by 0 program:

```
PROGRAM PLC_PRG
VAR
        dividend : INT;
        divisor: INT;
        division : INT;
END_VAR
```

For this example, we are going to have a dividend and a divisor, and the quotient of the two is going to be assigned to the variable division.

## The main program

This is the program that will go in the `PLC_PRG` file:

```
dividend := 5;
divisor := 0;
division := dividend / divisor;
```

The code in the file will attempt a division by 0. The code will assign a number of 5 to the dividend and 0 to the divisor. The final variable, `division`, is the quotient of the two. Since division by 0 is an illegal operation in any form of computer programming, the PLC program will crash, and an error similar to the one in *Figure 2.1* will be produced.

## The division by 0 error

The following screenshot is the output when the PLC attempts to divide a number by 0:

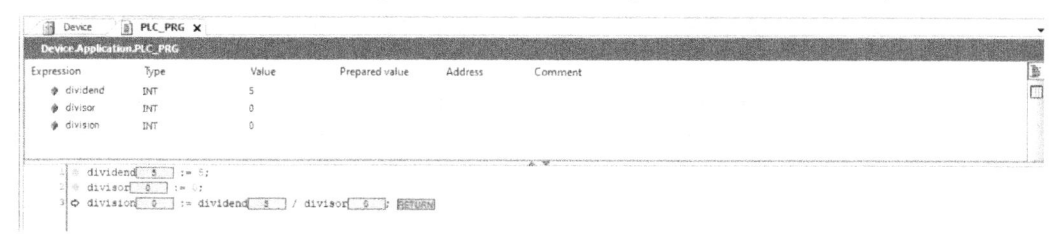

Figure 2.1 – Division by 0 error

After you run the program using the same procedure that was used in the previous chapter, you should notice two things. The first is that the line that does the computation is now yellow. This yellow highlight means there is an error present. The second thing you should notice is that the program automatically stops. If you watch the play button, it will automatically reenable.

If you try to change the number to a value that is not 0 and try to log back in with the default login selection, **Login with online change**, you will either be met with the pop-up error in *Figure 2.2* or the program will fail to run.

This is the **Download failed** popup:

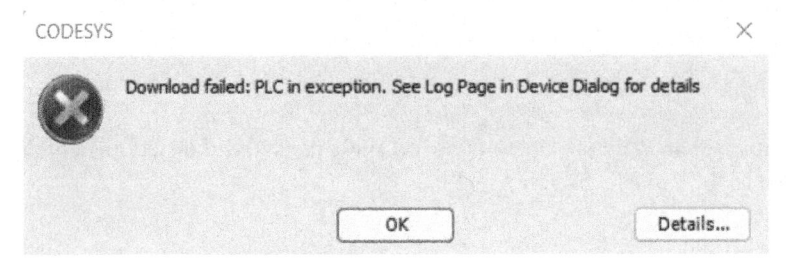

Figure 2.2 – Download failed popup

When an error such as a division by 0 error occurs, restarting the program can be problematic. The easiest way to fix the issue is to simply fix the error. In this case, change the 0 to any other non-zero value, then restart. Any code change will trigger the options in *Figure 2.3*. To restart the virtual hardware, you have to press the login button again, only instead of selecting **Login with online change**, you must select the **Login with download** option and ensure **Update boot application** is selected as well.

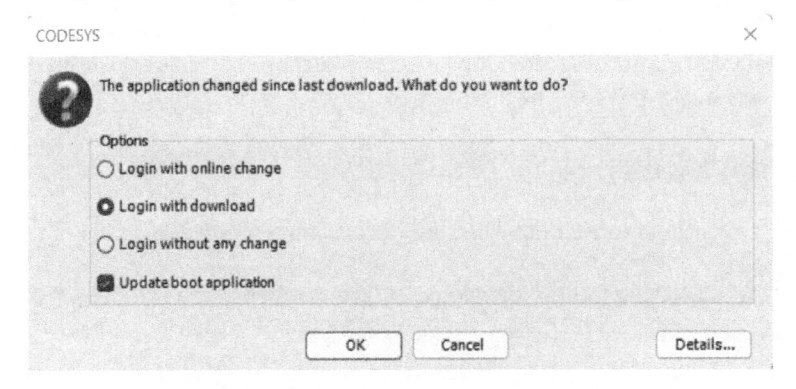

Figure 2.3 – Necessary selections to reset the PLC

Once **OK** is clicked, the application should be reset and you will be able to rerun your PLC program. As can be deduced, in a fast-paced production environment, having to perform these steps every time a value is set to 0 can easily become a major issue.

To remedy this problem, you could set the divisor in an `if` statement as in the following code.

## Checking for 0 code

This code ensures that the divisor is not 0:

```
dividend := 5;
divisor := 0;
IF divisor <> 0 THEN
    division := dividend / divisor;
END_IF
```

This code performs a simple check on the value of the divisor. If the value of the divisor variable is not 0, then it will perform the computation. However, if the value of the divisor is 0, it will not perform the computation.

This code is an applicable solution when there are only one or maybe two values that need to be checked. However, this is something that is easy to overlook during development, and when many equations are divided by 0, they can easily bloat the code. In other words, this isn't a very good solution.

## TRY-CATCH blocks

A better and more formal solution is to use a TRY-CATCH block. TRY-CATCH blocks are like safety nets. When code is in a TRY block, it is essentially tested for errors. If an error is found, the code in the CATCH block will execute. The basic syntax for setting up a TRY-CATCH block is as follows:

```
__TRY
    <code to test>
__CATCH
    <code to run when there is an error>
__ENDTRY
```

Three statements are required to implement a TRY-CATCH block. As was stated before, the TRY section will test the code and the CATCH block will run if there is an error. To end the TRY-CATCH block, the ENDTRY keyword is used. To demonstrate the TRY-CATCH block, let's look at an example. These are the variables that we're going to use to demonstrate TRY-CATCH:

```
PROGRAM PLC_PRG
VAR
    dividend : INT := 5;
    divisor : INT := 0;
    division : INT;
END_VAR
```

For this demonstration, we're going to preset the dividend and divisor in the variable section. The following is the TRY-CATCH code, a basic example of TRY-CATCH handling an exception:

```
__TRY
    division := dividend / divisor;
__CATCH
    division := -999;
__ENDTRY
```

The computation in the TRY block will throw a division by 0 error. When the error is thrown, the code in the CATCH block will run. As such, when the program is run, the division variable will be set to -999, as in *Figure 2.4*:

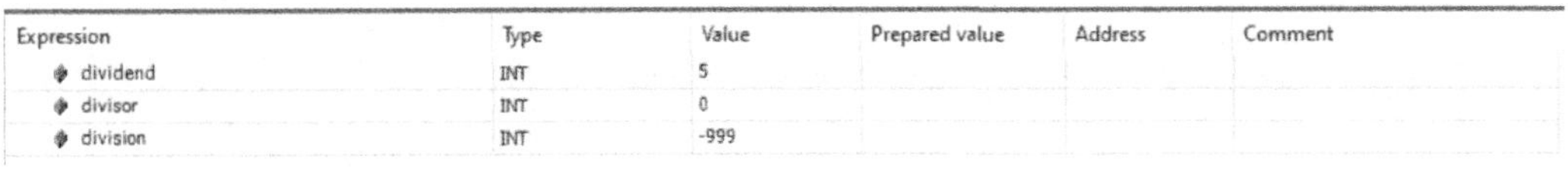

| Expression | Type | Value | Prepared value | Address | Comment |
|---|---|---|---|---|---|
| dividend | INT | 5 | | | |
| divisor | INT | 0 | | | |
| division | INT | -999 | | | |

Figure 2.4 – TRY-CATCH program output

As can be seen in *Figure 2.4*, the division variable is set to -999, which in turn means that the CATCH block ran and set the variable. If the divisor number is rewritten to be a value that will not cause a division by 0 error, the code will not need to be reset and the computation will execute without issues, as in *Figure 2.5*.

This is the output when the divisor value is set to 1:

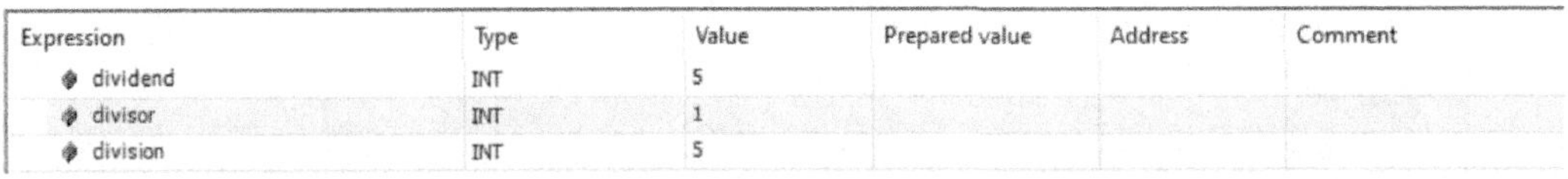

| Expression | Type | Value | Prepared value | Address | Comment |
|---|---|---|---|---|---|
| dividend | INT | 5 | | | |
| divisor | INT | 1 | | | |
| division | INT | 5 | | | |

Figure 2.5 – TRY-CATCH with no exception

When the divisor is set to 1 and the value is written, the division variable resets itself from -999 to 5. In short, the computation was executed without issues. The overall takeaway is that even if an exception occurs, the program will not crash. As such, when valid values are passed back in, the program will execute as normal without needing to restart the PLC.

The true power behind a TRY-CATCH block is that it can handle multiple errors. In other words, if you were trying to compute 20 different equations with a divisor that could possibly be set to 0, you wouldn't have to use 20 IF statements to check whether any of the divisors were 0. In all, you can test as much code as you need in a single TRY-CATCH block.

## FINALLY statements

There is one additional block that can be used with the `TRY-CATCH` statements, which is known as a `FINALLY` statement. A `FINALLY` block is an optional block that is used in conjunction with `TRY-CATCH`. The code in a `FINALLY` block will execute regardless of whether an exception occurs or not. Essentially, the code that goes into a `FINALLY` block is used to do things that must execute regardless of whether there is an error or not.

The following code is the syntax for a `TRY-CATCH-FINALLY` block:

```
__TRY
  <Code to test>
__CATCH
  <Code to run when there is an error>
__FINALLY
  <Code that will run whether there is an exception or not>
__ENDTRY
```

This code is `TRY-CATCH` with a `FINALLY` block. As can be seen, adding a `FINALLY` block is as simple as adding the extra keyword.

## Identifying and handling errors

The `TRY-CATCH` blocks that we have explored so far did not specify what the error was. In practice, this usually isn't the most preferred option. The type of `TRY-CATCH` block that we have explored so far can be called a generic except block. It is important to remember that many different things can throw an error. Therefore, if you want to address the issue, you will most likely need unique logic to handle it. For real-world applications, you generally do not want to use generic except blocks. In a real-world application, you want a `TRY-CATCH` block to have logic that can handle specific errors. For example, if you find yourself with an exception caused by division by 0, you may want to switch the dividend to 1 or conduct some other logic that will alleviate the situation so that the error does not occur again.

The first step in creating specific logic is setting up an `Exception` variable.

### Exception variables

This is how an `Exception` variable is declared:

```
PROGRAM PLC_PRG
VAR
    exc : __SYSTEM.ExceptionCode;
END_VAR
```

As can be seen, creating an `Exception` variable is as simple as creating a variable of any other data type. What `exc` is referring to is what is known as an enum, a concept we will explore in a bit. It has a reference to many of the different types of errors that can be caught. This means that when developed correctly, a developer can write unique code to handle each unique error specifically.

The first thing that we need to do is identify the error.

### *Variables for unique exceptions*

These are the necessary variables for `TRY-CATCH` with an `Exception` program:

```
PROGRAM PLC_PRG
VAR
    dividend : INT;
    divisor : INT;
    division : INT;
    exc : __SYSTEM.ExceptionCode;
END_VAR
```

These are all the needed variables for the program. As can be seen, these are the same programs that were used for the division by 0 programs, except the `exc` variables that will hold the exception.

### *Catching the exception*

This code will store the error in the `exc` variable:

```
__TRY
    division := dividend / divisor;
__CATCH(exc)
    division := -999;
__ENDTRY
```

This code is nearly the same as the code we used for the original `TRY-CATCH` program. The only difference between the programs is the `(exc)` code next to the `CATCH` statement. This variable will store what the exception is in the `exc` variable, as can be seen in *Figure 2.6*:

| Expression | Type | Value | Prepared value | Address | Comment |
|---|---|---|---|---|---|
| dividend | INT | 0 | | | |
| divisor | INT | 0 | | | |
| division | INT | -999 | | | |
| exc | EXCEPTIONCODE | RTSEXCPT_DIVIDEBYZERO | | | |

Figure 2.6 – Error output

*Figure 2.6* shows that the error the code picked up is `RTSEXCPT_DIVIDEBYZERO`. This means that the code picked up a division by 0 error.

As was stated before, it is usually considered a best practice and a good idea to implement custom logic to handle the error. For our purposes, we're going to set the division variable to `-999` only when a division by 0 exception is thrown.

### Handling custom exceptions

The following code is one way of implementing logic to respond to unique exceptions:

```
__TRY
    division := dividend / divisor;
__CATCH(exc)
    IF (exc = __SYSTEM.ExceptionCode.RTSEXCPT_DIVIDEBYZERO) THEN
        division := -999;
    END_IF
__ENDTRY
```

This code has an `if` statement that checks for a division by 0 exception. This code will only change the `division` variable to `-999` when a division by 0 exception occurs. This code serves two purposes. The first is that it protects from all errors. The second benefit this code provides stems from the fact that we are specifically handling division by 0.

All in all, many types of exceptions can be thrown. There is no magic bullet to determine when and where you should use a `TRY-CATCH` block. However, a good rule of thumb is to wrap things such as arrays, math equations, and so on in `TRY-CATCH` blocks. It is highly recommended that you explore all the errors that are in the `ExceptionCode` enum to become familiar with the various types of errors.

So, what is the purpose of doing it this way versus the way we did it before? In terms of functionality, there isn't much of a difference. However, this example demonstrates the use of logic to handle a specific error. In this case, we are reading the exception and performing a specific operation for the given error. By processing the exception, we are able to better handle the error with a specific operation, as opposed to just performing a generic operation for all errors.

As was mentioned before, pointers are often the case of fatal PLC program errors. However, what is a pointer? What does a pointer do? The following sections will explore pointers and references so you can understand how they work and how they can cause issues in a program.

## Understanding pointers

To understand a pointer, it is first necessary to understand the basics of how variables are stored in memory. For many PLC programmers, creating a variable or a tag is simply inputting a name and assigning it a data type. However, some mechanics go on under the hood. For starters, a variable is much more than just a name and a data type that holds a value. A variable is a dedicated memory block that the computer, in this case, the PLC, uses to hold a value of a specific data type. The memory block is generally not human-readable; as such, the variable name is just a human-readable facade that makes accessing and manipulating the data in the memory block easy.

## Representing PLC memory

*Figure 2.7* is a graphical representation of a PLC's memory. It is a simplified way of conceptualizing how the PLC sees its memory addresses:

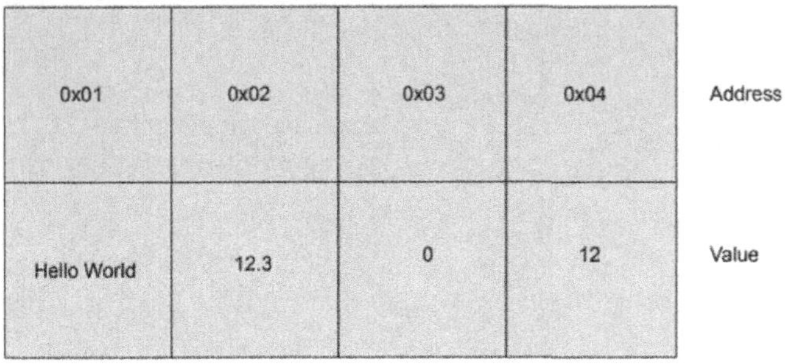

| 0x01 | 0x02 | 0x03 | 0x04 | Address |
|---|---|---|---|---|
| Hello World | 12.3 | 0 | 12 | Value |

Figure 2.7 – A graphical representation of computer memory

As you have probably deduced, working with the raw addresses would be very confusing and probably lead to bugs in the program. This is the reason why variable names are so important. In short, a variable name adds a layer of abstraction over the memory address.

Variables are not the only types of data that have a memory address. As we will see later, function blocks, methods, and more all have memory addresses when the program is running. A general rule of thumb is that if it has a name that you provide, it has a memory address.

More often than not, you'll want to work with a human-readable name over a memory block address. However, you can still directly access the memory address of a variable or anything else that has a memory address with what is known as a pointer. Pointers are declared similarly to regular data types; however, the value they hold is the memory address of a variable, function block, or whatever it might be.

## General syntax for pointers

This is the syntax that is used to declare a pointer:

```
PROGRAM PLC_PRG
VAR
    pt : POINTER TO <TYPE>;
```

```
END_VAR
```

This code is just declaring a variable with the POINTER and TO keywords. This variable declaration will be able to hold the address of a function, variable, or so on of any type. In short, this is all that is needed to declare a pointer.

Thus far, we have explored pointers. Although we have created a pointer, we haven't done anything with it. In short, we have created a pointer that points to nothing. For those of you who have a programming background, we have created the dreaded null pointer. As such, for a pointer to be of use, we need to explore the ADR operator.

## The ADR operator

The ADR operator will provide the address of whatever is passed into it. Many times, the ADR operator is used with pointers. In short, it is the main way to retrieve address information. So, it is usually assumed that if you're going to use a pointer, you're going to use the ADR operator as well.

To explore the ADR operator, we're going to create a small program that will display the memory address of a variable. The following are the variables we will need:

```
PROGRAM PLC_PRG
VAR
    pt : POINTER TO INT;
    testVal : INT := 10;
END_VAR
```

The pt variable is the variable that will hold the memory address. The testVal variable is the important variable for this program. This is a variable whose memory address we're going to read with the following code:

```
pt := ADR(testVal);
```

As can be seen in the code, the testVal variable is passed into the ADR operator, and that output is assigned to the pt variable. When the code is run, you should see an output similar to what is in *Figure 2.8*. It is the output of the pointer.

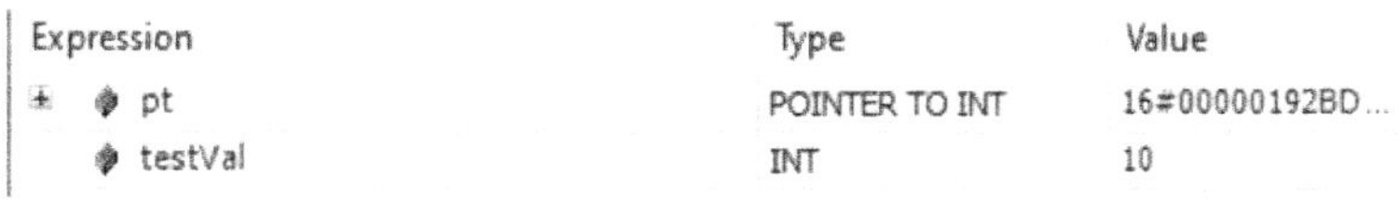

| Expression | Type | Value |
| --- | --- | --- |
| ± ● pt | POINTER TO INT | 16#00000192BD... |
| ● testVal | INT | 10 |

Figure 2.8 – Memory address output

The first row contains the memory output in the Value column. The memory address that you get when you run this program will probably be different from the one in the output screenshot.

What we have just done is get the memory address of a variable. This alone won't do much. To actually do something meaningful, we have to dereference the pointer.

## Dereferencing pointers

Obtaining a value out of a pointer is called dereferencing. You dereference by appending the ^ symbol to the pointer variable. The ^ symbol will allow you to access or manipulate the data in a pointer. The following program is a basic example of how to dereference a variable.

These are the variables that are going to be needed for assignments via pointers:

```
PROGRAM PLC_PRG
VAR
    testVal_pt : POINTER TO INT;
    testVal : INT := 10;
    testVal2 : INT;
END_VAR
```

This code creates a pointer variable to hold the memory address of `testVal`. The `testVal` variable is initialized with a value of `10`. The `testVal2` variable is not initialized and after the logic is run, it will be assigned the value of `testVal`.

This is the logic that is needed to assign values via pointers:

```
testVal_pt := ADR(testVal);
testVal2 := testVal_pt^;
```

As usual, the first line assigns the address of the `testVal` variable to the `testVal_pt` variable. The second line accesses the data in the `testVal_pt` variable and assigns it to the `testVal2` variable. As such, after the program runs, the value of `testVal2` should be `10`, similar to what is in *Figure 2.9*:

| Expression | Type | Value |
| --- | --- | --- |
| testVal_pt | POINTER TO INT | 16#00000192BD... |
| testVal | INT | 10 |
| testVal2 | INT | 10 |

Figure 2.9 – Dereferencing output

As can be seen, the value from the `testVal` variable was extracted and assigned to the `testVal2` variable via a pointer.

When a pointer is not properly configured, it can produce an invalid pointer. An invalid pointer is like a null pointer in a traditional language. Though they occur less frequently in PLC programming, you need to know how to handle them.

## Handling invalid pointers

If you've ever programmed in C/C++, Java, C#, or any traditional programming language, chances are you've run across a null pointer before. PLC programming is no different. If you're working with a pointer, you want to check that the pointer is pointing to something. For example, you may try to assign a value to a pointer variable, but, if the variable isn't pointing to anything, you have the PLC equivalent to a null pointer. Consider the following code.

The following code declares a pointer:

```
PROGRAM PLC_PRG
VAR
      testVal_pt : POINTER TO INT;
END_VAR
```

These are the variables we are going to use in the invalid pointer program. As can be seen, all we have is our standard pointer variable, `testVal_pt`.

The following is the logic that will be used to demonstrate the invalid or null pointer:

```
testVal_pt^ := 2;
```

As you can tell from the code, `testVal` doesn't point to anything. As it stands, the code is attempting to assign the number to a null pointer, as the pointer isn't pointing to a memory address at the moment. When the code is run, the output should match what is in the following figure:

Figure 2.10 – Null pointer

The program will instantly fail when someone tries to run it. Essentially, this code is trying to assign the value of 2 to an empty pointer. Since the pointer does not point to a memory address, the PLC will not know how to handle the situation and the program will crash.

### *Catching an invalid pointer*

To ensure that the pointer is pointing to something and a null pointer does not occur, you can either check to ensure that the pointer is pointing to a memory address or use a `TRY-CATCH` block.

Depending on what you're trying to accomplish with your code, the best way to check for an invalid pointer is to check the memory address. If the pointer is not pointing to anything, then the value will be 0. As such, an easy way to check whether a pointer is valid is to perform a simple check on it.

This logic will only try to assign a value to a pointer if the pointer is not null:

```
IF testVal_pt <> 0 THEN
    testVal_pt^ := 2;
END_IF
```

In short, this code will perform a simple `IF` check on the value of the memory address that is stored in the pointer variable. If the memory address is 0, then the program will ignore the assignment and not crash.

Compared to the other logic, this code did not crash. Essentially, the `IF` statement prevented the assignment that caused the previous code to crash by not allowing the assignment to be executed.

Using an `IF` statement to check for an invalid pointer is an excellent and very common way to detect invalid pointers. Many developers will always wrap their pointer code in a control statement as a best practice. However, as was mentioned before, you can also use a `TRY-CATCH` block.

### *TRY-CATCH for invalid pointer variables*

These are the necessary variables to demonstrate the `TRY-CATCH` invalid pointer program:

```
PROGRAM PLC_PRG
VAR
    testVal_pt : POINTER TO INT;
    exc : __SYSTEM.ExceptionCode;
END_VAR
```

Essentially, this is the same code that was used before, with the addition of the `exc` variable to catch the error.

The following is the logic to catch an invalid pointer:

```
__TRY
    testVal_pt := 2;
__CATCH(exc)
__ENDTRY
```

As can be seen, this is just a basic `TRY-CATCH` block. When this code is executed, you should see an output similar to the following figure:

| Expression | Type | Value |
|---|---|---|
| testVal_pt | POINTER TO INT | 16#0000000000000002 |
| exc | EXCEPTIONCODE | RTSEXCPT_NOEXCEPTION |

Figure 2.11 – TRY-CATCH invalid pointer output

The output in *Figure 2.11* is a little more descriptive than just wrapping the pointer in an `if` statement; however, the drawback to this method is that to remedy the underlining problem, you will still need custom `if` statements to handle the logic. On the other hand, the `TRY-CATCH` blocks will provide a blanket of protection for multiple pointers. Essentially, both code blocks will prevent the program from crashing. It will ultimately be up to you as the developer to choose which method is more appropriate.

Pointers are fine to use, and there are many codebases that still use them. However, modern PLC programming has introduced a more user-friendly way of working with pointers that requires less syntax. As such, for new codebases, it is usually a good idea to favor what are known as a reference over pointers.

## Understanding references

A reference is a type of pointer that is more user-friendly and requires less syntax than a traditional pointer. A few big advantages of using a reference are that you do not have to use the `^` symbol, you do not have to use the `ADR` operator, and finally, references are type-safe.

References share many similarities with pointers, including similar syntax. As such, much like pointers, a reference must be declared. Therefore, the first step in learning how to use pointers is to understand how to declare them.

### Declaring a reference variable

Declaring a reference is almost the same syntax as declaring a pointer. This can be thought of as a shorthand way of using a pointer. The only difference is that the `REFERENCE` keyword is used as opposed to the `POINTER` keyword.

This is the syntax to declare a `REFERENCE` variable:

```
<variable> : REFERENCE TO <data type>
```

Putting this syntax into practice, we can create a reference to an integer, as in the following code. These are the variables we will use for the reference demonstration:

```
PROGRAM PLC_PRG
VAR
    A : REFERENCE TO INT;
    B : INT := 3;
END_VAR
```

This example will only use two variables. The A variable is the `REFERENCE` variable that will be assigned the value that is in the B variable when the program is run.

## Example program

This is the code to demonstrate references:

```
A REF= B;
A := B;
```

The first line is equivalent to using the ADR operator; in this case, that variable would be A. The second line is equivalent to dereferencing using the ^ symbol. When the program is run, you should see an output similar to what is in the following figure:

| Expression | Type | Value |
|---|---|---|
| A | REFERENCE TO INT | 3 |
| B | INT | 3 |

Figure 2.12 – Reference program output

Essentially, the reference program does the same thing as the pointer program; however, the reference program uses a much simpler and more intuitive syntax. The first line is of vital importance. Since this line is equivalent to the ADR operator in the pointer program, if this line is neglected, you will get an invalid reference.

## Checking for invalid references

It is important to check for invalid references in the same way that it is important to check for invalid pointers. There is an easy operator that can be used to test whether a reference is valid or not. The most effective way to do this is to use the __ISVALIDREF operator. This operator will return TRUE if the reference is valid, or FALSE if it is not.

These are the bare minimum variables to test for an invalid REFERENCE variable:

```
PROGRAM PLC_PRG
VAR
    A : REFERENCE TO INT;
    B : INT := 3;
    valid : BOOL;
END_VAR
```

The `valid` variable isn't always necessary as the operator returns a TRUE or FALSE value. As such, it can often be embedded in a control statement. However, for this example, we are going to store the return to the `valid` variable. This is the logic we will use to check whether the reference is valid:

```
A REF= B;
valid := __ISVALIDREF(A);
```

As can be seen, all we are doing is passing the A reference variable into the operator and assigning the output of the operator to the `valid` variable. When the program is run, you should see the following output:

| Expression | Type | Value |
| --- | --- | --- |
| A | REFERENCE TO INT | 3 |
| B | INT | 3 |
| valid | BOOL | TRUE |

Figure 2.13 – The __ISVALIDREF output

As can be seen, since we have a `valid` reference, the operator's output is TRUE. Now, if you were to comment out the first line of the code and run the program again, the `valid` variable would be set to FALSE.

Generally, whether to use a reference or a pointer will boil down to the developer's preference. References offer more features and are safer to use. With that being said, it is still possible to see code that uses pointers.

Now that error handling, pointers, and references have been explored, it is time to transition to adding context to the code. Code is only good if someone else can read it. As such, proper documentation is the key to the longevity of a codebase.

## Understanding documentation

For many programmers in general, documentation is viewed as more of a petty pain than anything else. However, proper code documentation is vital to the longevity of a codebase. With that being said, it is often not clear what should be documented or, for that matter, how to properly write documentation.

The past code examples were easy enough to understand when the proper context was provided. However, if the context was taken away, chances are you would have a difficult time figuring out what the code did. In practice, this is very bad. Other developers that inherit your work should be able to open the project and understand the code with minimal effort.

Documentation also helps you personally. Let's assume that in 6 months when your customer wants an upgrade, you're going to be the one who is going to implement the upgrade. Chances are you're not going to remember how the code works. There is no silver bullet that'll make your codebase documentation perfect; however, the following are some tricks to guide you along the way to proper documentation.

### Self-documenting code

Proper code begins with self-documenting code. The term *self-documenting* means that the code should be as human-readable as possible. As such, the first place to start is to give your variables and code modules descriptive names. In automation, it is very important to name things descriptively. Consider the example variables and see which variables are more descriptive.

These are two variables with a descriptive name and a non-descriptive name:

```
PROGRAM PLC_PRG
VAR
     machineState : BOOL;
     isMachineOn : BOOL;
END_VAR
```

When you examine the preceding two variables, you can see that there is a `machineState` variable and an `isMachineOn` variable. Upon seeing this code for the first time, which variable name was more descriptive for knowing whether the machine is on or off? If you think about it, `machineState` can mean many different things: for example, the machine is in an error state, the machine is running, or the machine is in any number of other states. However, if you look at the `isMachineOn` variable and see that it returns a Boolean, it is pretty safe to assume that this variable is going to hold whether the machine is on or off.

A logically named variable is only useful when it is consumed by something. Generally, well-named variables will allow a program to be more logical and easier to follow when it is consumed; however, it is common for inexperienced programmers to try to code to actual values as opposed to variables. Coding to hard values can lead to issues when trying to troubleshoot bugs as it is common to forget what the values represent. As such, when developing a program, it is important to use easy-to-understand and descriptive names for variables.

## Code to variables

As a programming instructor, I often see inexperienced programmers using actual values in their code as opposed to using variables that are set to the intended values. This may seem trivial at first but, as stated before, coding to a variable as opposed to coding to a value adds clarity to a program and makes a program more modifiable. Consider the following code.

This code is an example of why coding to a variable is important:

```
IF speed >= 10000 THEN
     motorOff := TRUE;
ELSIF speed < 10000 THEN
     motorOff := FALSE;
END_IF
```

This code can be followed but, at first glance, `10000` has no context. Imagine that you're in the field trying to troubleshoot a problem that is linked to a motor speed issue. When you first see `10000`, do you think you're going to know what that value means, especially when the only instruction you have is that the motor speed needs to be modified?

Now, imagine you have to change that value. In this example, the value is only in two places, which means that altering that value won't be that difficult and chances are that you won't introduce a bug. However, suppose that value is in 10 places, and you need to change it. Chances are that if you must change that value in that many places, you're going to introduce some bugs. The moral of the story is it is better to declare a variable with a descriptive name and code to that variable.

These are the variables that we are going to use to demonstrate the concept of coding to a variable:

```
VAR
    speed: INT;
    motorOff : BOOL;
    motorSpeedCutOff : INT := 10000;
END_VAR
```

As can be deduced, the variable names are very descriptive. The `motorSpeedCutOff` variable gives a clear meaning to what the `10000` value stands for. In short, it gives context. To demonstrate this concept, let's take a look at what a program with a logically named variable looks like.

The following is the same code as before, with the exception of the `motorSpeedCutOff` variable instead of the raw `10000` value:

```
  IF speed >= motorSpeedCutOff THEN
      motorOff := TRUE;
  ELSIF speed < motorSpeedCutOff THEN
      motorOff := FALSE;
  END_IF
```

This code not only has much more context than the last example but it is also easier to change. Instead of changing `10000` in multiple places, you will only need to change it in one place. As such, you don't have to worry about accidentally setting the wrong value in one of the locations and introducing bugs.

These are just a couple of common self-documenting code practices that should be observed when developing new code. However, there are many other techniques to ensure the code documents itself. As a CIS instructor, these are the two that I often try to instill in my students the most, as they are the easiest and most powerful documenting practices to implement.

## Code commenting

Another practice that will help your codebase last the test of time is commenting. There are many programs, both PLC programs and not, that have no comments in the code. These programs will usually take extra time for developers new to the codebase to understand, and it can ultimately mean that the codebase will become unmaintainable after a while.

Consider our motor code block from the last example again. Though we discussed it and we can tell by the variable names that it is meant to turn a motor off and on, we don't know which motors it controls, why this code block is important, or anything beyond its basic functionality. This can be very dangerous in practice, as developers that are inexperienced in the codebase can remove code blocks, modify them without understanding them, and so on.

Logically named variables can only go so far. Many times, extra context must be given to explain code. This is where comments come into play. Comments are very important in the development of software; however, commenting is a bit of an art that must be practiced to master. As such, the following sections will explore proper and improper commenting.

### *Good comments*

Comments are notes in the source code to yourself and other programmers. Comments should be short, simple, to the point, and above all else, provide context. As was discussed before, the motor code from the past examples doesn't have a lot of context as to what it is meant for. A quality code comment can be found in the following code block:

```
//this code turns the conveyor motor on and off
IF speed >= motorSpeedCutOff THEN
    motorOff := TRUE;
ELSIF speed < motorSpeedCutOff THEN
    motorOff := FALSE;
END_IF
```

The comment in this code block is one line and indicates what the code does. It can be argued that using more descriptive names, such as `conveyorMotorSpeedCutOff`, would serve the same purpose, and it could. However, it is still wise to add comments to provide some context.

### *Bad comments*

Just because your code has a lot of comments does not mean that it is well documented. Too many comments can be as detrimental to a codebase as too few comments. Too many comments can clutter up the source code and overwhelm the reader. This is an example of a poorly documented program:

```
//this code turn the conveyor motor on and off
IF speed >= motorSpeedCutOff THEN
    //when the motor speed is greater than the cut off speed turn the motor off
    motorOff := TRUE;
ELSIF speed < motorSpeedCutOff THEN
    //when the motor speed is less than the motor cut off
```

```
speed turn the motor on
        motorOff := FALSE;
END_IF
```

The two comments in the `IF` statements are completely unnecessary. The self-documenting nature of the variable names provides enough information to the reader about what the `IF` statements do. The explanatory comments do little more than bloat the code file and possibly confuse the reader. In a situation like this, it is best to remove them.

Now that we have some more advanced tools at our disposal, it is time to employ them. One common pattern that is often used in PLC programming is the state machine. In the next section, we are going to use the concepts we learned in the previous section to build a state machine.

## Understanding state machines

As a software engineer, especially one that writes PLC code, you must understand what state machines are. State machines to PLC programmers are what the **Model-View-Controller** (**MVC**) pattern is for web developers. To be a quality PLC programmer, you must understand what a state machine is and how to implement one.

The most simplified way to think of a state machine is as a series of states that can transition from one state to another. A simple example of a state machine is a lightbulb connected to a switch. The following diagram represents the state of a lightbulb:

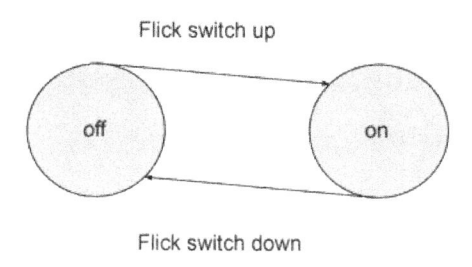

Figure 2.14 – Lightbulb state machine

As can be seen with the arrows, if the lightbulb is on, it can transition to off when the switch is flicked down. If the lightbulb is off, it can transition to an on state when the switch is flicked up.

The majority of state machines that you are going to work with as a PLC programmer are called **finite state machines** (**FSMs**). FSMs are state machines that have a limited number of states that the machine can be in. At most, the machine can be in exactly one state at a time. The lightbulb is an example of an FSM. There are only two states the light can be in at any given time: either off or on. To conclude the chapter, we are going to create a simple state machine that will have three states: on, off, and error.

## Variables for the state machine

These are the variables that will be used for the state machine:

```
PROGRAM PLC_PRG
VAR
     machineState            : INT := 1;
     motorSpeedCutOff        : INT := 10000;
     runTime                 : INT := 2;
     setSpeed                : REAL;
     numOfParts              : REAL := 8;
     motorOff                : BOOL;
     exc                     : __SYSTEM.ExceptionCode;
END_VAR
```

This program will have a number of preset values. The `machineState` variable is preset to 1, which means that the machine will automatically start in the off state. The `setSpeed` variable is the quotient of `numOfParts` divided by the `runTime` value. The `setSpeed` variable is a simulated motor speed value that will set the speed of a theoretical motor. These values simulate the number of parts that a line should produce in a given amount of time. If the operator accidentally inputs a 0 value for `runTime`, the line will transition to an error state, which will reset everything. Now that the variables for the state machine have been established, we can explore the logic that drives the state machine. As can be seen in the following section, the general structure of a state machine is very simple.

## Exploring state machine logic

This is the logic for the state machine:

```
CASE machineState OF
1:
     //machine off state
     motorOff := TRUE;
2:
     //machine run state
     __TRY
          //set motor speed
          setSpeed := numOfParts / runTime;
          IF setSpeed >= motorSpeedCutOff THEN
               motorOff := TRUE;
          ELSIF setSpeed < motorSpeedCutOff THEN
```

```
                motorOff := FALSE;
            END_IF
        __CATCH(exc)
            //throw machine into error state
            machineState := 3;
        __ENDTRY
 3:
    //error state
    runTime      := 0;
    setSpeed     := 0;
    machineState := 1;

END_CASE
```

As is the case with many state machines, the machine is built around a case statement. For this machine, Case 1 is the off state. In other words, when in Case 1, the machine is turned off. Case 2 is the machine running state. Case 2 computes and controls the motor speed. Since this is wrapped in a TRY-CATCH block, if there is an error, the machine will go into Case 3, which is an error state, and will then immediately transition into an off state.

## Case 1 – a non-running state machine

This is the state machine in what is considered an off state:

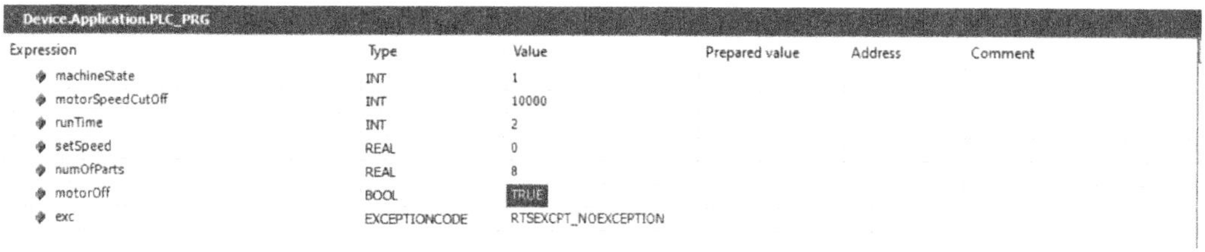

| Expression | Type | Value | Prepared value | Address | Comment |
|---|---|---|---|---|---|
| machineState | INT | 1 | | | |
| motorSpeedCutOff | INT | 10000 | | | |
| runTime | INT | 2 | | | |
| setSpeed | REAL | 0 | | | |
| numOfParts | REAL | 8 | | | |
| motorOff | BOOL | TRUE | | | |
| exc | EXCEPTIONCODE | RTSEXCPT_NOEXCEPTION | | | |

Figure 2.15 – State machine in an off state

The state machine is set to Case 1 by default, which means that the machine is in an off state. This can be seen by examining the setSpeed variable being set to 0 and the motorOff variable being in a TRUE state, which, in this case, means the motor is off.

## Case 2 – a running state machine

These are the variable outputs when the machine is running:

| Expression | Type | Value | Prepared value | Address | Comment |
|---|---|---|---|---|---|
| machineState | INT | 2 | | | |
| motorSpeedCutOff | INT | 10000 | | | |
| runTime | INT | 2 | | | |
| setSpeed | REAL | 4 | | | |
| numOfParts | REAL | 8 | | | |
| motorOff | BOOL | FALSE | | | |
| exc | EXCEPTIONCODE | RTSEXCPT_NOEXCEPTION | | | |

Figure 2.16 – Running state machine

When the `machineState` variable is set to 2, the machine will go into a running state, in which `setSpeed` is computed to 4 and the `motorOff` variable is set to `FALSE`. This means that the motor is running.

| Expression | Type | Value | Prepared value | Address | Comment |
|---|---|---|---|---|---|
| machineState | INT | 1 | | | |
| motorSpeedCutOff | INT | 10000 | | | |
| runTime | INT | 0 | | | |
| setSpeed | REAL | 0 | | | |
| numOfParts | REAL | 8 | | | |
| motorOff | BOOL | TRUE | | | |
| exc | EXCEPTIONCODE | RTSEXCPT_FPU_DIVIDEBYZERO | | | |

Figure 2.17 – Exception thrown

If you look at the value of the `machineState` variable, it is set to 1, which means off; however, the `runTime` and `numOfParts` variables are zeroed out, which only happens when the machine passed through the exception state. Essentially, the state machine transitioned states faster than you could notice it.

Compared to the other concepts we explored, state machines are an amalgamation of different concepts. A state machine is a pattern and, as such, how you implement the code will vary. Overall, you should now have a decent understanding of state machines and the core concepts explored.

# Summary

This chapter has explored some of the more complex features of the Structured Text language. Many of the features explored in this chapter can help improve the quality of your PLC code. In short, error handling can help catch unforeseen errors, proper code documentation can prolong the lifespan of your codebase, state machines are a vital part of any PLC program, and although pointers may not seem that important now, they will become more prevalent as the book progresses.

Of all the concepts explored, state machines, documentation, and error handling are going to be used the most in the day-to-day life of a PLC programmer.

Now that Structured Text has been explored, the next thing to explore is debugging.

## Questions

Answer the following questions based on what you've learned in this chapter. Cross-check your answers with those provided at the end of the book, under *Assessments*.

1. What is the difference between a pointer and a reference?
2. What does the ^ symbol do?
3. What are three keywords that can be used with a TRY-CATCH block?
4. What is self-documenting code?
5. What is the difference between a good comment and a bad comment?
6. Why should you code to a variable?

## Further reading

Have a look at the following resources to further your knowledge:

- CODESYS documentation for TRY-CATCH statements: `https://help.codesys.com/api-content/2/codesys/3.5.13.0/en/_cds_operator_try_catch_finally_endtry/`
- CODESYS documentation for pointers: `https://help.codesys.com/api-content/2/codesys/3.5.14.0/en/_cds_datatype_pointer/`

# 3
# Debugging — Making Your Code Work

Chances are you have never written a program of any significant size that worked as expected on the first go. In fact, chances are you hardly ever get a program to compile and run on the first go. Every software engineer knows that defects are a part of life. As such, debugging is a part of life as well.

Debugging is a skill. Just as a programmer must learn to write code, they must also learn to debug software. A developer can be the best developer in the world; however, if they cannot effectively debug their software, no one is going to consider them very effective. Just as there are techniques to develop code, there are techniques that can be used to debug software.

There are many different ways to debug software. Some methods are more sophisticated than others. It doesn't matter what method you choose to debug your software as long as the software is defect-free when you deploy it. Many tools can be used to troubleshoot code; however, the most common tool is known as a debugger. CODESYS comes packaged with debugging tools that can be used to find bugs. This chapter will explore the following concepts to help you become better at finding and removing defects in your code:

- What is debugging?
- Understanding debugging tools and techniques
- Troubleshooting – a practical example

Finally, to end this chapter, we're going to debug an actual program using the techniques we have learned. In short, we will troubleshoot a simulated real-world program to hammer in the concepts.

## Technical requirements

Unlike *Chapter 2*, this chapter will focus on debugging code as opposed to developing code. Code examples will be provided in the text; however, it is recommended that you pull down the code from

GitHub. The source code for this chapter can be downloaded at this link: `https://github.com/PacktPublishing/Mastering-PLC-programming/tree/master/Chapter%203`.

## What is debugging?

Debugging a program starts with understanding what a **bug** is. A bug is best thought of as a software defect. Bugs range in severity—some bugs might produce minor inconveniences in a program, such as producing the wrong text, while severe bugs will prevent a program from compiling. When a bug is detected, it is important to find the bug and repair it; this act is what is known as **debugging**.

Debugging is as much an art as it is a science. Debugging is the act of finding and eliminating defects in software. As was discussed earlier, defects are a given for a program of any significant size and functionality. As such, it is important for you, as a developer, to be able to troubleshoot defects. The following section is dedicated to understanding bugs and the debugging process.

### Types of bugs

Depending on who you ask and what article you read, there are many types of bugs. However, the following types of bugs are arguably the most common types you will run into as a PLC programmer:

- **Syntax errors**: This type of bug is triggered by invalid syntax. Syntax errors will usually prevent your program from compiling and running. Normally, CODESYS, as well as other **integrated development environments (IDEs)**, will produce a red squiggle line under the offending syntax.

- **Logic errors**: Similar to syntax errors, logic errors are very common. These are issues in your logic that will cause wrong outputs, wrong computations, and so on. These bugs can be a little more dangerous than syntax errors because you won't know about them until the program runs, and since these are caused by compliant syntax, you usually won't get a red squiggle line. These defects can cause infinite loops, crashes, and software failures, which in terms of automation will cause machine failures and other potentially dangerous situations. These are the bugs best suited to find via a debugger tool.

- **Functional errors**: Functional errors are bugs in the program that prevent the software from working as expected. For example, a functional error may come in the form of a button that does not turn on the correct assembly line when pressed. These errors are not the result of bad syntax, nor are they the result of logical errors. Depending on the severity of the bug, the issue can range from minor inconvenience to dangerous machine malfunctions. Normally, these bugs are discovered during the functional testing phase of the machine's development. In other words, these bugs are usually discovered when you're testing the machine to ensure it works as expected. Debugging tools can also help find and rout out these defects as well.

There are many more different types of bugs. However, these are the most common bugs that I encounter when programming PLCs. Depending on what you're working on, you may also have to contend with usability, security, performance, or any other type of bug that is negatively impacting your software. For this book, we are going to focus on fixing the three highlighted types of defects. Having explored the types of bugs that appear, we need to look at the difference between debugging and testing.

## Testing versus debugging

The terms *testing* and *debugging* are often used interchangeably, even among experienced programmers. However, there is a difference between the two terms. Testing is the act of finding bugs in software. As we will explore in *Chapter 9*, testing is a full-blown phase in the software development life cycle. When somebody refers to testing, they are referring to the act of ensuring that the software works as intended. In other words, they are referring to the act of finding defects in the software. Testing software has its own set of techniques and dedicated purpose.

Debugging is fixing defects that are found in the software. Unless the bug is a syntax error or some obvious logic error, many bugs—especially functional defects—are found during the testing phase. That is, many bugs are not uncovered until someone is verifying the software is working as intended. If a defect is found, whoever found the defect will report the bug to the developer, and they will use debugging tools and techniques to isolate and remedy the bug.

Let's summarize the differences between testing and debugging:

- **Testing**: Finding bugs
- **Debugging**: Fixing bugs

Testing and debugging are not the same thing, even though the two terms are often used interchangeably. Debugging can and should be carried out throughout the development process. Testing is not just the job of a dedicated person that is done at a specific point in the development phase of the software development life cycle. Though much of the major testing will be conducted after development is completed, a good developer will frequently run their code to ensure it compiles and behaves as expected as frequently as possible. A general rule of thumb is that the sooner a bug is caught, the easier it will be to remove.

As a developer, most of what you're going to be doing is debugging. Essentially, your QA team, customer, or project manager will give you a list of defects, and it'll be up to you to find the root cause of the defects and fix them. This can be a bit of a challenge; however, there are several steps that you can follow to help guide you along.

## Breaking down the debugging process

Debugging a program is a process. A good developer will never jump in and start modifying code without a clear understanding of the bug and a roadmap to a solution. Depending on which articles you read, the number of steps may vary, but the general steps do not. For this book, we are going to use the following steps to troubleshoot our defects. If you are familiar with the software development life cycle, a concept that we're going to explore in *Chapter 9*, this process may seem vaguely familiar.

A major pitfall you will see often as a PLC programmer is faulty hardware mimicking software defects. For mature systems, faulty hardware can often appear to be defects in the software. As such, if the system has been deployed for a while and there have been no issues or software updates, faulty hardware can oftentimes be confused with undiscovered bugs. However, regardless of whether or not the issue is related to software or hardware, the following steps can be used:

1. **Reproduce the problem**: The first step in troubleshooting a bug is reliably reproducing a bug. Before you can start troubleshooting the problem, you need to be able to trigger the defect on command. As a PLC programmer, this can often be difficult due to the hardware components. Often, a full machine setup is necessary to fully reproduce the bug. All things considered, it is of vital importance that you can reproduce the bug at will before proceeding. If you cannot reproduce the problem, the ultimate patch that you will create may not work as intended.

2. **Isolate the problem**: Assuming that the defect stems from the software, the next step is to isolate the problem. Isolating the problem in this context means figuring out the code that is causing the issue. This can be done using a variety of methods, such as putting in a `trace` statement (a statement that provides readable output) or using a debugger. Depending on your experience with the codebase and the defect, this can be a daunting task. This phase can easily take the most time in the debugging process as you may have to sort through the whole codebase to find the offending code, and the problem may be spread across multiple areas.

3. **Analyze the problem**: Once you find the problem, the next step is that you need to understand the problem. Much like the isolation step, this can also be a very daunting task, especially if what you're working on is a patch for an existing system. Often, you will be bouncing between the past steps to fully analyze and understand the issue. There are many ways to analyze the problem. Generally, this is what I refer to as playtime. Usually, a developer will have to play with the code by passing different values, triggering different conditions, and so on to fully understand the behavior of the offending code.

4. **Fix the issue**: Once you understand the problem and what is causing the issue, you can proceed to fix the defect. Developing a patch for a system requires an in-depth understanding of the system and how it is intended to work. In short, this is where you're going to implement your solution. If the problem is hardware-related, it is very tempting to try to compensate for the issue with a software patch. This is one of the worst things you can do as a PLC programmer. The program should assume it is working with properly working hardware. If you modify the source code to compensate for malfunctioning hardware, your program is now only compatible with that particular component. This means that once the faulty hardware is replaced, the source code will have to be restored to its original state, which can be challenging depending on how much code has been changed. Malfunctioning parts can be left in machines for many years; as such, when parts are replaced, bugs will be reintroduced into the system. In short, this type of modification should be avoided at all costs; it is never okay to compensate for

malfunctioning hardware such as broken encoders, poor motor or motor drives, sensors, and so on by changing the source code.

5. **Validate your solution**: The final phase in debugging your program is testing your fix. To ensure your solution fixed the problem, you need to verify that it works. This phase will require a working knowledge of the way the machine is intended to work, as well as the issue you were trying to fix and how to trigger the issue. If your patch does not work as intended, you should start over from *step 1*. If you have pre-written test cases, you should use those to test the patch. However, depending on where you're working and what you're working on, you may not have that luxury; as such, you should test it the best you can. When testing your patch, it is important to ensure that the patch didn't accidentally break something else. Therefore, it is important to thoroughly test not only the feature you're patching but also the whole system. If the machine is deployed or is functioning, it is a good idea to run a full cycle on the system. This means you should run at least one test production run on the system.

These steps will only provide a roadmap for troubleshooting a problem. The true trick in debugging is to find the bug. Depending on your experience with the codebase and the complexity of the codebase, this can be challenging. What we have covered thus far are merely theoretical concepts such as debugging steps, the difference between testing and debugging, and so on. On their own, these won't get you very far in actually routing out bugs. As such, it is important to understand some techniques to help find and eliminate bugs.

# Understanding debugging tools and techniques

There are many different tools and techniques that can be used to debug a program. As was discussed at the beginning of the chapter, some techniques are more sophisticated than others. It doesn't matter which technique you use as long as you debug the software and it works as intended. As such, the following section will explore some ways to track down problems in your code.

## Print debugging

The easiest way to debug a program is with `print` statements. Print debugging is used to isolate problems; in other words, this technique will help you find the offending code. The *IEC 61131-3* doesn't support a command that will output to a console or screen the same way languages such as Java or C++ do. However, this technique can still be used in PLC programming, and in some regards, it is a little easier to use. To demonstrate the use of print debugging, we are going to create a simple program that toggles a variable to TRUE when the ratio of two numbers *is not* less than 1 and FALSE when it *is* less than 1.

Since no command will print to the screen, to display a message in CODESYS or any other PLC environment, you create a string variable and assign your message to that string. To demonstrate this concept, let's look at an example. Suppose that we have a simple division program. The program itself has three variables, and the main logic consists of two variables being divided and the quotient being assigned to the third. If the third is less than 1, it will change a Boolean to TRUE. However, when we run the program, it crashes, and we're not sure why. Suppose our variables look like the following:

```
PROGRAM PLC_PRG
VAR
    division    : REAL;
    dividend    : REAL := 4;
    divisor     : REAL := 2;
    notLessThan1 : BOOL;
END_VAR
```

As can be seen, there are three variables related to the mathematical equations. The other variable, `notLessThan1`, is a Boolean variable that will turn TRUE when the quotient of the dividend and divisor is less than 1.

This is the main logic for the unstable program:

```
division := dividend / divisor;
IF division < 1 THEN
    notLessThan1 := TRUE;
ELSE
    notLessThan1 := FALSE;
END_IF
```

In short, the goal of this program is to turn the `notLessThan1` variable into a TRUE state when the `division` value is greater than 1. If the `division` variable is set to a value less than 1, the `notLessThan1` variable will be FALSE. If the program is run in its current state, we will get the following output:

| division | REAL | 2 |
| dividend | REAL | 4 |
| divisor | REAL | 2 |
| notLessThan1 | BOOL | FALSE |

Figure 3.1 – LessThan1 program output

As can be seen, `notLessThan1` is set to FALSE. This is not the expected output for the program. The expected output should be TRUE. As such, there is a defect in the software. For a problem such as this, print debugging can be a good technique.

To start the print debugging process, the first thing we need to do to use the technique is to create a string variable. So, we'll want to modify our variable list, as in the following code snippet.

This is the variable list with the print debugging variable:

```
PROGRAM PLC_PRG
VAR
     debugMsg    : STRING(20);
     division    : REAL;
     dividend    : REAL := 4;
     divisor     : REAL := 2;
     notLessThan1   : BOOL;
END_VAR
```

The size you make the debugging string will vary depending on the information you want to display. Generally, you don't need a lot of information; you just need enough to help you navigate the flow of the program. As such, 20 characters is usually overkill; however, it is nice to have the characters there if you need them. In all, you can make the string length any size you want. It is also a good idea to give the variable a name that is indicative of its purpose as a debugging variable and to put a reminder comment next to it that reminds both you and other developers to remove the variable when the debugging is done. You don't want to leave unused variables or code in your program as that'll become code rot, and it will do nothing other than clutter up your program and make it hard to maintain. So, it is generally a good habit to remove debugging variables as soon as you're done with them.

The next step in print debugging is setting up your `print` statements, or in the case of PLC programming, assigning your message to the `debug` string variable. This can be tricky, and a good technique is to place the print message or assignments at the top of and the bottom of the file or files that you think might be causing the issues. After you run the program, you want to start moving the messages around to different locations to see whether you can close in on the defect.

For our program, putting a message at the beginning and end of the program will let us know where our code is causing the issue or whether we have a more complex problem, such as a corrupted file. For this example, modify the code to match the following snippet:

```
debugMsg := 'start';
division := dividend / divisor;
IF division < 1 THEN
     notLessThan1 := TRUE;
ELSE
     notLessThan1 := FALSE;
END_IF
debugMsg := 'end';
```

This code has the debugging statement at the top and bottom of the program. It wraps our core logic around the two messages. If there is a fatal problem that is not related to our code, the `debugMsg` variable won't say anything, and chances are you won't be able to press the *play* button or run the program at all. If there is a problem with our logic and the program starts, the `debugMsg` variable will say start. Finally, if there are no problems and the program executes without error, `debugMsg` will say end.

The following is the print debugging message:

| Expression | Type | Value |
|---|---|---|
| debugMsg | STRING(20) | 'end' |
| division | REAL | 2 |
| dividend | REAL | 4 |
| divisor | REAL | 2 |
| notLessThan1 | BOOL | FALSE |

Figure 3.2 – Print debugging output

The `debugMsg` variable is set to end. This means that our code is executing; as such, we can deduce we have an issue with our logic.

The next step would be to move the end message to another place in the code. This is where strategy comes in. We know that our program is executing, so we need to move our end message higher up in the source code. This is where a little visual code analysis will go a long way. If you study the code, you will notice that the `notLessThan1` variable's state is changed in only one place—in the `if` statement. There are also two branches in that conditional statement—the main `if` statement and an `else` statement. For situations such as these, it is a good idea to put an output statement in both conditions to see the path the code is taking. As such, you should modify your code to match the following code:

```
debugMsg := 'start';
division := dividend / divisor;
IF division < 1 THEN
    notLessThan1 := TRUE;
    debugMsg := 'in main if';
ELSE
    notLessThan1 := FALSE;
    debugMsg := 'in else';
END_IF
```

The code will change the `debugMsg` variable to `in main if` when the `division` variable is less than 1 and `in else` when the `division` variable is greater than 1. When this code is run, you should get the following output:

| Expression | Type | Value |
|---|---|---|
| debugMsg | STRING(20) | 'in else' |
| division | REAL | 2 |
| dividend | REAL | 4 |
| divisor | REAL | 2 |
| notLessThan1 | BOOL | FALSE |

Figure 3.3 – debugMsg output

For this run, the `debugMsg` variable came out as `'in else'`. This means that the divisor is not getting set properly. Since there is only one place in the code where it is getting set, we can assume that we have values swapped in our variables. If you look at the variable list, you should see that by swapping the divisor and dividend, we'll get a value that is less than 1. After swapping the value, the output will be as follows:

| Expression | Type | Value |
|---|---|---|
| debugMsg | STRING(20) | 'in main if' |
| division | REAL | 0.5 |
| dividend | REAL | 2 |
| divisor | REAL | 4 |
| notLessThan1 | BOOL | TRUE |

Figure 3.4 – Corrected variable assignments

In this iteration of the program, you can see that the `debugMsg` variable is set to `'in main if'`, and the `notLessThan1` variable is set to TRUE. In other words, the program is working as intended.

Now, it is important to understand that this was a demonstration of print debugging. The true value of the `debugMsg` variable was shown in all the screenshots. Though the values are shown, when working with a non-trivial program or logic, it is not so straightforward. Print debugging is a very valuable technique to understand. In short, no matter what you're working on, you should know how to print debug. It is strongly recommended that you put together some trivial programs and debug them with the technique, as it is a vital skill to master.

## The CODESYS debugger tool

Now that a basic understanding of debugging and print debugging has been established, we can focus on the built-in debugger tool that CODESYS has to offer. Print debugging can be a very tedious process. Though print debugging is a very powerful way to find and remove errors in a program, it is sometimes not the best approach to debugging a program. When print debugging isn't an efficient approach, it is often necessary to use a debugger.

A debugger is a program that allows you to debug other programs. It should be thought of as a debugger tool for troubleshooting software. Most modern IDEs have built-in debugging tools, and CODESYS is no different. So, when you experience an issue that is too complex to troubleshoot with print debugging, it is vital to use this tool.

The first thing that you need to understand when using a debugger is breakpoints. Breakpoints are the backbone of most debugging tools. Without proper knowledge of how to use breakpoints, you won't be able to properly use a debugging tool. Breakpoints are pauses in your program. Essentially, you use a breakpoint to halt the execution of the program at a certain line without terminating the program.

Setting up a breakpoint in CODESYS is very easy. To create a new breakpoint, all you have to do is log in to (but not run) the application and right-click on the line you want your breakpoint to be at, and select **New Breakpoint**. This will open up a window where you create your breakpoint, as shown in *Figure 3.5*:

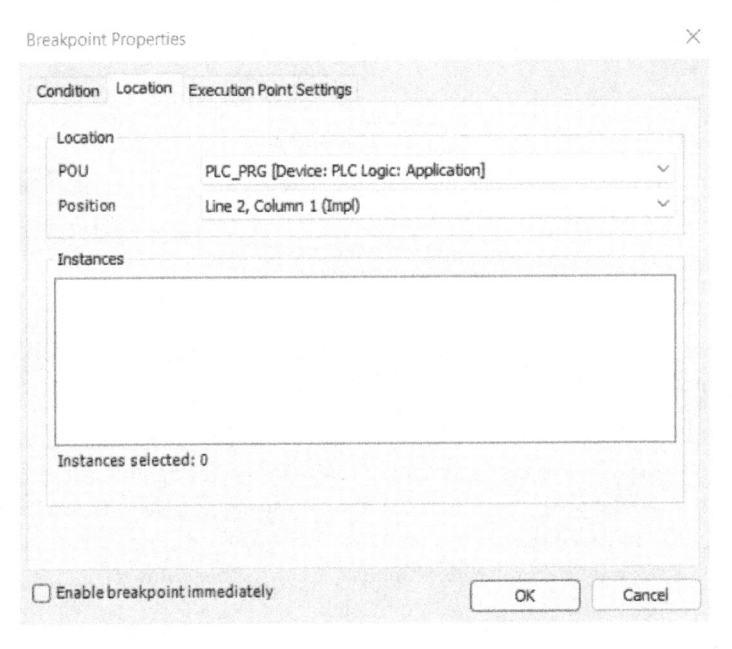

Figure 3.5 – Breakpoint Properties window

The purpose of this window is to set the properties of the breakpoint. The default **Location** tab is where you select where the breakpoint is going to be. The **POU** section is the file in which the breakpoint will be placed, and the **Position** section is the line where the breakpoint will be placed.

### Exploring breakpoints

To demonstrate and explore breakpoints, we will troubleshoot the program that we debugged with the print debugging technique. The goal is to debug the original program configuration so that we can get the same results as we did with the print debugging method.

Reset the variables in the program to their original state:

```
PROGRAM PLC_PRG
VAR
    division    : REAL;
    dividend    : REAL := 4;
    divisor     : REAL := 2;
    notLessThan1   : BOOL;
END_VAR
```

To demonstrate the debugging tool, we have reset the variables from the print debugging project to match the code snippet. Similar to the original error, we will have the dividend and divisor swapped. This is a very important step as you will want the same configuration that was used before so that it can be troubleshot.

For the breakpoint demo, we're also going to use the same logic as we did before. For this demonstration place, the breakpoint is on the code with the red box around it, which is the line that does the computation. So, you'll need to log in, right-click that line, select **New Breakpoint**, and set **Position** to Line 1. When you are finished, your code should look like the following.

```
division   0.5  := dividend   2   / divisor   4   ;

IF division   0.5   < 1 THEN
    notLessThan1 TRUE := TRUE;
ELSE
    notlessThan1 TRUE := FALSE;
END_IF
```

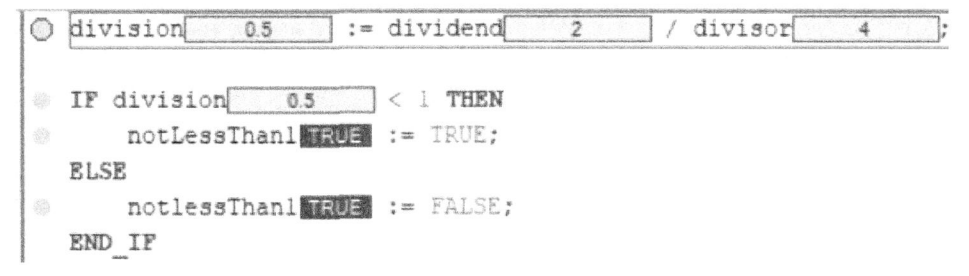

Figure 3.6 – Breakpoint on line 2

Notice the red outline around the second line. This red outline is an inactive breakpoint. Essentially, there is a breakpoint there, but it is turned off. This means that when the program is run, it will run as if there were no breakpoint present. For the breakpoint to pause the program, log in and right-click the line that is highlighted in red. You will see an option called **Toggle Breakpoint**, and depending on whether the breakpoint is enabled, you will either see **Enable Breakpoint** or **Disable Breakpoint**. The functionalities of these options are as follows:

- **Toggle Breakpoint**: This will either enable or disable the breakpoint. Essentially, if the breakpoint is enabled, it will disable the breakpoint, and if the breakpoint is disabled, it will enable it. No matter the state of the breakpoint, this option will always be available.
- **Enable Breakpoint**: This option will only be available when the breakpoint is disabled. When enabled, the breakpoint will act as a pause in the program. When a breakpoint has been enabled, the circle next to the line will be solid red.
- **Disable Breakpoint**: This option will only be available when the breakpoint is enabled. This option will disable the breakpoint, and when the program is run, it will be ignored. If the breakpoint is disabled, the circle next to it will be gray with a red circle around it.

To demonstrate a breakpoint, the first thing we're going to do is add a message variable and a message on *line 2*.

This is what the variable list should look like:

```
PROGRAM PLC_PRG
VAR
    message      : STRING(20);
    division     : REAL;
    dividend     : REAL := 4;
    divisor      : REAL := 2;
    notLessThan1    : BOOL;
END_VAR
```

For this code snippet, all we did was add a `message` variable. Unlike the `debugMsg` variable in the print debugging, this variable will serve as an output so that the debugging tool can be demonstrated as opposed to being used for debugging.

The following is the code to demonstrate the debugging tool:

```
message := 'before breakpoint';
division := dividend / divisor;
message := 'after breakpoint';
IF division < 1 THEN
    notLessThan1 := TRUE;
ELSE
    notLessThan1 := FALSE;
END_IF
```

Understanding debugging tools and techniques    55

This code is similar to the code we used for print debugging. However, to give a basic demonstration of the debugger, the `message` variable signals that the program is before the breakpoint, and a change in text signals that the program has moved past the breakpoint.

Once the code is modified, you will want to log in to the program, add a breakpoint on the second line that contains the division operation, and enable the breakpoint. When complete, your code should resemble the following screenshot:

```
1   message[ before bre ▶ ] := 'before breakpoint';
2 ● division[    0.5    ] := dividend[  2  ] / divisor[  4  ];
3   message[ before bre ▶ ] := 'after breakpoint';
4   IF division[   0.5   ] < 1 THEN
5       notLessThan1 TRUE := TRUE;
6   ELSE
7       notlessThan1 TRUE := FALSE;
8   END_IF
```

Figure 3.7 – Enabled breakpoint

The code in *Figure 3.7* shows an enabled breakpoint. Depending on the state of your code, it may be a red line instead of a yellow one. If your code does not match up exactly, don't worry. As can be seen, the whole line is solid red, as well as the circle next to the line. When you press *play*, you will get the following output for the variables and code:

| Expression | Type | Value | Prepa |
|---|---|---|---|
| ● message | STRING(20) | 'beforebreakpoint' | |
| ● debugMsg | STRING(20) | 'in main if' | |
| ● division | REAL | 0.5 | |
| ● dividend | REAL | 2 | |
| ● divisor | REAL | 4 | |
| ● notLessThan1 | BOOL | TRUE | |

```
1   message[ before bre ▶ ] := 'before breakpoint';
2 ● division[    0.5    ] := dividend[  2  ] / divisor[  4  ];
3   message[ before bre ▶ ] := 'after breakpoint';
4   IF division[   0.5   ] < 1 THEN
5       notLessThan1 TRUE := TRUE;
6   ELSE
7       notlessThan1 TRUE := FALSE;
8   END_IF
```

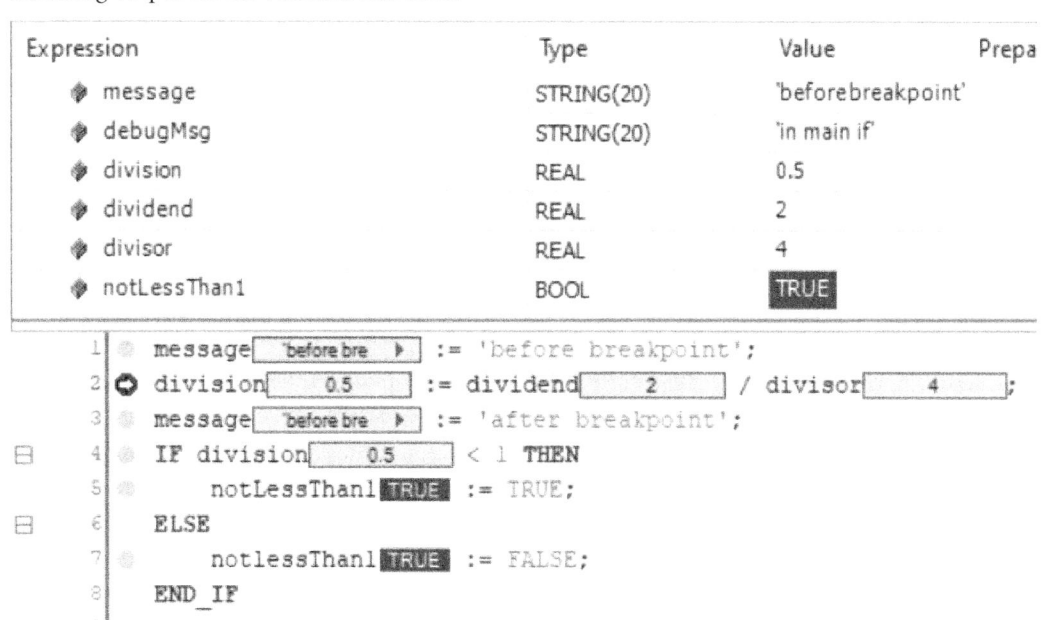

Figure 3.8 – Paused program

The `message` variable shows that the message is set to `before breakpoint`, and the second line is highlighted yellow in the code section. Also, notice that the `division` variable is set to 0. In other words, the computation did not run. It is important to remember that the line with the breakpoint will not run. Ultimately, this means that the program is paused at the second line and, as such, the breakpoint is working.

### Exploring stepping

Inserting breakpoints is only half the process of troubleshooting. As can be deduced from past examples, a breakpoint will virtually stop a program. In terms of troubleshooting, having a program that is effectively off at a certain line won't be that helpful. To remedy this, there is a concept called **stepping**.

Stepping is a tool used to manually control the flow of the program. There are a few stepping types, as follows:

- **Step Over**: The `Step Over` command allows the statement at the breakpoint to be executed and halts the execution again before the next command. A `Step Over` command can be called by pressing *F10* or the following button in CODESYS:

Figure 3.9 – Step Over button

This button will perform the same operation as pressing *F10*. One point to note is that if the next command is outside the current **Program Organizational Unit (POU)**—for example, a custom function block—that line will be executed as if it were one single command and not go into that code.

- **Step Into**: `Step Into` should be used when the next line of code is a POU, such as a subordinate POU, function block instance, function, method, or action. The `Step Over` command will treat the POU call as one command. This means that all the code in the POU will be executed as if it were one command. This is different from the behavior of `Step Into`, which will enter the POU and execute the first statement. After the first statement is executed, the program will then pause again. Essentially, you use `Step Into` when you're calling other POUs and you need to run those commands line by line. This is the **Step Into** button in CODESYS:

Figure 3.10 – Step Into button

The `Step Into` command can be called by pressing the button in *Figure 3.10* or by pressing *F8*. Unless you're trying to step into a specific POU, it is not necessary to use `Step Into`.

- **Step Out**: The `Step Out` command will run the POU code from the breakpoint to the end of the POU. Once the POU code has completed, the execution will return to the calling POU. This command is unique when compared to the other step commands because if this command is run in the main POU, it will execute to the end of the POU and will jump back to the first line of code in the POU. Once at the first line, it will pause there.

This is the **Step Out** button in CODESYS:

Figure 3.11 – Step Out button

The `Step Out` command can be invoked by pressing the button or by pressing *Shift + F10*.

Other debugging commands exist in CODESYS. Each of these commands has its benefits and is used for different things. Since the examples thus far have only contained a single POU, we will focus solely on troubleshooting using the `Step Out` command. For now, you can experiment with the breakpoint program and `Step Over` command.

No matter how you're debugging, you will usually have to manually change values. As you debug, you will want to supply the program with different inputs to explore the outputs, similar to the way we've been doing it. When manually changing values, you can either write a value as we've been doing or you can force a value.

## Forcing variables

Thus far in the book, we have only written variables in the program. Writing values is a very common way of altering variables during runtime; however, it is not the only way to alter the value of a variable. Another way to alter a variable is known as *forcing*. Forcing is a common way to manually alter values. In terms of troubleshooting, knowing the differences between the two writing techniques is very important as there are differences between the two techniques.

### *Forcing versus writing values*

On the surface, writing and forcing are very similar, but there are differences between the techniques. As you have probably noticed, when you write a variable, the program is able to change it. For example, in the program that we are using to test breakpoints, if we write the `bool` variable to a FALSE state when the dividend is changed to 1, the variable will change to TRUE. However, if you force a variable, the program will not be able to alter it. As such, the best way to summarize the difference between writing and forcing is as follows:

- **Writing**: Manually changing a value that can be altered by the program during runtime
- **Forcing**: Manually changing a value, but the program will not overwrite the value

Forcing a value is much like writing a value. To force a variable, you will set the value in the variable window and right-click the value. However, instead of clicking **Write Value**, you will click **Force Value**. When you correctly force the value, as opposed to writing it, you will see an output like the one shown in the following screenshot:

| Expression | Type | Value |
| --- | --- | --- |
| message | STRING(20) | 'beforebreakpoir |
| debugMsg | STRING(20) | 'in main if' |
| division | REAL | 0.5 |
| dividend | REAL | 2 |
| divisor | REAL | 4 |
| notLessThan1 | BOOL | F TRUE |

Figure 3.12 – Forced Boolean value

As can be seen in *Figure 3.12*, when the `notLessThan1` variable was forced, it not only changed the value to TRUE—it added the red circle with **F** in it. This circle means that the value has been forced.

Similar to the way a forced variable is manually set, to change the value, the variable must be manually unset. It is very important to remember that if you force a variable, the program cannot change that value. The only way to allow the program to execute as normal is to unforce the value. Essentially, the process of unforcing is the same as forcing. All you have to do is right-click in the variable area and select **Unforce Value**. When you perform this operation, the red circle will disappear, which means the value can be altered as appropriate by the program.

It is common to write and force variables when connected to a live machine to troubleshoot problems. With that being said, any time you manually manipulate a variable, you are running the risk of abnormal behavior in the machine. This can be very dangerous and can lead to machine damage, damage to the surrounding area, and—most importantly—injuries or, in extreme cases, deaths.

Generally, forcing values is considered to be more dangerous than writing. A forced value will not be altered by the program once that value is set. As such, if you accidentally set a value for a motor to continuously turn, or disable safety switches, the program cannot react as it normally would. For example, if someone walked into the path of a moving machine, they may get injured. Sometimes there is no way around troubleshooting issues that can be dangerous; however, you must be fully aware of what you are doing and of the people around you when manually manipulating values, especially when forcing values.

### *Forcing versus writing troubleshooting*

Now that a definition of forcing and writing has been established, a logical question is: *When should you use forcing or writing?* The answer to this question depends on what you're trying to accomplish. Generally, this is a good rule to follow:

- **Writing**: Troubleshooting the code's behavior under a certain condition
- **Forcing**: Troubleshooting a block of code under a certain condition to check whether something such as a solenoid is working properly or a motor encoder is working

To expand on this, you write a variable when you want to test how the code behaves under different values. You may want to check whether a certain block of code is going to run under a certain condition, similar to what we've been doing. On the other hand, you will force a variable when testing a block of code. For example, consider the following code snippet:

```
IF division < 1 THEN
    notLessThan1 := TRUE;
ELSE
    notLessThan1 := FALSE;
END_IF
```

This is a dummy program that sets `notLessThan1` to TRUE or FALSE. As we have seen with this code in the past, for the `notLessThan1` variable to be TRUE, `division` will need to be less than 1. So, if we wanted to make sure that `notLessThan1` toggles between TRUE and FALSE, we could set the values that assign the value of `division` to different values and see whether `notLessThan1` toggles. On the other hand, if we wanted to always ensure that `notLessThan1` was either TRUE or FALSE, we would force `notLessThan1` to the necessary value. By forcing, we can keep the `division` variable assigned to a value that is less than 1 or greater than 1, depending on what we're trying to accomplish, and then, we can observe the behavior of the program without worrying about those values changing.

At this point, we have explored all the necessary information to troubleshoot a problem of our own. To practice our troubleshooting skills, let's modify the state machine that we made in the last chapter and add a new case that will result in slowly stopping a motor. State machines are excellent tools to practice debugging skills as it is common to mix up states and confuse logic that assists with the transitions.

## Troubleshooting – a practical example

When working with motors, it is sometimes necessary to incrementally stop a motor. Sometimes this is due to the process, while other times it is due to the motor or component. To demonstrate practical troubleshooting, recreate the state machine from *Chapter 2* with the following modifications, which are the necessary variables to power the new iteration of the state machine:

```
PROGRAM PLC_PRG
VAR
    machineState            : INT := 1;
    motorSpeedCutOff        : INT := 10000;
```

```
            runTime                 : INT := 2;
            setSpeed                : REAL;
            numOfParts              : REAL := 8;
            motorOff                : BOOL;
            exc                     : __SYSTEM.ExceptionCode;
            motorSlowDown           : INT := 100;
            speed                   : INT;
END_VAR
```

These are the variables that we used in *Chapter 2* with the addition of the motorSlowDown and speed variables. The speed variable will be used in a for loop that will be used to incrementally slow down the motor by subtracting speed from it. The motorSlowDown variable will be used as the value that will determine the slowdown rate. In this case, 100 will represent 100 RPMs per loop iteration.

This is the modified state machine from the previous chapter:

```
CASE machineState OF
1:
    //machine off state
    motorOff := TRUE;
2:
    //machine run state
    __TRY
        //set motor speed
        setSpeed := numOfParts / runTime;
        IF setSpeed >= motorSpeedCutOff THEN
            motorOff := TRUE;
        ELSIF setSpeed < motorSpeedCutOff THEN
            motorOff := FALSE;
        END_IF
    __CATCH(exc)
        //throw machine into error state
        machineState := 3;
    __ENDTRY
3:
    //error state
    runTime      := 0;
    machineState := 4;// go into motor slow down
```

```
5:
    //Motor Wind Down
    IF setSpeed <= 500 THEN
         FOR speed := 100 TO 500 BY motorSlowDown DO
              setSpeed := setSpeed - speed;
         END_FOR;
    END_IF;
    machineState := 1;
END_CASE
```

The purpose of this code is to gracefully stop the motor from running. Essentially, this is the same state machine from the previous chapter except with the additional case. However, this code block has a particular problem. According to the customer, the motor is not slowing down.

Now that we have a reported problem, let's start the debugging process. The first thing that we need to do is reproduce the problem. We know that the only time the motor should be turned off is when there is an error of some kind. As such, what we can do is throw a division by 0 error.

The first thing we are going to want to do is run the application as normal. Since we know that the normal step to run the state machine is to set the `machineState` variable to case 2, we are going to write that variable, similar to what is seen in the following screenshot, which shows the motor in a normal state:

| Expression | Type | Value |
|---|---|---|
| machineState | INT | 2 |
| motorSpeedCutOff | INT | 10000 |
| runTime | INT | 2 |
| setSpeed | REAL | 4 |
| numOfParts | REAL | 8 |
| motorOff | BOOL | FALSE |
| exc | EXCEPTIONCODE | RTSEXCPT_NOE... |
| motorSlowDown | INT | 100 |
| speed | INT | 0 |

Figure 3.13 – Normal motor operations

As can be seen in *Figure 3.13*, the motor appears to be on and operating as would be expected. Since the motor will only turn off when an error is thrown, let's set the `runTime` variable to 0.

After setting `runTime` to 0, we should be met with the following error:

| Expression | Type | Value |
|---|---|---|
| machineState | INT | 4 |
| motorSpeedCutOff | INT | 10000 |
| runTime | INT | 0 |
| setSpeed | REAL | 4 |
| numOfParts | REAL | 8 |
| motorOff | BOOL | FALSE |
| exc | EXCEPTIONCODE | RTSEXCPT_FPU_DIVIDEBYZERO |
| motorSlowDown | INT | 100 |
| speed | INT | 0 |

Figure 3.14 – Abnormal behavior in motor

*Figure 3.14* shows that we have an error; however, our motor did not shut off. Upon examining the variable output, we can see that everything looks as if it should work. From here, we can either start with breakpoints or we can use print debugging. This is ultimately a matter of preference; however, for issues such as these, where we're not sure exactly why a case isn't transitioning, print debugging can often provide enough information for the amount of effort.

To carry out the print debugging process, we're going to add a variable called `msg` to the variable list. Essentially, just add the following code somewhere in the variable list:

```
msg: STRING(20);
```

This variable will be a temporary variable that will only be utilized to display which case the program lands inside.

In cases such as this, it is important to follow the flow of the program to get to the point where it should be executed as expected. In this case, we would want to have a `trace` statement in the `catch` block, in the `error` block, and in the new `case` statement, as with the code in the Chapter 3 project on GitHub. The code can be pulled down by following the link in the *Technical requirements* section.

After you pull down the code, notice that all the `msg` variables are set at the top of the statement. We do this so that no other logic will interfere with its execution. We need to display where we're at as soon as we get into the block. After we run the program and set the value to the values in *Figure 3.15*, we should get the same values in the **Value** column:

Troubleshooting – a practical example    63

| Expression | Type | Value |
|---|---|---|
| machineState | INT | 4 |
| motorSpeedCutOff | INT | 10000 |
| runTime | INT | 0 |
| setSpeed | REAL | 40000 |
| numOfParts | REAL | 80000 |
| motorOff | BOOL | TRUE |
| exc | EXCEPTIONCODE | RTSEXCPT_FPU_DIVIDEBYZERO |
| motorSlowDown | INT | 100 |
| speed | INT | 0 |
| msg | STRING(20) | 'in error case' |

Figure 3.15 – Program output

Here, we have something a little strange. We are held up in the error case, so we're in case 3. However, our `machineState` variable is set to 4. This means that we either have our case mislabeled or we have our state set to the wrong case. If we look at the source code for the state machine, we can see that our `motorSlowDown` case is labeled as 5. This would cause the error as the case is trying to go to is case 4. As such, there is no case for it to go to, so it is hung up. We can fix this bug by either changing the case number or the case itself. Since it makes little sense to have case 3 followed by case 5, we will just relabel case 5 as case 4. Upon making the code change and running it, we will be met with the following proper case transition output:

| Expression | Type | Value |
|---|---|---|
| machineState | INT | 1 |
| motorSpeedCutOff | INT | 10000 |
| runTime | INT | 0 |
| setSpeed | REAL | -42000 |
| numOfParts | REAL | 80000 |
| motorOff | BOOL | TRUE |
| exc | EXCEPTIONCODE | RTSEXCPT_FPU_DIVIDEBYZERO |
| motorSlowDown | INT | 100 |
| speed | INT | 4100 |
| msg | STRING(20) | 'in off state' |

Figure 3.16 – Properly transitioning program

In the preceding screenshot, we can see that the motor is in the correct case.

Now, if we look at *Figure 3.16*, we can see something odd. The set speed is still way off. We have a negative number. When the `setSpeed` variable hits 0, it should simply cut off, so we should never have a value less than 0. This means we have a bug.

This bug can be found and remedied simply by looking at the code. If we have a `for` loop, it is going to run for a given number of intervals. For our program, this is not as desired. As soon as our variable is less than or equal to 0, we want the loop to break so that we can move on to the next statement. As such, a more appropriate loop would be a `while` loop. We can modify the code from *Chapter 2*'s *Exploring state machine logic* section to include a fourth case to match the code shown next.

### Case 4 – a while loop

This is the modified case 4 code:

```
//Motor Wind Down
msg := 'in motor case';t
IF setSpeed >= 500 THEN
        WHILE setSpeed >= 0 DO
                setSpeed := setSpeed - speed;
        END_WHILE;
        IF setSpeed <= 0 THEN
            setSpeed := 0;
        END_IF
END_IF;
machineState := 1;
```

You can simply replace the code in case 4 with this code. Essentially, the `while` loop will execute until the `setSpeed` variable—the variable that controls the motor speed—is either 0 or less than 0. Similar to the `for` loop, the `while` loop can also produce a value less than 0; as such, we'll include an `if` statement that will set the variable to 0 when the value is less than or equal to 0. Essentially, this is just a sanity check to ensure the value is set to 0.

To test the code, force the `machineState` value to 4 so that it will not leave the case, and you should be met with the following output:

| Expression | Type | Value |
|---|---|---|
| machineState | INT | 4 |
| motorSpeedCutOff | INT | 10000 |
| runTime | INT | 2 |
| setSpeed | REAL | 0 |
| numOfParts | REAL | 80000 |
| motorOff | BOOL | TRUE |
| exc | EXCEPTIONCODE | RTSEXCPT_NOEXCEPTION |
| motorSlowDown | INT | 100 |
| speed | INT | 4100 |
| msg | STRING(20) | 'in motor case' |

Figure 3.17 – while loop output

As can be seen, when we force the `machineState` variable, the `setSpeed` variable will always be 0. This is what we want. As such, we have fixed multiple bugs using forcing and print debugging, and the state machine is now working.

## Summary

In this chapter, we explored debugging. Debugging is something that is overlooked by developing programmers and is a skill that is often learned by trial and error. The main takeaway from this chapter is that debugging is a skill much like coding and you have to practice it to hone it. As we have seen, much like the way software development is a process, so is debugging.

Many different tools and techniques can be used to help debug programs. However, the greatest tool that a developer has at their disposal is critical thinking. As we practiced with the state machine, you don't always need tools such as debuggers. Though they are widely used tools that you should master, it is oftentimes just as effective to use print debugging and deductive reasoning to troubleshoot code. In short, as we move forward with an in-depth look at variables, it is important to note that you will continue to use debugging.

## Questions

Answer the following questions based on what you've learned in this chapter. Cross-check your answers with those provided at the end of the book, under *Assessments*.

1. Define print debugging.
2. Define interactive debugging.
3. Define the debugging process.
4. What types of bugs were found in the practical example?

## Further reading

Have a look at the following resources to further your knowledge:

- CODESYS testing and debugging documentation: `https://help.codesys.com/api-content/2/codesys/3.5.15.0/en/_cds_struct_test_application/`

# 4
# Complex Variable Declaration — Using Variables to Their Fullest

Variables are the backbone of any program. As you can guess, any program of significant size and functionality will use variables. Thus far, we have used variables extensively. However, in terms of the raw power of variables, what we have explored so far has hardly scratched the surface. For systems such as CODESYS, variables are a very rich concept, and unlike traditional languages such as C++ or Java, there are many tools available to help you easily implement variables.

If you've ever programmed in a traditional programming language, such as C++, C#, Java, or the like, you'll notice many similarities between those languages and the concepts in this chapter. Though we have used variables till now, we have not taken full advantage of all the attributes that are offered by variables. As we transition into object-oriented programming in the coming chapters, it is important to understand how to organize variables and add a level of protection to them so that we do not accidentally alter something. As such, this chapter will cover the following topics:

- Auto declaring variables
- Constants
- Arrays
- Global variable list
- Structs
- Enums
- Persistent variables
- Final project – motor control program

## Technical requirements

As usual, to follow along with this chapter, all you will need is CODESYS installed and working. Similar to past projects, the source code can be found at the following URL: `https://github.com/PacktPublishing/Mastering-PLC-programming/tree/master/Chapter%204`. All code examples presented in this chapter will be presented in the GitHub directory. As usual, for the optimal learning experience, it is best to explore the examples and modify them.

## Auto declaring variables

In the previous chapters, we manually created variables. Manually creating variables is a common way of creating variables, but it is not the only way. CODESYS offers a simple tool for declaring variables called an **Auto Declare** tool. The tool is useful for declaring variables with data types that you may not be familiar with, creating a variable in a different file, and so on. The tool is also handy for assigning actual outputs, comments, and more for the variable.

The purpose of this tool is to help assist you in the creation of variables. However, this tool can easily be more trouble than it is worth. For the most part, it will be easier to declare a regular variable as we have been doing so far – that is, simply using the editor to write the variable's code manually. However, if something is complex, in a different file, or you simply forgot the syntax, the tool can be of great value. In short, use the tool wisely and when necessary.

The Auto Declare tool isn't as visible as many of the other tools that are used to create data structures, such as structs, or variable lists that are used to group variables of different types. With that being said, the tool for auto declaring variables can be opened in two ways. It can be opened by pressing *Shift + F2* or by navigating to **Edit | Auto Declare**. On using either of these methods, you should be met with the **Auto Declare** window. This tool can now be used to automatically declare variables.

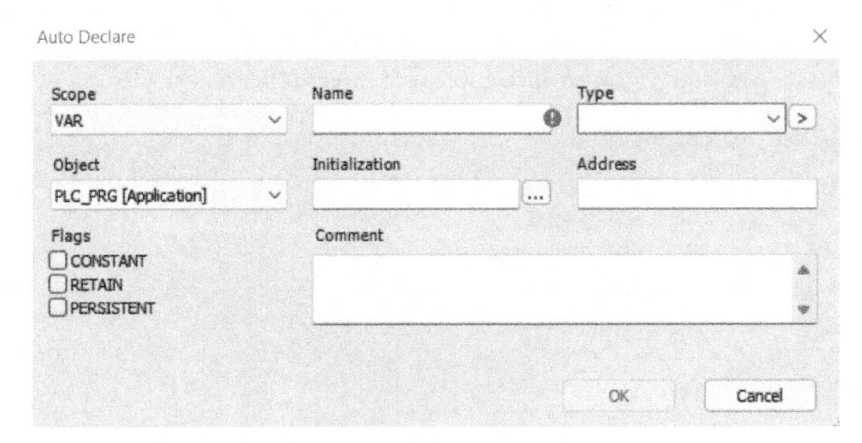

Figure 4.1 – Auto Declare tool

The tool utilizes the fields to format the variable data. To see the functionality of the tool, consider *Figure 4.2*. The data in the fields will create a variable called `test` of the `INT` type.

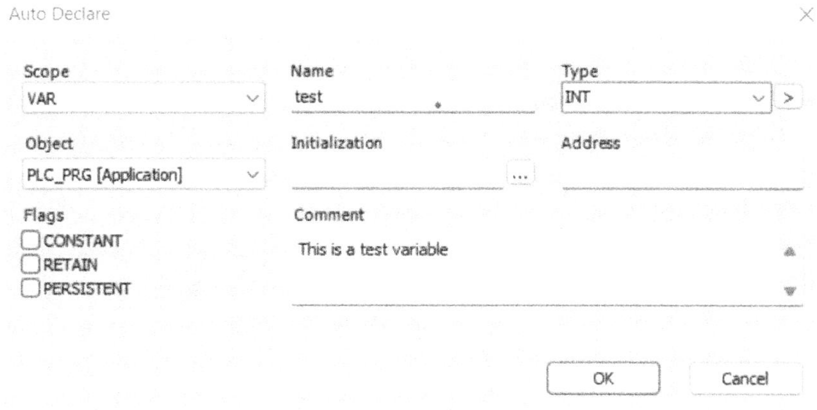

Figure 4.2 – Creating a variable

When the **OK** button is pressed, this code should be created:

```
PROGRAM PLC_PRG
VAR
// This is a test variable
    test: INT;
END_VAR
```

Essentially, this is the final output for the Auto Declare tool. As can be seen, this is the same code we've been generating manually.

In all, the tool is there to make life easier for you as a developer. It is also important to note that this particular tool is unique to CODESYS and not to the IEC 61131-3 standard. As such, it may be different, if it is available at all, in other systems. If you take a look at the checkboxes in the Auto Declare tool, you will notice **CONSTANT**, among other boxes. Constants are very important in programming, especially in object-oriented programming.

## Understanding constants

It is often necessary to have an immutable variable. In other words, it is often necessary to declare a variable, assign it a value, and ensure that the value never changes. Possible examples of required constants are as follows:

- Mathematical constants such as pi
- Motor speeds that never change
- Machine part sizes for calculation (things such as gear ratios)

This list is by no means comprehensive, nor will you always need to declare constants for the preceding bullet items. Whether or not you declare a constant is up to you and the application that you're developing. In short, you will declare a constant when you want to add a level of protection so that a variable's value never changes.

Declaring a constant is very simple. You can either use the Auto Declare tool and simply check the **CONSTANT** box, or you can declare one manually with the following syntax:

```
VAR CONSTANT
    const: INT := 23;
END_VAR
```

This code will declare a variable called `const` with an initial value of 23. Essentially, the syntax is the same as declaring a regular variable with the additional CONSTANT keyword after the VAR keyword.

The CONSTANT block will go in the same tab that regular variables are declared in. For example, the following code is valid:

```
PROGRAM PLC_PRG
VAR
    // This is a test variable
    test: INT;
END_VAR

VAR CONSTANT
    const: INT := 23;
END_VAR
```

As can be seen, the `test` variable is the variable we declared with the Auto Declare tool in the last section, while the block under it is the CONSTANT block.

Let's say you attempt to change the value of the CONSTANT variable to 5, as in the following code:

```
const := 6;
```

Then, you will get a compile error when you try to run the program. *Figure 4.3* shows what will happen when you attempt to mutate the value of a CONSTANT variable.

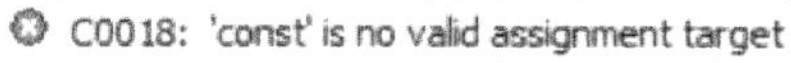

> C0018: 'const' is no valid assignment target

Figure 4.3 – Compile error

The red line means that there is an error that will prevent the code from compiling. If you hover over it with your mouse, you will see that the issue stems from us trying to modify the const variable, which is declared to be a constant.

Constants are one of the most important concepts to understand when it comes to properly implementing variables. No matter what you're programming, a general rule of thumb is to have your variables change as little as possible. As such, it is a good practice to declare whatever you can as a constant. With that being explored, the next concept we'll look at is arrays. Much like constants, arrays are another vital concept that every software engineer must know.

## Investigating arrays

**Arrays** are pivotal to any program. Arrays are data structures that allow you to declare one variable, which can hold many different values of a certain data type. The general syntax for declaring an array is as follows:

```
name: ARRAY[<start_element>..<ending_element>] OF <TYPE>;
```

Therefore, if we wanted to declare an array of Boolean values called TestArray with elements ranging from 1 to 10, we would use the following syntax:

```
TestArray: ARRAY[1..10] OF BOOL;
```

Much like regular variables or constants, arrays can be declared manually or with the Auto Declare tool. The declaration of an array is different from traditional languages such as Java or C++. So, if your background is primarily in a traditional language, you may find the declaration of arrays to be awkward, especially when it comes to a concept known as **multidimensional arrays**. If you come from a traditional computer science background, you will probably be familiar with this term; if not, we will cover the concept in a bit. Whatever your background is, if you find the syntax awkward, you may want to simply use the Auto Declare tool.

## Complex Variable Declaration — Using Variables to Their Fullest

By clicking the arrow to the right of the **Type** drop-down box, an array declaration wizard will appear.

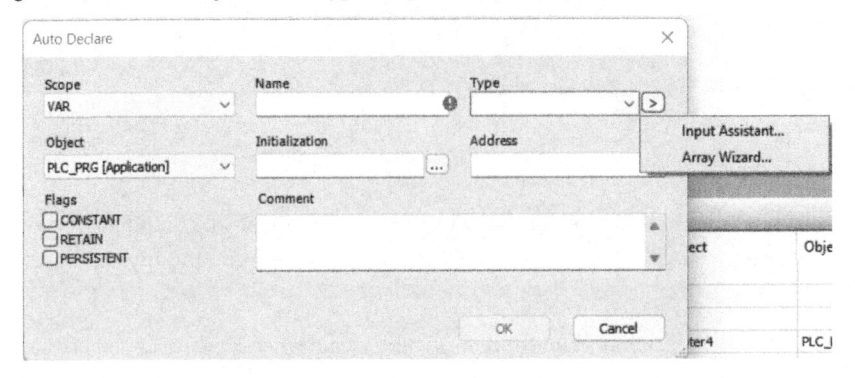

Figure 4.4 – Auto array generation

To declare the array, fill out the fields as you would normally do. Once you click on **Array Wizard...**, you should get the following screen, which is the array generation wizard:

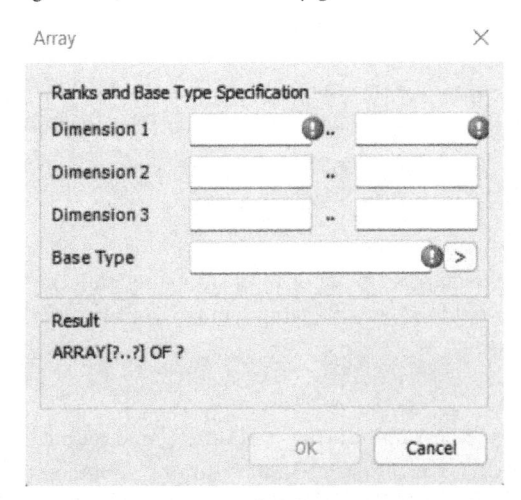

Figure 4.5 – Array wizard

As seen in *Figure 4.5*, the wizard will allow you to create an array with up to three dimensions. For now, we will only create an array with elements ranging from 1 to 10. The following are the necessary fields to generate an array:

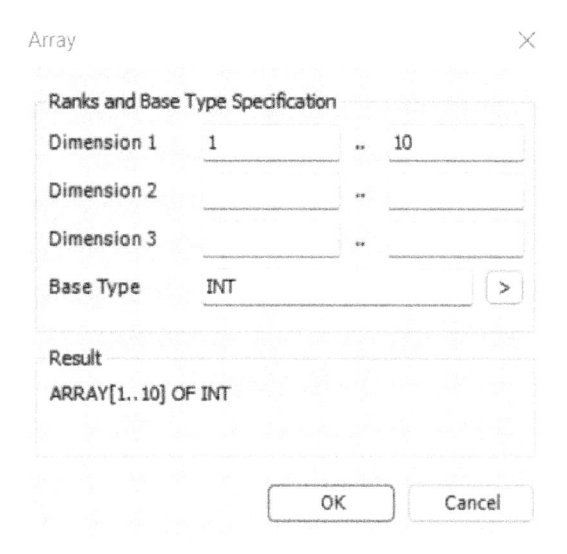

Figure 4.6 – Array declaration via the wizard

With these fields filled out, the wizard should produce a single-dimensional array with elements 1 to 10 of the INT type. After you click **OK**, you may get a pop-up warning. If you do get the popup, simply acknowledge it. Once you click **OK**, you should see the following code generated:

```
VAR
     testArray: ARRAY[1..10] OF INT;
END_VAR
```

This means that the wizard generated an unutilized array. This, in turn, means that the values in the array have no values. In many instances, this is not a wanted outcome. Sometimes, you want to add values to the elements when you declare them. This is known as initializing.

## Initialized arrays

Similar to traditional programming languages, arrays can be initialized at declaration. If you plan on initializing an array, you can either do it manually or with the tool. Generally, if you initialize an array at declaration, it is easier to use the tool, especially if you're going to assign multiple different values. Assigning values is very easy with the tool; all you have to do is click the button with the three dots next to the initialization field. This window will allow you to assign values to the individual elements of the array.

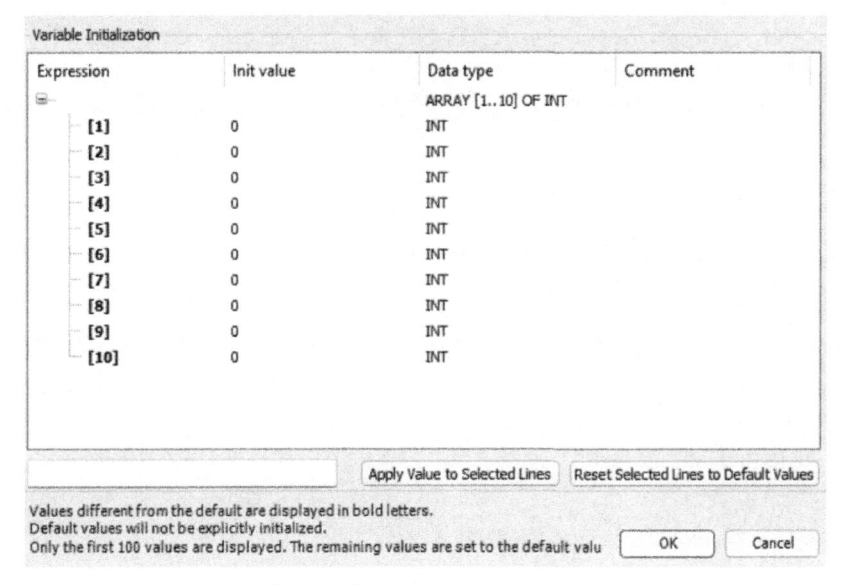

## Complex Variable Declaration — Using Variables to Their Fullest

Figure 4.7 – Initialization window

The values in the **Init value** column are the values that the individual elements will be assigned to when the program is run. To change a value, all you have to do is double-click the value you want to change and enter the new value. Once you have changed all the values, you can click **OK**.

If you wanted to change all the values to 1, you would change all the values in the **Init value** column to 1, as in the following figure:

Figure 4.8 – Array elements set to 1

When you press **OK** on the Auto Declare screen, the following code should be generated:

```
testArray: ARRAY[1..10] OF INT := [10(1)];
```

The code that the tool generates is an abbreviated syntax. Essentially, `10(1)` means that all 10 elements are set to the value of 1. If we were to assign the values 1, 2, 3, and 4 to elements 1-4, the syntax would look like the following:

```
testArray: ARRAY[1..10] OF INT := [1, 2, 3, 4, 6(0)];
```

In this syntax, 1, 2, 3, and 4 are assigned to the first four elements, while the remaining six elements share the value of 0.

## Multidimensional arrays

Until now, we have only used single- or one-dimensional arrays. However, multidimensional arrays are also supported. Multidimensional arrays can best be thought of as arrays embedded in arrays. In terms of automation, a multidimensional array is used for many things. One common application is to use multidimensional arrays to control devices in clusters. For example, you may have three banks of motors with each bank containing five motors. The syntax for declaring a multidimensional array is as follows:

```
multidimensionalArray: ARRAY[<outer>, <inner>)] OF INT;
```

As such, to declare a regular array, we would use the following syntax:

```
multidimensionalArray: ARRAY[1..10, 1..5] OF INT;
```

In this example, there will be an array with 10 elements and each element is an array that contains 5 elements.

If you look at *Figure 4.6*, you can see that the tool is able to generate up to 3D arrays; however, you can manually create as many embedded arrays as you want. Though there is no limit to the number of dimensions you can have, it should be noted that the more dimensions you have, the more convoluted the array will be to work with.

Thus far, we have only added values in the array in some form. We have also looked at declaring and initializing an array. Adding values to the array is only half the effort; however, adding values without consuming the values is useless. For that reason, we now need to explore accessing array elements.

### *Accessing elements*

Accessing elements in a multidimensional array is quite simple. To access an element in a multidimensional array, we would use the following syntax:

```
multidimensionalArray[<outer>, <inner>]
```

Depending on how this syntax is used, we can assign a value to that element or retrieve a value. For example, if we wanted to assign the value 23 to the fourth element in the inner array in the third outer array, we would use the following code:

```
multidimensionalArray[3, 4] := 23;
```

If we run this code and view the variable, we will get the output as seen in the following figure:

| ◆ multidimensionalArray[3, 4] | INT | 23 |

Figure 4.9 – Multidimensional array output

As can be seen in the preceding screenshot, the [3,4] row has a value of 23 in it.

Arrays are a very important concept in any type of programming. As has been said previously, variables are a very rich concept in IEC 61131-3 programming. Hence, the next concept that needs to be explored is the **global variable list** (**GVL**).

## Exploring global variable lists

In traditional programming, global variables are usually considered dangerous and bad practice. However, the philosophy in PLC programming is a little different. Global variables can be dangerous; however, it is common that many processes depend on the same values. As we'll explore later, there are ways to encapsulate and pass data around, but when there are many different code blocks that consume the values, it can often be inefficient to pass the data around.

Global values are often placed in special files called **GVLs**. Variables in a GVL can be accessed and manipulated by any file. Consequently, GVLs can be kind of dangerous to use, and code that utilizes variables from a GVL can be difficult to troubleshoot. Since a variable can be altered by any block of code from any file, it can be very difficult to figure out where a defective value stems from. Also, if a value is forced, it can trigger a response in many processes. This means that it is very important to have an in-depth understanding of the code, as damage or injury can occur if the wrong variable is forced.

### Creating a GVL

GVLs are special files that contain variables. They are special files that must be manually added to a project. The process for adding a GVL is as follows:

1. Right-click **Application** in the PLC project tree.
2. Hover over **Add Object**.
3. Click **Global Variable List**.

Once those steps have been completed, you should see the following window appear:

Figure 4.10 – Global variable list wizard

This is the wizard that will generate the GVL. In the case of the GVL file, all you have to do is input a unique name in the **Name** field and press **Add**. For this example, input `testGVL` into the name field. Once you press **Add**, `testGVL` should be added to the project tree. The code in the file should resemble the following code snippet:

```
{attribute 'qualified_only'}
VAR_GLOBAL
END_VAR
```

This is the code that powers a GVL. Similar to the way we have been declaring variables before, the variable is declared between the `VAR_GLOBAL` and `END_VAR` blocks.

## Demonstrating a GVL

The first step in demonstrating a GVL is to add a variable to the GVL file that we just created. In this case, add a variable called `gvlVar` to the GVL file and give it an `INT` data type. When you are done, your GVL file should look like the following:

```
{attribute 'qualified_only'}
VAR_GLOBAL
     gvlVar : INT;
END_VAR
```

This code creates a global variable. When properly called, this variable can be used in any file in the program. In this case, the `gvlList` variable is referencing the `testGVL` file. As such, if you need to manipulate a variable outside of the `testGVL` list, you would use something akin to the following code snippet:

```
gvlList.gvlVar := 3;
```

If we combined all these snippets and ran the program, we should see an output similar to what is in *Figure 4.11*:

| Expression | Type | Value |
|---|---|---|
| gvlVar | INT | 3 |

Figure 4.11 – GVL output

This example can be used in any file. If you want to try this, you can input the code in the `PLC_PRG` file.

GVLs are a way to group variables but not the only way to do so. Like the C, C++, and C# programming languages, IEC 61131-3 supports structs.

## Understanding structs

Structs are special data structures that allow you to group logically related data into a single data structure. Structs in IEC 61131-3 work very similarly to a struct in a C-like language. They are custom data types that contain variables of different data types in a singular data structure. If you've never programmed in a C-like language, structs may seem a lot like classes, a concept that will be covered later.

### Declaring a struct

Creating a struct is very similar to creating a GVL. Similar to a GVL, you create a struct with the following steps:

1. Right-click **Application**.
2. Hover over **Add Object**.
3. Click **DUT**.

When you finish these steps, you should see a wizard that is very similar to the wizard used to create a GVL, except that it has a few more options. The wizard can be viewed in *Figure 4.12*. For now, the only thing that you will need to do is change DUT in the **Name** field to `motorStruct` and click **Add**. Once you click **Add**, a new file will appear in the application tree under **Application**.

## DUT wizard

The following screenshot is of the **Data Unit Type (DUT)** wizard. As stated earlier, this wizard will be used to create many different types of data structures and will determine whether a struct will inherit from another struct. If you are not familiar with inheritance, it will be covered in *Chapter 6*. For now, do not check the **Extends** box.

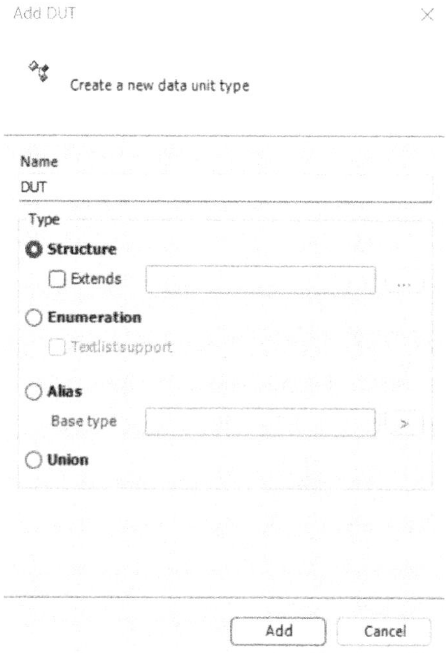

Figure 4.12 – DUT wizard

After you change the name and press **Add**, the struct file that will be added will be similar to the code in the following code snippet:

```
TYPE motorStruct :
STRUCT
END_STRUCT
END_TYPE
```

Similar to the way we added variables to a GVL, we add variables to a struct by simply declaring them.

Consider this example: suppose we have a motor and we want a specific struct to manage the motor's current speed, maximum RPM, and minimum RPM. We could use the following code in the struct:

```
TYPE motorStruct   :
```

```
STRUCT
    motorSpeed  : INT;
    maxRPM      : INT;
    minRPM      : INT;
END_STRUCT
END_TYPE
```

You have to explicitly state whether a file can use a struct or not. To do this, you create a variable in the variables list of the file that needs to manipulate the struct:

```
PROGRAM PLC_PRG
VAR
    motor1 : motorStruct;
END_VAR
```

In this case, `motor1` is a reference to the `motorStruct` data type. In short, `motor1` will have attributes called `motorSpeed`, `maxRPM`, and `minRPM`. Similar to other data types, we can have as many variables as we want. Consider the following code and its output in *Figure 4.13*:

```
PROGRAM PLC_PRG
VAR
    motor1 : motorStruct;
    motor2 : motorStruct;
END_VAR
```

The values for `motor1` and `motor2` can be set with the following logic in the `PLC_PRG` file's logic section:

```
motor1.maxRPM      := 3000;
motor1.minRPM      := 1000;
motor1.motorSpeed  := 1500;

motor2.maxRPM      := 4000;
motor2.minRPM      := 1000;
motor2.motorSpeed  := 2000;
```

These are the outputs when the code is run:

| Expression | Type | Value |
|---|---|---|
| ⊟ ● motor1 | motorStruct | |
|     ● motorSpeed | INT | 1500 |
|     ● maxRPM | INT | 3000 |
|     ● minRPM | INT | 1000 |
| ⊟ ● motor2 | motorStruct | |
|     ● motorSpeed | INT | 2000 |
|     ● maxRPM | INT | 4000 |
|     ● minRPM | INT | 1000 |

Figure 4.13 – Motor struct output

The output shows that we have two different motors and each motor has different values.

Structs can be used anywhere. You can use them in any type of variable file, even a GVL. However, another important data type other than structs that you should know is an **enum**.

## Getting to know enums

Similar to a struct, an enumeration is also a user-defined data type that is composed of comma-separated values. These values are predefined constants. Essentially, when a value in an enumeration is set, it cannot be changed. As such, enumerations are excellent tools for defining threshold limits, motor speeds, temperature values, and more. You declare an enumeration with the same wizard that we used to declare a struct, so be sure to view *Figure 4.12*.

For this example, create an enum name, `motorSpeeds`, using the same DUT wizard as before but by checking **Enumeration** as opposed to **Structure**, and leaving **Textlistsupport** unchecked. Once the code is generated, you can remove the `enum_member` attribute that is auto generated. Once that is done, modify the code to match the following:

```
{attribute 'qualified_only'}
{attribute 'strict'}
TYPE motorSpeeds :
(
    maxSpeed := 2000,
    minSpeed := 500

);
END_TYPE
```

Notice that the values in the enum end with a comma, except for the last entry. This is because values in an enum are separated with a comma; this is how the system knows when one value ends and the next begins. All entries will have a comma at the end except for the last entry – in this case, `minSpeed`.

After completing this, modify the code in the `PLC_PRG` file to match the following:

```
motor1.maxRPM      := motorSpeeds.maxSpeed;
motor1.minRPM      := motorSpeeds.minSpeed;
motor1.motorSpeed  := 1500;
```

When you run this program, you should get the following output, which is the result of setting the motor's speeds with an enum:

| Expression    | Type        | Value |
|---------------|-------------|-------|
| motor1        | motorStruct |       |
| motorSpeed    | INT         | 1500  |
| maxRPM        | INT         | 2000  |
| minRPM        | INT         | 500   |

Figure 4.14 – Motor speeds set with an enum

Notice how the `maxRPM` and `minRPM` fields now reflect the values set in the enum.

Enums are great tools to use for declaring constants as we did with the motor speeds. They are very robust, as you can either assign a value to them or not. If you opt not to include a value, remember that the first value declared will be set to 0 by default, and each subsequent entry will be set as the previous value plus 1. In our case, if we didn't have the values set, `maxSpeed` would be 0 and `minSpeed` would be 1. Now that we have a grasp on enums, we need to shift our attention to persistent variables.

## Exploring persistent variables

Compared to many of the other concepts, such as enums and structs, that have been explored, **persistent values** are most similar to constants. However, there is a difference between constant and persistent variables. A persistent variable will hold its value in case of a cold system reset, a warm reset, or a repeated download. In other words, the value in the variable won't be lost during a hard shutdown, but the value can still be changed during runtime. According to the documentation, applications for persistent variables range from counters to hour meters. Essentially, use persistent variables for values that must be preserved in case of things such as power failures.

Declaring a persistent variable is quite easy. You can manually insert the following code in any file in which you want to create a persistent variable:

```
VAR PERSISTENT
END_VAR
```

You can also declare a persistent variable using the tool and check the **PERSISTENT** box. Once the variable is declared, you can use it as you would any normal variable.

Variables are a very rich concept in PLC programming. Concepts such as arrays, structs, enums, GVLs, constants, persistent variables, and so on are attributes that are usually only touched on but can drastically enhance your code.

## Persistent variable list

Much like a GVL, persistent variables have their own list. They are created in the same way that a GVL is; however, you can only have one. The only difference from creating a GVL is selecting **Persistent Variable List** as opposed to **Global Variable List**. Persistent variable lists will also follow the same logic to access variables. Hence, the only main difference is that a variable in a persistent variable list will be persistent and retain the value.

## Final project – motor control program

To demonstrate all the concepts we have covered so far, let's build a motor control program. The program will simulate five motors. The motors will be in an array and the program will set the speed of the motors based on a persistent variable. To begin, let us create a motor structure:

```
TYPE motorStruct :
STRUCT
    motorStateMsg : STRING[20];
    motorState    : BOOL;
    motorSpeed    : INT;
END_STRUCT
END_TYPE
```

This code will create a structure that will dictate whether the motor is on with a Boolean variable, the motor speed (which will be set with an enum value), and a string that will tell which state the motor is in. After this structure is created, add an enum named `motorSpeeds`. Once you create the enum, add the code to match the snippet:

```
{attribute 'qualified_only'}
{attribute 'strict'}
TYPE motorSpeeds :
(
    maxSpeed := 4000,
    minSpeed := 3000,
    avgSpeed := 2000,
    offSpeed := 0
);
END_TYPE
```

Here, `motorSpeeds` will be declared. Since this is an enum, the values are constant. Next, we need to add the persistent variable list called `motorState`. You can do this by simply adding a GVL and adding the `PERSISTENT RETAIN` command. After you create the list, add the variable to match the code, as shown in the following snippet:

```
{attribute 'qualified_only'}
VAR_GLOBAL PERSISTENT RETAIN
    maxSpeed : BOOL;
    minSpeed : BOOL;
    avgSpeed : BOOL;
END_VAR
```

Now, in the `PLC_PRG` file, we will create an array of `motorStruct` and a counter variable, like so:

```
PROGRAM PLC_PRG
VAR
    motors: ARRAY[1..5] OF motorStruct;
    count : INT;
END_VAR
```

What this code will do is create an array of five motors of the `motorStruct` type. Essentially, each motor will have each of the attributes we declared in the structs. The main logic of the file should match the following:

```
IF motorState.avgSpeed = TRUE THEN
    FOR count := 1 TO 5 DO
        motors[count].motorStateMsg := 'avg speed';
        motors[count].motorState := TRUE;
        motors[count].motorSpeed := motorSpeeds.avgSpeed;
        motorState.minSpeed := FALSE;
        motorState.maxSpeed := FALSE;
    END_FOR
ELSIF motorState.maxSpeed = TRUE THEN
    FOR count := 1 TO 5 DO
motors[count].motorStateMsg := 'max speed';
motors[count].motorState := TRUE;
motors[count].motorSpeed := motorSpeeds.maxSpeed;
motorState.avgSpeed := FALSE;
        motorState.minSpeed := FALSE;
```

```
        END_FOR
    ELSIF motorState.minSpeed = TRUE THEN
        FOR count := 1 TO 5 DO
            motors[count].motorStateMsg := 'min speed';
            motors[count].motorState := TRUE;
            motors[count].motorSpeed := motorSpeeds.minSpeed;
            motorState.avgSpeed := FALSE;
            motorState.maxSpeed := FALSE;
        END_FOR
    END_IF
```

This code will check to see which setting the motors are set to and loop through the motor array to determine which speed from the enum to set the motors to. Set one of the variables to TRUE – for example, set `avgSpeed` in the `motorState` GVL to TRUE – and you should be met with the following output in *Figure 4.15*:

| | | |
|---|---|---|
| motors | ARRAY [1..5] OF mo... | |
| motors[1] | motorStruct | |
| motorStateMsg | STRING(20) | 'avg speed' |
| motorState | BOOL | TRUE |
| motorSpeed | INT | 2000 |
| motors[2] | motorStruct | |
| motors[3] | motorStruct | |
| motors[4] | motorStruct | |
| motors[5] | motorStruct | |

Figure 4.15 – Motor array state

Now, turn whichever variable you set to TRUE back to FALSE, set another variable (such as `minSpeed`) to TRUE, and view your output.

There are many ways to structure a project like this. However, it is common to have a motor structure that serves a purpose, such as speed and state control, while it is also common to have an enum provide constant motor speeds. The persistent variable list was chosen so the state would not be lost in a power outage, and we chose an array for the motor bank to make it easy to loop through and assign a value to each motor. This is by no means the only or best way of accomplishing our goal; however, it is a good way to see each of the concepts working in unison.

## Summary

In this chapter, we have explored many types of variables such as GVLs, enums, constants, structs, and more. In traditional PLC programming, concepts such as these are rarely used. However, as technology advances, these concepts are going to become more ingrained into the development of automation equipment. The concepts we have explored in this chapter will allow you to better organize your code. They will also serve as a way to better encapsulate data.

As we continue our journey in PLC programming, variables will play a vital role. In the next chapter, we are going to explore functions. As such, understanding variables will be pivotal in the exploration of function arguments.

## Questions

Answer the following questions based on what you've learned in this chapter. Cross-check your answers with those provided at the end of the book, under *Assessments*.

1. How many dimensions can be included in an array that is declared using the tool?

    A. 1
    B. 2
    C. 3
    D. 4
    E. 5

2. What is the difference between a GVL and a struct?
3. Describe the difference between an enum and a constant.
4. What is a common error you may have with an array?
5. Declare a 2D array of dimensions 5 and 6.

## Further reading

Have a look at the following resources to further your knowledge:

- CODESYS documentation for persistent variables: `https://help.codesys.com/api-content/2/codesys/3.5.14.0/en/_cds_var_persistent/`
- CODESYS documentation for enums: `https://help.codesys.com/api-content/2/codesys/3.5.14.0/en/_cds_datatype_enum/`
- CODESYS documentation for structs: `https://help.codesys.com/api-content/2/codesys/3.5.14.0/en/_cds_datatype_structure/`

# Part 2 – Modularity and Objects

This part explores code modularity. It starts off with developing and properly utilizing functions. After you have a grasp of function and code modularity, the section will transition into the exotic topic of **object-oriented programming (OOP)**. Due to the novelty of OOP in the IEC 61131-3 standard, this section will introduce you to the cutting edge of PLC programming.

This part includes the following chapters:

- *Chapter 5, Functions — Making Code Modular and Maintainable*
- *Chapter 6, OOP — Reducing, Reusing, and Recycling Code*
- *Chapter 7, OOP — The Power of Objects*

# 5
# Functions — Making Code Modular and Maintainable

As a college-level programming instructor, the first thing I like to teach after teaching the basics, such as loops and flow control, is functions. For many new and non-classically trained programmers, the purpose of functions often makes little sense. For the most part, new and inexperienced software developers see functions as a useless code organization technique that convolutes code. However, I usually counter this logic by stating that programmers should be like sewists. When a sewist creates a quilt, they take individual patches and sew them together to create the quilt. When it's time to create the quilt, there is little concern about creating a patch. The only thing they are worried about is integrating the patch into the quilt as a whole.

For the most part, a programmer should consider themselves to be a sewist of software and the patch of choice should be functions. As we will explore in this chapter, the code should be as modular as possible. A well-written program will be modular. In a well-written program, adding or removing functionality will be as simple as adding or removing function calls. We will explore functions by exploring the following concepts:

- What modular code is and reasons for using it
- Functions
- Return types
- Arguments

To combine everything, we are going to build a temperature conversion function. The temperature converter is a common implementation of a method due to the fact that the code will need to be run many times. Moreover, keeping in line with best practices, we only want the code implemented in one place for easy maintenance.

## Technical requirements

For this chapter, a working installation of CODESYS will be needed. The code for the examples can be found here: `https://github.com/PacktPublishing/Mastering-PLC-programming/tree/master/Chapter%205`.

## What is modular code?

To understand the importance of modular code, consider a car. A car is not created of a single piece of material. Instead, a car is the amalgamation of individual components that, when combined, form a fully working vehicle. For example, a car has an engine, a transmission, brakes, and so on. By creating a car in components, the designers can swap out broken parts, upgrade individual components, use certain parts in other cars, and so on without redesigning the whole car.

A program should follow the same logic. Code needs to be placed into logical containers that organize code. The most basic of these containers is called a function. A function is analogous to a patch and a program is analogous to a quilt. A well-written program is a program that is composed of functions that are stitched together to form a fully working program.

With that in mind, what is the underlying purpose of a function, or module? The answer to that question is simple but will usually take a little practice to fully understand. A function is a block of code that performs a single task. Essentially, a module exists to do something. A good rule of thumb is to have a module be responsible for one thing; for example, the brakes on a car. In terms of PLC programming, a function may be responsible for starting or stopping a machine after a given amount of time, a data logger, or any other component. In short, a function is just like a cog in a machine. Just as cogs are responsible for the smooth execution of a machine, functions are used for the smooth execution of a program.

With all that being said, many of my students often ask me why we should modularize code. Why not just write one big program to handle the task? This is a very logical question and is one that we will explore next.

## Why use modular code?

Modular code is vital for the longevity of a program. Consider the car example again; it would be inefficient and expensive for a person to have to buy a new car every time they need to change their brake pads. A program is not different. A program needs to be structured in such a way that if a defect is found, you or another developer can easily navigate to the block of code where the defect occurs.

Though the organized structure of a program is a very important concept in code modularization, it is not the only reason to modularize code. There are many other reasons why you would want to modularize your code. It could be for the sake of code reusability. In other words, when your function is properly written, you can use it in another project without having to modify it. This comes back to the patch mentality. If you're making a quilt, a quality patch can be used in any type of quilt; it is just a matter of sewing or integrating it into the project.

However, the biggest reason to modularize your code has to do with reducing the amount of redundant code. In programming, when things need to change, you want to change them in one place and one place only. If you have redundant code in multiple places, what'll usually happen is bugs will be introduced whenever a change is made. However, if you modularize your code properly and follow the rules that are laid down in the next section, you'll generally have fewer bugs introduced when modifying your code and reduce the overall complexity of the program.

In all, there are many reasons to modularize your code. With that being said, it is now time to explore how code modularization works in PLC programming. As such, the remainder of this chapter will be the exploration of functions, how they are used, and how they work.

# Exploring functions

The most fundamental module of any program in any programming language is the function. We've thrown around the term quite a bit, but what is a function? In the most basic sense, a function is a named, callable block of code that performs a task. Essentially, a well-written program will be a collection of functions.

With that being said, it is common for entry-level students and non-formally trained professionals to assume that a function is merely allocating functionality to separate files. This, however, is a major fallacy that can lead to a poorly written program that will not last. Though it is very common to split functions into separate files in PLC programming, a separate file *is not* a function. As such, it is important to remember that there is a major difference between code in separate files and functions.

## What goes into a function?

Before we start writing functions, we need to determine what goes into a function and when a function needs to be broken out. Generally, the second question is easier to answer. A good rule of thumb that I use and teach my entry-level students is that if you see two or more lines of code appear in two or more places, it is usually a good idea to figure out what those lines are doing and put them into a function of their own. This ties into the rule of thumb we discussed in the previous sections. If you see the same code appearing multiple times in a program, you'll usually want to abstract that code out into a function and simply call the function. As we explored in the last section, when we do this, the risk of introducing bugs into the program, as well as the overall complexity of the program, is lowered.

Another general rule of thumb is what I like to call the one-sentence rule. I honestly don't remember where I discovered this trick; however, when applied properly, it can drastically improve the quality of your code. The gist of the rule is to summarize your function in a single sentence. If the word *and* appears in the sentence, then the statement after the word *and* should be broken out into a function of its own. For example, consider these two sentences:

- This function turns on the assembly line
- This function turns on the assembly line and hopper

The definition of the first function is correct. This sentence describes a function that does one thing. On the other hand, the second function does way too much. The second function turns on both the assembly line and the hopper. As such, if a modification ever has to be made to either one of those operations, or a situation occurs where only one of the operations needs to be turned on, you're going to risk either introducing defects into the other process or having to create redundant code to control the targeted operation.

Now that we have explored the fundamental concepts of a function, such as why we use them, when we use them, and what should go into a function, we can move on to exploring how to create one. Compared to other languages, creating a function in IEC 61131-3 and, more specifically, CODESYS, is a bit more in-depth. In IEC 61131-3, a function is a **Program Organization Unit** (**POU**). As such, you use a wizard to create one.

## Creating a function

Unlike in many traditional languages, a function in a system such as CODESYS or TwinCat lives in a file of its own. For the most part, the misconception of another file being a function stems from this.

To create a function, the first thing you should do is create a new structured text program, then right-click **Application**, navigate to **Add Object**, and click **POU**. When you do this, you should be met with the following popup:

Exploring functions 93

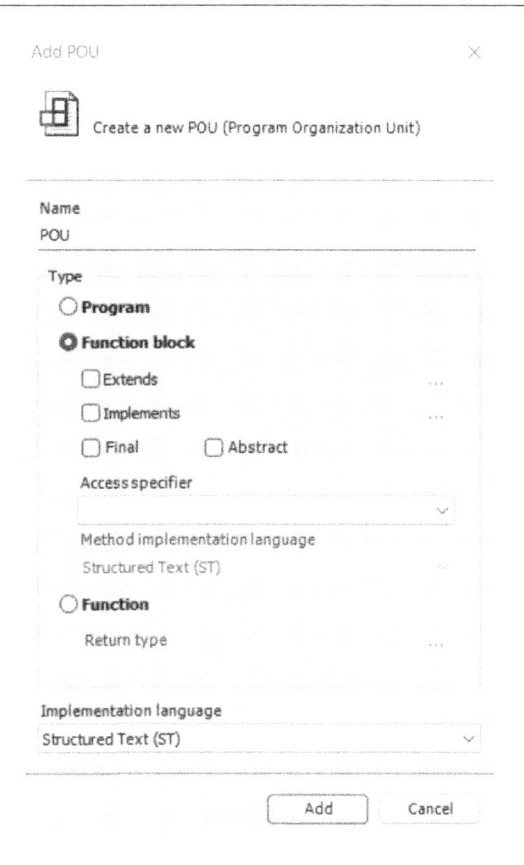

Figure 5.1 – POU wizard

In my opinion, *Figure 5.1* shows one of the most used windows for any object-oriented programmer that is developing with CODESYS. At first glance, we see options called **Function** and **Function block**. There is a major difference between these two options and they should not be confused. A function block is akin to a class in C++, Java, or other traditional programming languages. For this chapter, we are interested in the **Function** option, which will create one — if not the smallest—code modules in IEC 61131-3. The POU that the **Function** option will create is what you would recognize as a method in C++, Java, and so on. Essentially, these are the sewist patches that were alluded to earlier.

For our example, we are going to select the **Function** option and input `Addition` into the **Name** field. For **Return type**, click the button with the three dots, and select **INT**. The return type is very important for functions. The return type specifies the data type of the value that the function will ultimately output. For now, just ensure that your POU creation wizard matches *Figure 5.2*.

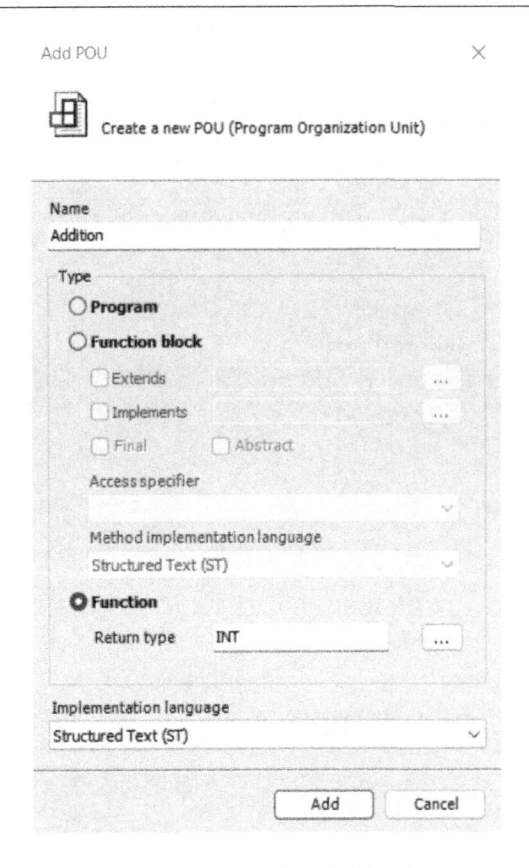

Figure 5.2 – POU setup for Addition function

Notice in the preceding figure that the language can also be selected. This is very important to remember as each function can be written in a programming interface that best suits it. For example, for simple programs, ladder logic might be more appropriate; for functions that require a heavy feedback loop, you can opt to use Function Block Diagram. Regardless, for this example, we're going to keep the language as **Structured Text (ST)**.

The function we are going to make is going to add two hardcoded numbers and return the value. If you're not sure what return values are, we're going to explore that in the *Return type* section. For now, your main focus should be on simply understanding how a function operates.

After filling out the wizard and clicking the **Add** button, a file with the name `Addition` should be generated in the file tree under **Application**. Navigate to the file and open it. You should see the following code:

```
FUNCTION Addition : INT
VAR_INPUT
END_VAR
```

```
VAR
END_VAR
```

Notice that there are two variable sections. The VAR_INPUT section is used for variables that will be used for what are called arguments or parameters. This is a concept that we will explore in the section *Arguments*. The VAR section is used to declare variables that are internal to the function. This means that the variable cannot be accessed from outside the function and cannot be used for arguments. For this example, let's add two variables, a and b, of type INT to the VAR section. We're going to assign the values 3 and 4 to the variables, respectively. In other words, your code should match the following code snippet:

```
FUNCTION Addition : INT
VAR_INPUT
END_VAR
VAR
    a : INT := 3;
    b : INT := 4;
END_VAR
```

In the logic section of the Application file, input the following code:

```
Addition := a + b;
```

Essentially, this line of code means that the output of the function is the sum of variables a and b.

Now, the code in a function block will not execute until it is called. Generally, this is the purpose of the PLC_PRG file. In most well-written programs, this file is equivalent to the main function or entry point for a program. Its main job is to only invoke the functions that are needed to kickstart and run the PLC program. For the most part, you want this file to be as short as possible.

With that in mind, a function is invoked by calling its name and passing in the necessary arguments. You can call a function from another function or any other file that is allowed to call the function. For our example, we're going to call the function from the PLC_PRG file. As such, navigate to that file, open it, and add the following code:

```
PROGRAM PLC_PRG
VAR
    x : INT;
END_VAR
```

This variable will be used to hold the return value from the function. Since our function's return value is of type integer, it is important to declare the variable x as an integer.

This code snippet is how a function is invoked:

```
x := Addition();
```

The code boils down to invoking the `Addition` function and assigning the return value to the `x` variable. When the code is run, you should see an output that is congruent to the screenshot in *Figure 5.3*.

| ♦ x | INT | 7 |

Figure 5.3 – Output from the Addition function

## The PLC_PRG file

A function can be called from any file. Essentially, a function similar to the example is a global function. This ultimately means that you can call a function from any POU, such as another function, function block, method, or any other POU. When it comes to using patterns, it is quite common to have functions invoke other functions.

The most common place to invoke a function is usually in the `PLC_PRG` file. Now, no rule says that you must call a function in the file. However, as we touched on earlier, this file is usually used for invoking blocks of code. In many traditional programming languages, there will usually be the main function or entry point that will start the logic. In the case of CODESYS, this is the `PLC_PRG` file. Usually, this file is used just to call functions and function block methods. Thus far, we have been using the `PLC_PRG` file to do pretty much everything. This is fine for practice, but in a real-world application, this file should be thought of as an orchestra conductor with all the functions and function blocks playing the role of musicians.

What goes into the main function or in the case of CODESYS, the `PLC_PRG` file, is often a difficult concept to grasp. Many programming students and entry-level programmers often want to group way too much logic in the program's entry point. Up until this point, we have used this file for all of our examples. However, in a real-world application, the only logic that should go in this file is the logic needed to coordinate the program's operation. Common elements that go into this file are things such as state machines that control the overall state of the machine and so on.

In all, this section has simply explored calling a function. Thus far, we have only seen return types in action. Return types, in practice, are relatively simple concepts; however, many new programmers are often confused by return types. As such, the next section is going to be dedicated to exploring the concepts.

## Examining return types

Return types can often be very confusing for new programmers. The main hang-up for many of the students that I have taught is that they often have a difficult time understanding what a return type is. As we have seen, a return type is simply a value that a function returns. In very simple terms, the returned value is simply the output of a function.

Each function must be declared with a return type. This return type can be any data type that is supported by IEC 61131-3; for example, the integer data type from the `Addition` function. In all, a function can return exactly one value of the type the function was declared with. So, if you declared a function with a return type of `INT`, you must return an integer similar to what we did with the `Addition` function. As we saw with the `Addition` function, returning a value is as simple as assigning the function name to a statement, as we did in the preceding code snippet for invoking a function.

This is a simplistic definition of return types. If this concept is still unclear, you will understand it as the book progresses, as this will be a concept that will be used from here on out. The key takeaways about return types are as follows:

- A return type is a function's output
- There can be exactly one return type per function
- One function can return exactly one value
- A return type can be any supported data type

Even though a function does return a value, sometimes it shouldn't. Sometimes we simply want a function to terminate before the value can be returned. For this, there is a special command known as `RETURN`, which will terminate a function before it completes its execution cycle.

## The RETURN statement

A function does not always have to return a value. In certain cases, it might be more appropriate to simply terminate a function as opposed to returning a value. For example, an otherwise fatal error that won't return a valid value or something along those lines will usually benefit from using a `RETURN` statement.

Compared to traditional languages such as Java or C++, the `RETURN` statement does not return a value. However, as stated before, it does terminate a function. This means that it is very common to use the `RETURN` statement in some type of control statement such as in an `IF` statement or inside a `CATCH` block. Consider the following scenario. Suppose we have a system where the RPMs of a motor are input as a multiple of 1,000. As such, if the operator wanted to program the machine for 4,000 RPMs, they would enter the value as 4 and the function would multiply the value by 1,000.

To build this function, we would normally use arguments; however, for now, we will hardcode the value for simplicity. The function should return the converted RPM values; however, if an invalid entry is input, such as a negative number, the function will simply terminate. For this example, let's create a new function called `RPMs` with a return value of type `INT`, and once the file is generated, add a variable called `rpmsInput` in the `VAR` section. Once done, your code should look like the following:

```
FUNCTION RPMs : INT
VAR_INPUT
END_VAR
VAR
    rpmsInput : INT := 4;
END_VAR
```

The following code snippet is the function's main logic, which is responsible for converting the user input into the proper RPM value or executing a RETURN command for an invalid value.

This is the logic that represents the RPMs function:

```
IF rpmsInput < 1 THEN
    RETURN;
ELSE
    RPMs := rpmsInput * 1000;
END_IF
```

Essentially, this code will return the converted RPMs as long as rpmsInput is greater than or equal to 1. If the value is less than 1, then the function will simply terminate. To run this code, PLC_PRG will also need to be modified to the following, which will invoke the RPMs function:

```
x := RPMs();
```

When the code is run with all the values as shown, you should see the output in *Figure 5.4*.

| x | INT | 4000 |

Figure 5.4 – Successful RPM conversion

Now, if rpmsInput is changed to -2, you should be met with an output similar to *Figure 5.5*.

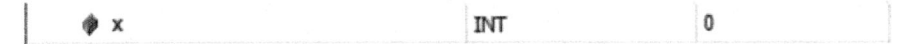

| x | INT | 0 |

Figure 5.5 – Invalid RPM conversion

Essentially, the output in *Figure 5.5* is 0 because the value was not changed as such; it defaulted to 0. As we can see, the RETURN statement resulted in the function terminating before a value was returned.

With all this being said, there are four important takeaways regarding the RETURN statement. They are as follows:

- The RETURN statement will terminate a function usually before a value is returned. This means that the RETURN statement is the last command executed in the function.

- A function can contain many different RETURN statements when they are used in a control statement but only one will execute per function call.

- The RETURN statement is usually wrapped in some type of control statement such as an IF statement or CATCH block.

- Unlike many traditional programming languages, the RETURN statement does not return a value; it simply terminates the function.

As we saw with both the Addition function and the RPMs function, hardcoding values can limit the usefulness and usability of functions. In other words, we need to add some flexibility so that other values can be used. With that being said, it is time to explore function arguments.

## Understanding arguments

Where return types are a function's output, arguments are a function's input(s). Arguments are optional, as we saw with the Addition function, where no inputs were required. However, for many functions, especially for functions that do math like our Addition function, arguments are usually necessary to provide reusability to a function. For example, our Addition function only added two hardcoded values. For the most part, this function is useless unless we want to add 4 and 3 every time the function is called. As such, a better approach would be to modify our function to take values as inputs.

The first step in creating functions with arguments is declaring variables in the VAR_INPUT section of the file that was automatically generated by the POU wizard. For our modified Addition function, we are going to have the function take two inputs, a and b. As such, we're going to modify that section of code to match the following. These are the necessary variables for the function:

```
FUNCTION Addition : INT
VAR_INPUT
    a : INT;
    b : INT;
END_VAR
VAR
END_VAR
```

As can be seen, we simply have two variables labeled a and b. These variables will hold the values that we input into the function. For this example, we are going to keep the same addition logic that we used before, so that logic should match the following:

```
Addition := a + b;
```

The key takeaway here is that no actual values are being assigned to any of the variables in the function. In this case, all values are supplied when the function is called.

Now, when we call the function, we have to supply the values. To demonstrate this, we are going to modify our code in the `PLC_PRG` file. The first thing we are going to do is modify the variable list to match the following. This is the necessary logic for invoking and passing values to the `Addition` function:

```
PROGRAM PLC_PRG
VAR
    x : INT;
    input1 : INT;
    input2 : INT;
END_VAR
```

In this case, we still have our x variable, which will hold the sum of the `input1` and `input2` variables, which will serve as our function inputs. To run the function, we use the following code:

```
x := Addition(input1, input2);
```

Notice that we have `input1` and `input2` separated by a comma in the parentheses. This is how we supply the function with the arguments. In this case, `input1` will be assigned to variable a in the `Addition` function and `input2` will be assigned to variable b. In short, for this way of passing in variables, the first argument goes to the first variable declared in the `VAR_INPUT` block.

Though this is a very common technique for passing arguments to a function, it is not the only way. Depending on what you're trying to accomplish, it is sometimes better to explicitly state which variables get which value. This technique is known as named parameters.

## Named parameters

In computer science, the concept of named parameters allows developers to explicitly state which variable gets which value. As such, by using named parameters, we are not bound to the traditional one-to-one approach of argument assignment that we have explored thus far. Named parameters allow us to assign the value to a specific variable by assigning it in the argument list. To demonstrate this, create a function named `Subtraction` with a return type of `INT`. Once the file is generated, add the variables a and b to the `VAR_INPUT` list as in the following code block:

```
FUNCTION Subtraction : INT
VAR_INPUT
    a : INT;
    b : INT;
END_VAR
```

```
VAR
END_VAR
```

Similar to the `Addition` function, the logic for the `Subtraction` function is as follows:

```
Subtraction := a - b;
```

Now that we have the functions set up, we can work on invoking the code. However, there are two different ways of passing arguments to the `Subtraction` function:

- **Variables for function call**: These variables will hold the values of the different subtraction calls:

    ```
    PROGRAM PLC_PRG
    VAR
          diff1 : INT;
          diff2 : INT;
    END_VAR
    ```

- **Logic to invoke the function**: The following code snippet represents two different ways to pass arguments to the `Subtraction` function:

    ```
    diff1 := Subtraction(3, 2);
    diff2 := Subtraction(b:=3, a:=2);
    ```

    When this code is run, you will get an output similar to *Figure 5.6*.

| Expression | Type | Value |
|---|---|---|
| diff1 | INT | 1 |
| diff2 | INT | -1 |

Figure 5.6 – Different argument order

As can be seen, we passed in the variables in a different order and got two different sets of results. Essentially, on the first line, we traditionally pass in the variables, and a gets assigned 3 and b gets assigned 2. As such, the difference is 1. On the other hand, in the second line, we declare the value of the b variable first and then the a variable with the following line:

```
diff2 := Subtraction(b:=3, a:=2);
```

In this code snippet, we are physically assigning values to the variables. When we use this methodology, we can pass the arguments in any order we want.

Passing arguments, though common, can be very cumbersome. Suppose you have many arguments, but for the most part, the values never change? It would be very inefficient to have to pass those values every time the function is called. Much like in many other languages, there is a solution for this. This special technique of argument assignment is often referred to as default arguments.

## Default arguments

Default arguments, are a way of pre-assigning a set of values to a function's signature, that is, setting variable values in the function itself. Essentially, a default argument is a pre-set value for an argument.

In terms of use cases for named parameters, default arguments and named parameters often walk hand in hand. In short, if you opt to use default arguments, you are, for the most part, forced to use named parameters.

A logical question that I often get asked by my entry-level students is why should we use default arguments? As we have touched on before, if we have arguments that hardly change, it is often better to simply assign them a default value. These default values should not be thought of as constants. Default parameters simply provide an overridable value for the function to use in the case that a value is not explicitly assigned when the function is invoked.

Now that we have a little background information on default parameters, let's take a look at an example of this concept in action. For this example, let's revisit our `RPMs` function. As it stands, we use a hardcoded conversion constant of 1,000. For the most part, this is fine; however, for whatever reason, suppose we need to use a different conversion value. If we decide that value needs to change, we will need to add a traditional argument similar to what we've been using, which means that we will always have to pass that value when we call the function. But a better solution would be to use a default parameter.

The magic of default parameters resides in the `VAR_INPUT` section. You create a default input by simply assigning a value in that block. With that being said, modify the `RPMs` function `VAR_INPUT` to match the following code snippet, which will set the `rpmsConversion` variable to `1000` by default:

```
FUNCTION RPMs : INT
VAR_INPUT
    rpmsInput       : INT;
    rpmsConversion  : INT := 1000;
END_VAR
VAR
END_VAR
```

As can be seen in the code snippet, all we did was simply assign the value `1000` to the variable. This means that this value will not necessarily have to have a value assigned to it when we invoke the function.

For this example, we're going to modify the function's logic to use the variable; as such, modify the function logic to match the following:

```
IF rpmsInput < 1 THEN
    RETURN;
ELSE
```

```
        RPMs := rpmsInput * rpmsConversion;
END_IF
```

To invoke this function, we're going to change the `PLC_PRG` file to match the following code snippet. These are the variables that we will use to demonstrate the `RPMs` function:

```
PROGRAM PLC_PRG
VAR
    convertedRpms : INT;
    motorRpms     : INT := 4;
END_VAR
```

The logic for calling the function is as follows:

```
convertedRpms := RPMs(rpmsInput := motorRpms);
```

When the program is run, the following will be output:

| convertedRpms | INT | 4000 |
| motorRpms | INT | 4 |

Figure 5.7 – RPMs conversion output

Now, as it stands, we will multiply the input by 1,000. However, for whatever reason, if we need to multiply by 100, it is still possible to do so without changing any code. We can simply pass the extra arguments when we call the `RPMs` function in the `PLC_PRG` file. As such, we can simply modify the `RPMs` call in the `PLC_PRG` file in the following manner:

```
convertedRpms := RPMs(rpmsInput := motorRpms, rpmsConversion := 100);
```

When this line of code is run, it will produce output similar to *Figure 5.8*.

| convertedRpms | INT | 400 |
| motorRpms | INT | 4 |

Figure 5.8 – Overridden RPMs conversion

As can be seen, default arguments are a very powerful concept in computer programming. As was demonstrated in the preceding code example, default arguments are excellent to use when you have a value that may only need to change sometimes. In short, by simply giving an argument variable a value, you can free up yourself or other developers from needlessly passing in an extra value. In turn, this means that your code will become more robust and as such will be easier to maintain and modify in the long run. Now that we know about functions, return types, and various forms of passing arguments around, let's build an actual function that can be used in a real-world project.

# Final project – temperature unit converter

Often, as PLC programmers, we are asked for our software to monitor temperatures. These temperatures could be inside the housing of a control panel, the temperature of a part we are fabricating, or many other applications. Now, it is quite common to need to be able to convert between temperature units, especially when the program is deployed to places around the world.

Temperature converters are prime examples of functions as they can be used across multiple projects and the code never has to change. As such, we want this code to be able to be inserted into multiple projects with minimal effort. For our function, we are going to create a state machine to trigger our conversion from one unit to another.

Our program will need to perform the following operations:

1. F -> C
2. F -> K
3. C -> F
4. C -> K
5. K -> F
6. K -> C

Our state machine will have six states. Therefore, create a function called `tempConverter` with a return type of `REAL` and match the code to the following snippets.

These are the variables that will be used for the temperature conversion:

```
FUNCTION tempConverter : REAL
VAR_INPUT
    state : INT;
    temp  : REAL;
END_VAR
VAR
END_VAR
```

Once you have these variables in place in the function file, add the following logic:

```
CASE state OF
1:
    //F -> C
    tempConverter := ((temp - 32) * 5) / 9;
2:
```

```
        //F -> K
        tempConverter := (((temp - 32) * 5 ) / 9) + 273.15;
    3:
        //C -> F
        tempConverter := ((temp * 9) /5) + 32;
    4:
    //C -> K
    tempConverter := temp + 273.15;
    5:
    //K -> F
    tempConverter := (((temp - 273.15) * 9)/5) + 32;
    6:
    // K -> C
    tempConverter := temp - 273.15;
ELSE
    RETURN;
END_CASE;
```

As can be seen, this code is simply a state machine where each conversion is a state. Once you have the function file squared away, modify the `PLC_PRG` file to match the following variables:

```
PROGRAM PLC_PRG
VAR
    convertedTemp    : REAL;
    state            : INT;
    temperature      : REAL;
END_VAR
```

Once you have the variables in place, you can call the conversion function with the following line:

```
convertedTemp := tempConverter(state, temperature);
```

For a simple test, we will convert 100°F to Celsius. To do this, we will input `1` for `state` and `100` for `temperature`. When we write the value, we will get the following output:

| | | | |
|---|---|---|---|
| convertedTemp | | REAL | 37.77778 |
| state | | INT | 1 |
| temperature | | REAL | 100 |

Figure 5.9 – Fahrenheit to Celsius conversion

As we can see, it correctly converted Fahrenheit to Celsius. Now, you can input different values and states to test the code.

At this point, we have had an in-depth look at functions. As the book progresses, we will be using functions and their cousins, known as methods, more and more. This chapter is a foundational chapter for the rest of the book, so, if you're not comfortable with the material, it is recommended that you go back and re-read this chapter and play with some of the examples or create some of your own.

## Summary

In this chapter, we explored functions, return types, arguments, and more. The goal of this chapter was to demonstrate how to modularize code. In short, this chapter was an introduction on how to become a programming sewist. For programs to survive, they must be modular.

The key takeaway is that a well-written program is a modular one. The functions that we have explored are the backbone of modularity. Essentially, what we have covered in this chapter is the foundation that will help you create code that can be easily modified without breaking other sections of the program. In other words, what we have covered in this chapter will help you build more durable code.

In the next chapter, we are going to expand on this concept a bit more and look at object-oriented programming and explore how we can modularize programmatic blueprints.

## Questions

Answer the following questions based on what you've learned in this chapter. Cross-check your answers with those provided at the end of the book, under *Assessments*.

1. What is a function?
2. What are default arguments?
3. What are named parameters?
4. In which order are arguments received in a function?
5. What is a function signature?
6. What goes into a function?
7. What is a return type?
8. Can the return type `INT` be used with a variable of type `REAL`?

## Further reading

Have a look at the following resources to further your knowledge:

- CODESYS documentation for functions: `https://help.codesys.com/api-content/2/codesys/3.5.12.0/en/_cds_obj_function/`

# 6
# Object-Oriented Programming — Reducing, Reusing, and Recycling Code

Almost all modern software applications utilize **object-oriented programming** (OOP) in some fashion. The most popular programming languages, such as Python, Java, C#, and C++ (among many others), are all object-oriented. Even most languages that are not traditionally object-oriented such as Microsoft's F# will usually have object-oriented features. In short, OOP is a staple of modern-day software development.

Until recently, PLC applications were one of the only forms of programming that did not utilize OOP in some fashion. This is mainly due to the nature of PLC applications. For many older applications, it was not necessary to use OOP, as many PLC applications were relatively simple. For the most part, separated files and ladder logic were enough for most applications. However, with the new sophistication of automation systems that are encompassing ever more complexity, a more robust and logical way of organizing, designing, and implementing PLC code is needed.

To accomplish this, the *IEC 61131-3* standard has introduced a unique way of programming PLCs, OOP, into its arsenal. By adopting OOP, you will be able to quickly and easily model machine behavior, port code, and reuse code. OOP is best used for complex projects. Using OOP for simple projects such as traditional PLC applications may be overkill. OOP is best used for complex applications, especially those that involve complex movements such as robotics. To explore OOP in this chapter, we will cover the following topics:

- What is OOP?
- Why use OOP?
- Understanding function blocks
- Getting to know objects

- Getting to know methods
- Getting to know properties
- Understanding the purpose of a getter and setter
- Understanding recursion and the `THIS` keyword
- Final project – creating a unit converter

For the final project, we are going to pull everything together and build a unit conversion function block that will have multiple methods responsible for converting measurements.

## Technical requirements

OOP may seem like an exotic concept, and if you're not familiar with it, you may already be assuming that you need to install extra plugins. However, as long as you have CODESYS installed, you're ready to adopt OOP. The code for this chapter can be found here: `https://github.com/PacktPublishing/Mastering-PLC-programming/tree/master/Chapter%206`.

## What is OOP?

OOP is widely misunderstood in the automation field. It is often confusing, as the support of OOP features varies from one PLC brand to another. However, new and increasingly popular PLCs, such as those produced by Beckhoff or Wago, support a very pure form of OOP. There is also prejudice from many in the field to adopt the usage of OOP due to people not understanding the paradigm and the benefits that it offers. Much of the prejudice and misunderstanding stems from the novelty of OOP in the PLC programming realm.

With all that being said, what is OOP? The first step in understanding OOP is to understand what OOP isn't. Many non-formally trained developers think of OOP as either breaking programs into files, similar to functions, or programming with classes—or, as they are known in *IEC 61131-3*, **function blocks**. However, this is a gross simplification and an inaccurate definition of what OOP is. It can be argued that this is where much of the prejudice comes from, as many automation programmers merely see OOP as a means of breaking a program into different files.

The most effective way of conceptualizing OOP is by considering the engineering process of a car. A car is not built out of a single piece of material. Instead, a car is very modular. A car is made of many components that work in unison. Engineers that are building cars will focus on combining the individual parts to form the larger whole that is the car. The car model that the blueprints describe is then put into production, and the result is a collection of cars that share the same parts and functionality with only minor differences such as color.

OOP works similarly. The concept of OOP revolves around what are known as **objects**. Normally, objects are referred to as things. In terms of the car example, an object is a car. Each object that is instantiated will share similar functionality—for example, our car object will have code that describes how the engine will rev, as well as data such as the miles per gallon the car will consume.

In the same way a car must have a blueprint, so, too, must an object. Objects are derived from digital blueprints that come in the form of function blocks that will be explored later. Similar to the way a car has parts, such as an engine and transmission that are described by a set of blueprints, objects contain code and data fields that do the same thing. Also, similar to how any number of cars can be built from a single blueprint, any number of objects can be instantiated or built from a single function block. In all, the key takeaway is that the backbone of OOP stems from objects that are created from digital blueprints.

## Why use OOP?

OOP is a staple of the modern programming world. As stated before, almost all modern applications utilize the object-oriented paradigm. When designed properly, object-oriented code will provide, but is not limited to, the following benefits:

- **Reusability**: The ability to move code modules from one application to another.
- **Code maintenance**: The ability to quickly fix issues that arise in the software.
- **Reduces redundant code**: Code is in one place and one place only.
- **Reduces the memory usage on the controller**: Since OOP reduces redundant code, it will cut down on the total memory usage.
- **Leverages design**: Function blocks are usually stitched together to form complex system architectures that cannot be accomplished without the use of objects.
- **Increased productivity**: Object-oriented code generally produces an overall better-quality product faster and cheaper than a non-object-oriented software system. When designed properly, the modules can be ported to another project (as will be explored later), bugs will be easier to find and fix, and code is less likely to be accidentally broken.

The listed reasons are high-level reasons as to why an organization would want to adopt OOP. In the next chapter, the programmatic benefits—known as the four pillars of OOP—will be explored in detail. However, from a purely organizational point of view, OOP encourages cleaner code that will have fewer bugs and will be easier to maintain in the future. In short, an object-oriented codebase will have a longer shelf life than a PLC codebase programmed traditionally. When designed properly, many of the objects that make up the object-oriented codebase can be easily ported over to other machines.

In all, if a PLC system does support the *IEC 61131-3* standard for OOP, there is no reason not to adopt it. Even if there is an initial investment to get a team up to date with OOP, it will ultimately be worth it in the long run. Without implementing OOP, an organization will lose out on the ability to easily modify, debug, produce, and ultimately reuse code.

### The four pillars – A preview

The main benefits of OOP stem from what is known as the four pillars. These pillars of OOP will reduce the amount of code used in a program while allowing for safer and more restricted access to other attributes. Typically, the four pillars are generalized as the following:

- Encapsulation
- Abstraction
- Inheritance
- Polymorphism

For this chapter, we are not going to focus on the pillars in practice (we will in the next chapter), but instead, we will use public access specifiers, which will allow us to access function block attributes from anywhere in the program. In other words, we will be using abstraction to a degree.

## Understanding function blocks

The term *function block* can be confusing. Unlike ladder logic, where a function block is merely a pre-built operation that carries out a specific task, when digging into OOP, that idea can be greatly expanded upon. In terms of OOP, function blocks are the code structures that allow developers to blueprint their objects. For readers with knowledge of languages such as C++, C#, Java, or the like, a function block is similar to a class. Generally, PLC programmers that have adopted OOP usually consider function blocks to be the equivalent of classes, and many will even refer to them as *classes*. In *IEC 61131-3*, a function block can hold data and code similar to the way a class in a traditional object-oriented language can. As will be explored later, a function block can also inherit from other function blocks and be inherited from, similar to classes in traditional object-oriented languages.

Classes are the backbone of any modern object-oriented language. *IEC 61131-3* does not support classes per se. In *IEC 61131-3*, the structure that is almost like a class is what is known as a function block. Function blocks are the *IEC 61131-3* standard's version of classes. For the most part, anything you can do with a class in a language such as C++, C#, or Java you can do with a function block in an *IEC 61131-3*-compliant PLC programming system.

As was mentioned before, the purpose of a function block is to provide a digital blueprint for an object. Function blocks are analogous to blueprint paper. That is, function blocks allow us to create digital blueprints. The overall purpose of function blocks is to describe how an object will behave or function.

In PLC programming systems such as CODESYS, declaring a function block is a very simple task. Function blocks are generated via the same wizard we used to create a function. As such, if you are unfamiliar with that process, see *Chapter 5*. To create a function block, right-click on **Application**, then navigate to **Add Object**, select **POU**, check **Function block**, and set the fields to match the following screenshot:

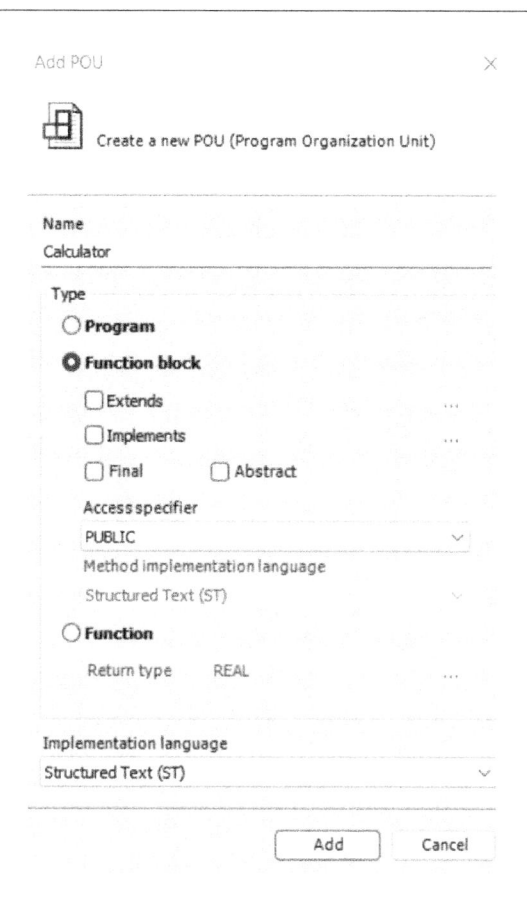

Figure 6.1 – Calculator function block

This function block is going to be a `Calculator` function block. When we are done with this function block, we are going to have four methods that are responsible for adding, subtracting, multiplying, and dividing. After you create the file, you should see the following code in the variable block of the file:

```
FUNCTION_BLOCK PUBLIC Calculator
VAR_INPUT
END_VAR
VAR_OUTPUT
END_VAR
VAR
END_VAR
```

The file that was generated when you created the function block and the variables that it contains are similar to a constructor method in a language such as C++, C#, or Java.

To demonstrate the initialization, first match the code in your function block to the following snippet:

```
FUNCTION_BLOCK PUBLIC Calculator
VAR_INPUT
END_VAR
VAR_OUTPUT
    msg : STRING(20);
END_VAR
VAR
END_VAR
```

Once you have completed this, insert the following code:

```
msg := 'Hello world!';
```

If you have any experience with OOP in a language such as C# or Java, you will know that we now need to create a variable to reference the class—or in this case, the function block—and initialize it.

The reference variable is declared in the variable block of the `PLC_PRG` file, like so:

```
PROGRAM PLC_PRG
VAR
    c1 : Calculator;
END_VAR
```

In this case, the `c1` variable is a reference to the `Calculator` function block. As with any other OOP language, once a reference variable has been declared, we have to initialize it to have it do anything. Initializing a reference variable is very easy, as it follows the snippet:

```
Reference_variable(args);
```

As such, to initialize the `c1` variable, we would use the following code:

```
c1();
```

When we run the code, we should get the following output:

| Expression | Type | Value |
|---|---|---|
| ⊟ ◆ c1 | Calculator | |
|     ◆ msg | STRING(20) | 'Hello world!' |

Figure 6.2 – Calculator function block output

As can be seen in the preceding output, the code in the function block body was run. As it stands now, we have just called a method in a very convoluted way. However, the power here lies in our ability to create many different instances of c1 with their own data. We'll explore this more in the next section. For now, objects are the backbone of OOP, and now that we've seen them in action, let's explore them in more depth.

## Getting to know objects

The root term in OOP stems from objects, which are things. Essentially, the c1 and c2 variables in the PLC_PRG file are objects; they are different instances of the Calculator function block. In other words, the variables are compact copies of the function block. This is a very powerful concept because though the variables reference the same code in the function block, they can hold different data. To demonstrate this, match the following code snippet to your Calculator function block:

```
FUNCTION_BLOCK PUBLIC Calculator
VAR_INPUT
    input : INT;
END_VAR
VAR_OUTPUT
    output : INT;
END_VAR
VAR
END_VAR
```

In this demonstration, we have simple input and output variables. This code will require an input variable to be provided when the object variable is initialized. All the logic will do is assign the input to the output, as in the following snippet:

```
output := input;
```

To provide an argument to the function block, we have to use named parameters. As such, we are going to create object reference variables in the PLC_PRG file. Therefore, match your PLC_PRG file to the following:

```
PROGRAM PLC_PRG
```

```
VAR
    c1 : Calculator;
    c2 : Calculator;
    out1 : INT;
    out2 : INT;
END_VAR
```

The variables are simple enough. We have two references to the `Calculator` class and two `out` variables that will hold the returned values from the function block. Initializing the variables and assigning the outputs is accomplished with the following code:

```
c1(input := 3);
c2(input := 4);
out1 := c1.output;
out2 := c2.output;
```

When this code is run, we get the following output:

| | | |
|---|---|---|
| c1 | Calculator | |
| c2 | Calculator | |
| out1 | INT | 3 |
| out2 | INT | 4 |

Figure 6.3 – Calculator references

The preceding screenshot shows that we can have two different variables reference the same code with different data. In short, the variables are like cars; they both are built using the same blueprints, but at the end of the day, they are still two different cars.

Function blocks can be used for so much more than just holding data. It is usually the norm to have special operations associated with the function block. In other words, these operations are special functions known as methods.

## Getting to know methods

The term *method* has been thrown around a few times thus far in the book. A method is a special type of function that belongs to a function block. Methods are members of function blocks that consist of blocks of code that are executed when called. Without methods, a function block is mostly useless. To conceptualize a method, consider the blueprint example before. If the function block is a car, then the methods are the brakes and engine of said car.

Getting to know methods 115

Unlike the functions we explored in *Chapter 5*, methods are not global. This means that they cannot be called from anywhere like the functions we previously explored. Essentially, the only place that can call these functions is a file with an object to the function block or from somewhere inside the function block, such as another method.

To create a method, you must first have a function block. Since we already have a `Calculator` function block, we are going to add four methods to it that will handle addition, subtraction, multiplication, and division.

Adding a method is relatively simple. To add a method, all you have to do is right-click the `Calculator` function block, hover over **Add Object**, and click **Method…**, similar to what is shown in *Figure 6.4*.

## Adding a method

This is the navigation path to add a method:

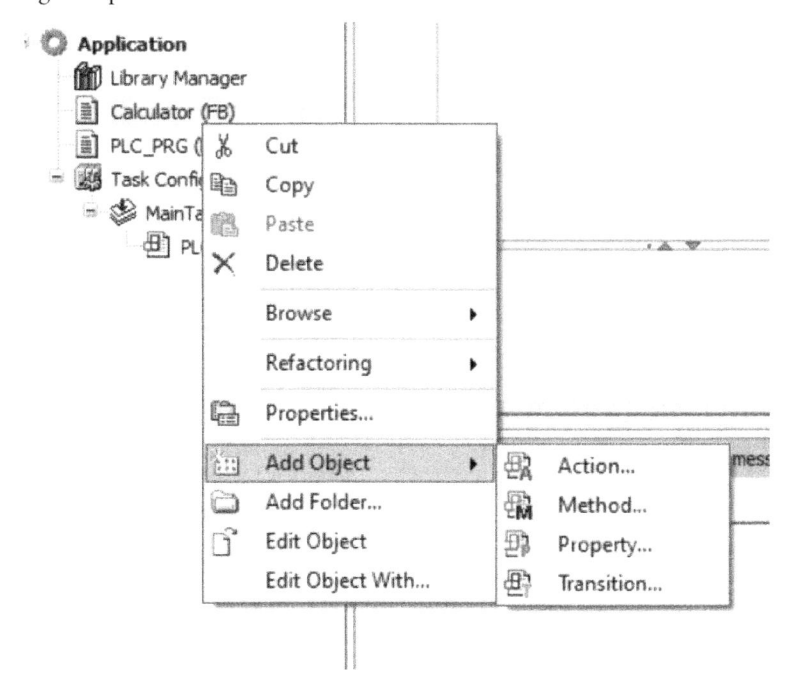

Figure 6.4 – Adding a method

When you follow these steps, you will be met with a wizard that will generate a method similar to what is shown in *Figure 6.5*. For this particular method, we're going to name it `AddNumbers`, set the **Return type** value to **REAL**, and the **Access specifier** value to **PUBLIC**:

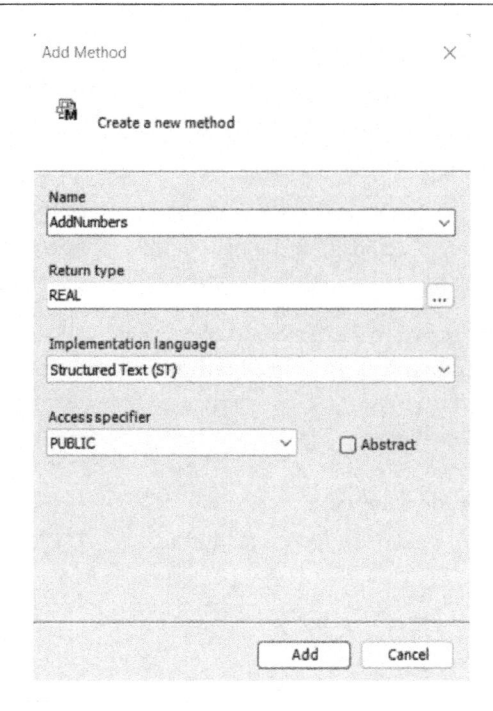

116  Object-Oriented Programming — Reducing, Reusing, and Recycling Code

Figure 6.5 – Method wizard

The return type in methods is the same concept that we explored with functions. The only different option is the **Access specifier** field. This is a special concept that will be explored in *Chapter 7*. For now, just ensure that you set that field to **PUBLIC**.

After you finish creating this method, create methods called `SubNumbers`, `MulNumbers`, and `DivNumbers` with the same parameters as `AddNumbers`. These will be the four functions of the calculator. When you are done, your function block should look like the one shown in *Figure 6.6*. These are the methods when added to the function block:

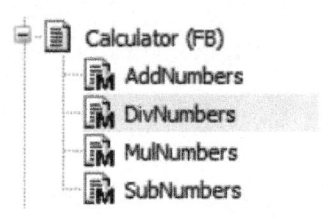

Figure 6.6 – Function block methods

In each of the methods, you should see a variable block, as in the following code snippet:

```
METHOD PUBLIC AddNumbers : REAL
```

```
VAR_INPUT
END_VAR
```

Each of these methods will take two arguments that we will call `val1` and `val2`. Similar to how we used inputs with functions, we will declare these variables in the `VAR_INPUT` block. With that being said, the variable block of each of the methods should resemble the following:

```
VAR_INPUT
    val1 : REAL;
    val2 : REAL;
END_VAR
```

Notice that these variables are REAL. This is so the calculator can have decimal value inputs. The logic in each of the methods should match the following:

- `AddNumbers` method:

    ```
    AddNumbers := val1 + val2;
    ```

- `SubNumbers` method:

    ```
    SubNumbers := val1 - val2;
    ```

- `MulNumbers` method:

    ```
    MulNumbers := val1 * val2;
    ```

- `DivNumbers` method:

    ```
    IF val2 <> 0 THEN
        DivNumbers := val1 / val2;
    ELSE
        DivNumbers := 0;
    END_IF
    ```

Notice that the `DivNumbers` method has a little extra logic compared to the other methods. This is so that if a division-by-zero situation occurs, the code will not crash. Essentially, the method will return 0 if the bottom number is inputted as 0.

Once you have the method code assembled, you can move on to preparing the `PLC_PRG` file. These are the variables that will be needed for the calculator program:

```
PROGRAM PLC_PRG
VAR
```

```
        c   : Calculator;
        sum : REAL;
        dif : REAL;
        pro : REAL;
        rat : REAL;
END_VAR
```

Here, the `c` variable is a reference to the `Calculator` class. The other variables that are of type `REAL` are the holders for the return values.

The following `PLC_PRG` file code will invoke the methods:

```
sum := c.AddNumbers(1, 3);
dif := c.SubNumbers(3, 2);
pro := c.MulNumbers(5, 5);
rat := c.DivNumbers(8, 2);
```

The logic in this `PLC_PRG` file calls the methods. Essentially, we use the same syntax as we did when we were accessing variable values from the constructor. We also pass arguments the same way we did when we were calling functions.

Once you have the `PLC_PRG` file set up and you run the code, you should be met with the following output:

| | | |
|---|---|---|
| c | Calculator | |
| sum | REAL | 4 |
| dif | REAL | 1 |
| pro | REAL | 25 |
| rat | REAL | 4 |

Figure 6.7 – Calculator output

As can be seen, all the methods were correctly called and are computing the correct values.

In OOP, it is generally frowned upon to directly manipulate data. As we will see in the next chapter, most variables—and, for that matter, methods—will become private. The reason we make attributes private is to limit access to them, but sometimes we have to read and write the values. As such, to accomplish this, we use what is called a **property**.

## Getting to know properties

Properties are extensions of the *IEC 61131-3* standard. Properties are special methods that are used to manipulate encapsulated data. When you create a property, you will get two files named `get` and `set`. These are your getter and setter methods. The getter method will be used to read data, while the

Getting to know properties | 119

setter method is used to write data to an attribute. The true value of properties will be explored in the next chapter; however, for now, we're going to explore them to a limited degree.

## Adding a property

Adding a property is a lot like adding a method. You will essentially follow the same flow of clicks that can be found in *Figure 6.4*, with the only exception being that you will select **Property...** instead of **Method...**. When you click **Property...**, you should be met with a wizard similar to the one shown in *Figure 6.8*. The wizard is very similar to the wizard that is used to create methods. To follow along, add the property in *Figure 6.8* to the `Calculator` function block:

Figure 6.8 – Property creation wizard

For this example, we're going to name the property `Prop1` and give it a return type of `REAL`. As usual with this chapter, we will give the property an **Access specifier** type with a value of `PUBLIC`. Once you click the **Add** button, you should see a property file generated with two methods, similar to what is shown in *Figure 6.9*:

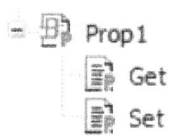

Figure 6.9 – Property-generated getter and setter

The preceding screenshot shows a getter file named `Get` and a setter file named `Set`.

## Understanding the purpose of a getter and setter

Now that we have a property set up, we need to answer a logical question: what are they used for? As stated before, if you have a variable that belongs to a function block, you never want to directly access it. At first glance, this may seem like a convoluted way of doing things; however, the power of a getter method comes in the form of the logic that it contains. Essentially, both getter and setter methods can support logic that can vet how and what's reading or writing the variable.

### Getter method

A **getter method** is used to read a variable in a function block. For the most part, getters usually have very simple logic. In short, many of the getter methods that I have written in the past (regardless of the language or project that I am working on) are usually methods that simply return a class, function block, or variable.

A basic demonstration for a getter method is reading a variable from the function block we have set up. To do this, set the code in the `Get` file to match the following:

```
Prop1 := defaultVal;
```

For the `Calculator` file, just create and set the `defaultVal` value to 33, as in this example:

```
FUNCTION_BLOCK Calculator
VAR_INPUT
END_VAR
VAR_OUTPUT
END_VAR
VAR
    defaultVal : REAL := 33;
END_VAR
```

The final file to modify is the `PLC_PRG` file. We're going to add the following lines of code to the file. When your code is modified from the last example, it should look like the following:

```
PROGRAM PLC_PRG
VAR
    c   : Calculator;
    sum : REAL;
    dif : REAL;
```

```
    pro : REAL;
    rat : REAL;
    def : REAL;
END_VAR
```

The following is the code that will call the methods from the `PLC_PRG` file:

```
sum := c.AddNumbers(1, 3);
dif := c.SubNumbers(3, 2);
pro := c.MulNumbers(5, 5);
rat := c.DivNumbers(8, 2);
//getter call
def := c.Prop1;
```

The last line calls the `Get` method. Calling this method is a little different from calling a normal method. Notice that all we did was use the property name, and CODESYS figured out that we want to use the getter method.

When the code is run, you should see the following output:

| | | |
|---|---|---|
| c | Calculator | |
| sum | REAL | 4 |
| dif | REAL | 1 |
| pro | REAL | 25 |
| rat | REAL | 4 |
| def | REAL | 33 |

Figure 6.10 – Default value output

As you can see in the last row, the `def` variable is set to `33`, which is the value of the `defaultVal` variable in the `Calculator` class. More logic can be added to vet data; however, this is a common implementation of a getter method. The total gist behind getter methods is that these methods are specifically designed to get values from classes. This is especially true for getting variables that are *encapsulated*, which is a concept that'll be explored in the next chapter. Though getters are usually simple, **setter methods** often have more logic and are more complex.

## Setter method

Setter methods write to variables in function blocks. These methods often require more logic to properly vet values that are being written to variables. For the most part, the syntax for a setter method is just the inverse of the syntax for a getter method. To demonstrate this, we are going to set up a very simple setter method that only assigns a value with no complex logic, similar to what we did with the getter method.

For this example, simply add the following line of code to the setter method:

```
defaultVal := Prop1;
```

In this case, we are only going to add one line of code. When you are finished, your `PLC_PRG` file should look like the following:

```
sum := c.AddNumbers(1, 3);
dif := c.SubNumbers(3, 2);
pro := c.MulNumbers(5, 5);
rat := c.DivNumbers(8, 2);
//setter call
c.Prop1 := 89;
//getter call
def := c.Prop1;
```

In this code, the setter code will simply call the setter method and assign the value 89 to the `defaultVal` variable in the `Calculator` function block. When you run the code, you should see the following output:

| c   | Calculator |    |
|-----|------------|----|
| sum | REAL       | 4  |
| dif | REAL       | 1  |
| pro | REAL       | 25 |
| rat | REAL       | 4  |
| def | REAL       | 89 |

Figure 6.11 – Setter method

In this output, the `def` variable is now set to 89. In this case, the setter method assigned the value 89 to `defaultVal` variables, and the getter method is now pulling the value that was assigned to `defaultVal`, which is 89.

As stated earlier in this section, setters and getters are simply methods and can have complex logic to properly vet values being read and written. The past getter and setter examples are simply the barebones of how to use them. It is recommended that you play around with getters and setters. Now that getter and setter have been explored, it is time to move on to another concept, known as **recursion**.

## Understanding recursion and the THIS keyword

Recursion is a looping concept that isn't used much in today's world. However, it is a concept that often pops up in interviews and is something that all software engineers need to understand. In a nutshell, recursion is where a method calls itself. Recursion is a valid concept and is an important concept to know; however, for many applications, some type of loop will be more appropriate.

If you do opt to use recursion, exercise great caution. Recursion is generally considered resource-heavy, and in the automation world, where many PLCs have traditionally limited computing resources compared to full-fledged computers, it can consume precious resources.

Recursion is also somewhat dangerous as it is easy to create what is known as an infinite recursive loop. These loops are recursive loops that continuously call themselves. Many modern compilers do check for this and will usually throw a compile error before the code is run. However, you should be aware of this and need to look out for the issue.

## THIS keyword

To understand recursion, you must first know what the THIS **keyword** is. The CODESYS documentation states that THIS keyword is a pointer of a function block to its own function block instance. In other words, THIS is a keyword for a function block pointer that points to itself. The general syntax for the THIS keyword is as follows:

```
THIS^.method()
```

## Recursion in action

To demonstrate recursion, let's implement a very common recursive function that calculates the factorial of a number. To do this, add a new method to the Calculator function block, name it Factorial, and give it the return type of INT:

```
METHOD Factorial : INT
VAR_INPUT
    x : INT;
END_VAR
```

This logic is the actual implementation of the Factorial function:

```
IF x > 1 THEN
    //method calls itself
    Factorial := THIS^.Factorial;
ELSE
    Factorial := x;
END_IF
```

## 124　Object-Oriented Programming — Reducing, Reusing, and Recycling Code

The line in the `if` statement is what calls the method. The method takes an argument by default, called x. When the method is called, it is supplied an initial value, and that value has 1 subtracted during each iteration and the value is multiplied by the current value of x. For example, if the initial value supplied is 4, the method will compute the following:

$$4 * 3 * 2 * 1 = 24$$

To demonstrate the code in action, modify the `PLC_PRG` file to match the following:

```
PROGRAM PLC_PRG
VAR
        c   : Calculator;
        sum : REAL;
        dif : REAL;
        pro : REAL;
        rat : REAL;
        def : REAL;
        fac : INT;
END_VAR
```

In the case of the variables, all we did was add `fac`:

```
sum := c.AddNumbers(1, 3);
dif := c.SubNumbers(3, 2);
pro := c.MulNumbers(5, 5);
rat := c.DivNumbers(8, 2);
fac := c.Factorial(4);// Frac method
//setter call
c.Prop1 := 89;
//getter call
def := c.Prop1;
```

When the code is run, you should see the following output:

| | | |
|---|---|---|
| c | Calculator | |
| sum | REAL | 4 |
| dif | REAL | 1 |
| pro | REAL | 25 |
| rat | REAL | 4 |
| def | REAL | 89 |
| fac | INT | 24 |

Figure 6.12 – fac output

As can be seen, `fac` is at value `24`, which means the factorial function works!

At this point, you should have a decent understanding of methods. Methods are, without a doubt, the backbone of any well-written object-oriented program. Now that we are starting to get a grasp on OOP, let's move on to our final project.

# Final project – creating a unit converter

In automation programming, it is very common to have to convert between different units of measurement to support clients around the world. This is especially true if you have a single codebase that supports a specific machine that is deployed to many different regions. To accommodate the different units of measurement, it is common to create a function block that can do this. For our final project, we are going to create a very simple function block that can convert the following units:

- Lbs -> kgs and kgs -> lbs
- Feet -> meters and meters -> feet

Depending on what you're working on, there will probably be many more units; however, this is just an example.

The first thing we need to do is create a function block called `UnitConversion` and add two methods called `weight` and `length` to it. Both methods should have a **Return type** value of **REAL** and an **Access specifier** value of **PUBLIC**. When you're done, your function block should look like the following:

UnitConversion (FB)
  length
  weight

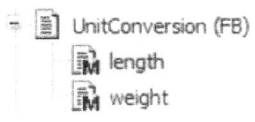

Figure 6.13 – Unit convertor with methods

For this project, we do not have to make any changes to the `UnitConversion` function block. The only changes made will be to the `length` and `weight` files, as follows:

- `length` file variables:

  ```
  METHOD PUBLIC length : REAL
  VAR_INPUT
      lengthInput : REAL;
      metric      : BOOL;
  END_VAR
  ```

- `length` file logic:

  ```
  IF metric = TRUE THEN
      //feet to meters
      length := lengthInput / 3.281;
  ELSE
      //meters to feet
      length := lengthInput * 3.281;
  END_IF
  ```

- `weight` file variables:

  ```
  METHOD PUBLIC weight : REAL
  VAR_INPUT
      weightInput : REAL;
      metric      : BOOL;
  END_VAR
  ```

- `weight` file logic:

  ```
  IF metric = TRUE THEN
      //lb to kg
      weight := weightInput * 0.4536;
  ELSE
      //kg to lb
      weight := weightInput / 0.4536;
  END_IF
  ```

Essentially, both methods will work off a Boolean value. If the value is `true`, it will convert the numerical argument to its metric counterpart; if it is `false`, it will convert to a standard value.

To call these methods, we will add the following lines of code to the PLC_PRG file:

- PLC_PRG variables:

```
PROGRAM PLC_PRG
VAR
//final project
    convert : UnitConversion;
    meters  : REAL;
    feet    : REAL;
    pounds  : REAL;
    grams   : REAL;
END_VAR
```

- PLC_PRG logic:

```
meters := convert.length(32, TRUE);
feet   := convert.length(32, FALSE);
pounds := convert.weight(100, TRUE);
grams  := convert.weight(100, FALSE);
```

When the code is run, you should get the following output:

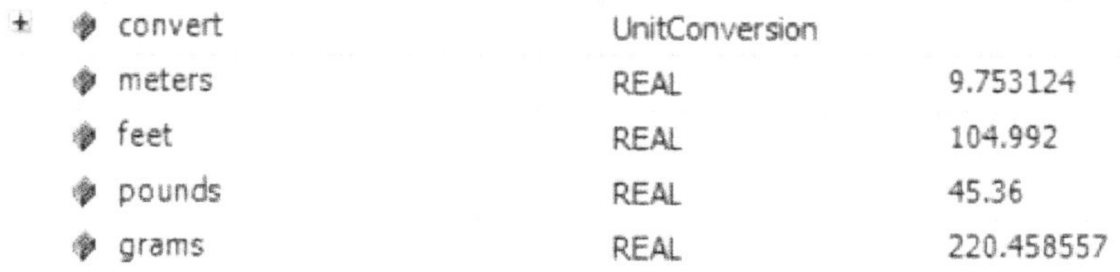

| convert | UnitConversion | |
| meters | REAL | 9.753124 |
| feet | REAL | 104.992 |
| pounds | REAL | 45.36 |
| grams | REAL | 220.458557 |

Figure 6.14 – Unit conversion

This is an example of a real-world function block that can be put into a production machine. However, if you do decide to put a unit converter into your machine, you may want to add some more conversion methods. By now, you should have a basic understanding of function blocks, methods, and recursion.

## Summary

OOP is the backbone of all modern programs. OOP is so ingrained into the IT world that you can't function as a programmer without an in-depth knowledge of the concept. The days of being able to get away with simply programming machines, in a procedural sense, with ladder logic are quickly fading.

This chapter was simply a soft introduction to OOP. OOP is way more than just organizing your code into function blocks, as many concepts govern the paradigm. Now that we have a grasp of function blocks, methods, properties, and recursion, we can learn to leverage them to reduce redundant code, create cleaner code, and apply actual architecture to programs.

## Questions

Answer the following questions based on what you've learned in this chapter. Cross-check your answers with those provided at the end of the book, under *Assessments*.

- What is a function block called in a traditional programming language?
- What is recursion?
- What is the purpose of the THIS keyword?
- What are the two methods that make up a property?
- What is the difference between a getter and a setter?

## Further reading

Have a look at the following resources to further your knowledge:

- CODESYS documentation function blocks:
  https://help.codesys.com/api-content/2/codesys/3.5.12.0/en/_cds_obj_function_block/

- CODESYS documentation calling methods:
  https://help.codesys.com/api-content/2/codesys/3.5.13.0/en/_cds_method_call/

- THIS keyword CODESYS documentation:
  https://help.codesys.com/api-content/2/codesys/3.5.13.0/en/_cds_method_call/

# 7
# OOP — The Power of Objects

For some, including developers that are developing traditional applications with traditional languages, **Object-Oriented Programming (OOP)** is just programming with classes, or in our case, function blocks. OOP can be described as a paradigm. OOP is a way of doing things, not just programming with classes. Though classes or function blocks are the backbone of object-oriented programs, there are many principles and features that govern the paradigm.

Compared to the last few chapters, the concepts in this chapter are going to be much more abstract. If you are a traditional PLC programmer who has only worked with basic structured text and ladder logic, the concepts in this chapter will seem difficult to understand and, at times, counterproductive. However, these concepts will help you produce quality, maintainable code. Whereas the last few chapters have been about organizing code, this chapter is about the concepts that govern the design of properly written software. This chapter will also introduce the important concept of design patterns, which are code designs that can be used to solve common problems that developers will regularly encounter.

To explore the nature and concepts of OOP, the following will be explored:

- Access specifiers
- The pillars of OOP
- Inheritance and composition in practice
- Interfaces
- Patterns

To round out the chapter, we will create a simulated assembly line using the concepts that we will explore.

## Technical requirements

Though we are dealing with complex programming, there are no extra plugins needed to follow along with the examples. As usual, to follow along, you will need a copy of CODEYSYS installed on your computer. The code for the examples can be found at the following GitHub URL: https://github.com/PacktPublishing/Mastering-PLC-programming/tree/master/Chapter%207.

## Understanding access specifiers

Until this point, all of our methods have used a public access specifier. Multiple other access specifiers can be used that allow different levels of access to function block attributes. In terms of what we have been using thus far, the public access specifier means that any file from anywhere in the program can access the attribute as long as it has access to an object variable that references the function block.

Generally, you want as few public attributes as possible. The only reason we have been setting our methods to *public* is for the sole sake of example. In OOP, you want your attributes to be as hidden as possible. In other words, the fewer files that can access a function block, the better off your program will be. Essentially, by properly hiding attributes, your program will be easier to maintain, with less possibility of code corruption. This is a concept that we will explore when we look at abstraction in the following section.

For now, we are going to explore the private access specifier. To do this, assume we are creating a PLC program that is designed to calculate both the area and the perimeter of a square. This type of program is very common in the automation world, as oftentimes multiple parts have to be fabricated based on a single dimension. In practice, creating two public methods can lead to over-complexity issues. For example, if the calculation needs to be executed multiple times in the program due to the size of a part changing during fabrication (or something along those lines), the programmer can forget to call one of the methods or, at the very least, the program will become more bloated with the method calls. In this case, it is better to create three methods: two private methods would be used to perform the calculations, and one public method would be used to provide a level of abstraction and hide complexity.

## Calculation program

This program will have multiple variables across multiple files. As stated before, we will need to create the following files:

- `calculation` (function block)
- `area` (private method with a return type of `REAL`)
- `perimeter` (private method a return type of `REAL`)
- `calculate` (public method with a return type of `REAL`)

When you're done creating the files, you should have the following tree:

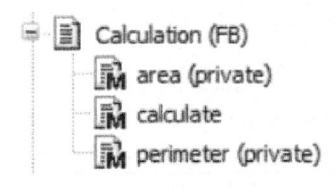

Figure 7.1 – Calculation tree

Note in the preceding figure that all the private methods are marked as such in the tree automatically. The `perimeter` function will have a variable block like the following:

```
METHOD PRIVATE perimeter : REAL
VAR_INPUT
     input : REAL;
END_VAR
```

The `area` function will have a variable block like the following:

```
METHOD PRIVATE area : REAL
VAR_INPUT
     input : REAL;
END_VAR
```

As can be seen, both methods will only take a single input variable. The logic for the two methods is as follows:

- `perimeter` method logic:

    ```
    perimeter := 4 * input;
    ```

- `area` method logic:

    ```
    area := input * input;
    ```

Once you have these methods completed, you will need to set up the `calculate` method. This method is a public method that will call the other two methods. As such, you will set up the variable block like the following:

```
METHOD PUBLIC calculate : REAL
VAR_INPUT
     input : REAL;
END_VAR
```

The logic for this method will be as follows:

```
are := area(input);
per := perimeter(input);
```

As can be seen, the `calculate` method will only serve as a facade to call other methods. There is a design pattern that serves a similar function that will be explored in the *Getting to know design patterns* section of this chapter.

The final file that must be modified is the `Calculation` function block. This code will simply be two variables that will be accessed by the `PLC_PRG` file. The code is as follows:

```
FUNCTION_BLOCK Calculation
VAR_INPUT
END_VAR
VAR_OUTPUT
    per : REAL;
    are : REAL;
END_VAR
VAR
END_VAR
```

The next file we will modify is the `PLC_PRG` file. This file will be responsible for calling the `Calculation` function block and kick-starting the process:

- These are the variables for the `PLC_PRG` file:

    ```
    PROGRAM PLC_PRG
    VAR
        c   : Calculation;
        are : REAL;
        per : REAL;
    END_VAR
    ```

- This is the logic for the `PLC_PRG` file:

    ```
    c.calculate(2);
    are := c.are;
    per := c.per;
    ```

When you have modified all the files and run the code, you should be met with an output that is similar to what is shown in the following figure:

| c   | Calculation |      |
|-----|-------------|------|
| are | REAL        | 4    |
| per | REAL        | 8    |

Figure 7.2 – Program output

When you set up your method, you probably noticed that there are multiple access specifiers. Each one adds a different level of accessibility for the attribute. However, the two that you usually use the most are public and private. Now that we have some background of what access specifiers are and what they do, we're going to move on to the four pillars of OOP.

## Exploring the pillars of OOP

Depending on who you talk to, OOP is governed by four pillars: *encapsulation*, *abstraction*, *inheritance*, and *polymorphism*. Some sources will cite only three pillars due to some developers grouping abstraction and encapsulation as a singular concept. Academia usually teaches that there are four pillars, and it is more common to hear about four pillars as opposed to three. For this book, we will explore the four pillars.

### Encapsulation versus abstraction

In OOP, we want to hide as many of the attributes as possible. We do this so attributes outside of the function block can't accidentally use them and cause issues. This will make the program easier to troubleshoot and maintain in the long run. However, there are some attributes that do have to be used by outside attributes. In this case, we need to provide the bare minimum to the outside function blocks, that is, we need to expose only as much as is needed of the process outside of the function block. To do this, we need to understand encapsulation and abstraction.

Essentially, encapsulation and abstraction are very similar concepts and they walk hand in hand. In practice, abstraction and encapsulation are inseparable, especially in languages such as Java or C#. In terms of PLC programming, encapsulation is implemented using function blocks and using the private access specifier, and ensuring that as many variables as possible are as hidden from other files as possible:

- **Abstraction**: In OOP, the fewer elements that can access a variable or method, the better. Generally, you want to hide as many of your function block attributes as you can. The concept of hiding attributes and only having the necessary attributes visible to other files is what is known as abstraction. In short, abstraction is the concept of hiding data from unauthorized files.
- **Encapsulation**: Encapsulation is a concept that deals with binding data into logical units. Function blocks are the normal modular unit in which logically related attributes are bound. The goal of encapsulation is to group related data and hide the complexity of the unit. This is often why both encapsulation and abstraction are confused and many times grouped as one pillar.

An easy way to think of abstraction and encapsulation is as follows:

- **Abstraction**: Hiding attributes from other files and abstracting the complexity.
- **Encapsulation**: Grouping logical attributes such as methods together and exposing only the necessary attributes to other files.

One example that I like to use to help visualize abstraction and encapsulation is to think of a car. A car engine is a very complex piece of machinery with many moving parts that have to operate flawlessly;

however, a driver does not have to know how to operate every single component for the engine to move the car. In fact, for the average driver, trying to manipulate the inner workings of an engine can be detrimental. All the average driver needs to know is how to press their foot on the gas pedal to make the car go. Many of the engine components are encased in plastics and metals to hide the complexity of the engine from the average driver.

For all the complexity of a car engine, without an operator pressing a gas pedal, the engine will not move the car. A car has basic features that the driver needs to know how to operate so they can properly drive the vehicle. This is a prime example of encapsulation. The complexity of moving the car is hidden; however, since the driver still needs to be able to operate the car, engineers added a gas pedal to drive the vehicle but hid the complexity of operating the engine. The gas pedal adds a level of abstraction over operating the engine.

If you think about the example in the *Understanding access specifiers* section, we had a function block with three methods. Two of the methods were private and one was public. Our program needs both calculations; therefore, we encapsulated the methods together and abstracted the complexity. In short, we abstracted the method so we only need to call one method to get both calculations.

Abstraction and encapsulation are the first two pillars of OOP. However, as stated before, there are a total of four pillars. The next two pillars we are going to explore are known as inheritance and polymorphism.

## Inheritance

**Inheritance** mostly conjures up thoughts of receiving material possessions from the dearly departed. In terms of programming, the concept is similar without anyone needing to pass away. In programming, special relationships can be formed between function blocks. This relationship allows one function block to use certain attributes of another function block. In other words, inheritance allows you to cut down on redundant code. Inheritance will let you reuse reliable code that exists in one function block in a totally different one. As such, inheritance will help you write code that can be used in many different places but will allow you to keep the code in one central location.

In all types of programming, the concept of inheritance is often abused, especially among inexperienced programmers. Inheritance is not meant to be a way of circumnavigating the private access specifier and using attributes! Instead, you use inheritance when there exists an *is a* relationship between function blocks. In other words, if you think of the animal kingdom, a cat, a tiger, and a lion are all felines. The animals have common attributes: they all have four legs, have a tail, and in one form or the other, meow. This means we can create a function block called `feline` and create other function blocks that are known as base or child function blocks that represent the cat, tiger, and lion. To demonstrate this, we are first going to create a base function block called `Felion`. There will be nothing special about this function block. We will create it as we have done any of the other function blocks we have created thus far.

After you create the `Felion` function block, you will need to add a `Lion` block, a `Cat` block, and a `Tiger` block. However, when you create these blocks, you need to check the **Extends** checkbox as in the following screenshot.

Figure 7.3 – Extends selection

Once you check this block and click the button with three dots on it, you will be met with a screen like the following:

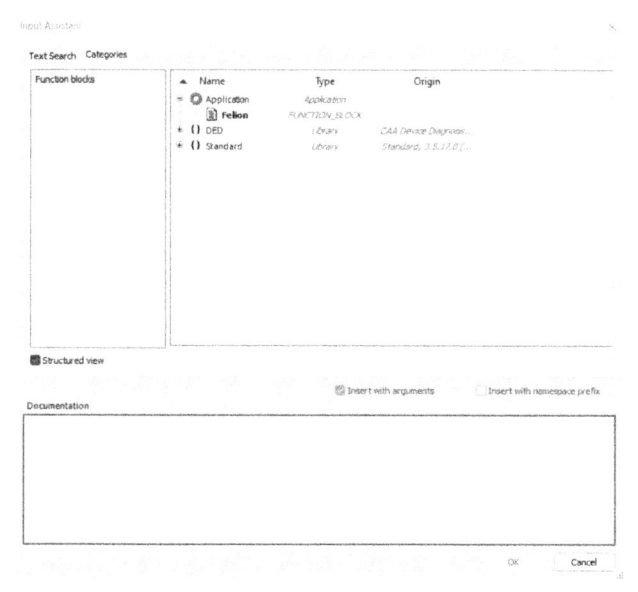

Figure 7.4 – Input Assistant block

This operation will set up the necessary code for the child function block, in this case, the `Lion` block, and the ability to use certain attributes in the base function block, in this case, the `Felion` block.

You will need to perform this operation for all three of the child function blocks. When you are done, you should see code similar to the following in the `Cat`, `Tiger`, and `Lion` function blocks:

```
FUNCTION_BLOCK Lion EXTENDS Felion
VAR_INPUT
END_VAR
VAR_OUTPUT
END_VAR
VAR
END_VAR
```

As can be seen in the code, the `EXTENDS Felion` keyword is added to the function block code. This is key to inheritance. The EXTENDS keyword is essentially telling the PLC to use visible attributes from the `Felion` function block in the `Lion` function block. Until now, we have used the terms child or sub-function block and parent or base function block. In normal OOP, the term function block would be replaced with the term class. However, the meaning is the same. A child or subclass is a class that inherits from a base or parent class. All the private or protected attributes from the base or parent class get passed down to the child class for use. These are very important terms to understand as the terms are often used in the OOP design phase and it is important to understand the difference between the terms.

With that in mind, let's set up a function in the `Felion` class called `speak`. To implement this method, you will want the variables section to match the following:

```
METHOD PUBLIC speak : WSTRING
VAR_INPUT
END_VAR
```

The main logic of the method is very simple as it will just return `meow`:

```
speak := "meow";
```

To consume this method, we are going to set up three object variables in the `PLC_PRG` file:

```
PROGRAM PLC_PRG
VAR
    cat         : Cat;
    lion        : Lion;
    tiger       : Tiger;
```

```
    catS     : WSTRING;
    lionS    : WSTRING;
    tigerS   : WSTRING;
END_VAR
```

In this case, all we have are references to the function blocks and string variables to hold the return values from `speak`.

The logic that will call the `speak` method is as follows:

```
catS   := cat.speak();
lionS  := lion.speak();
tigerS := tiger.speak();
```

If you recall, we only declared the `speak` method in the `Felion` class and not in any of the child function blocks. However, when we run the code, we will get the following output:

| Expression | Type | Value |
|---|---|---|
| cat | Cat | |
| lion | Lion | |
| tiger | Tiger | |
| catS | WSTRING | "meow" |
| lionS | WSTRING | "meow" |
| tigerS | WSTRING | "meow" |

Figure 7.5 – Inherited outputs

The preceding screenshot represents inheritance. We declared an attribute in one function block and we can use it in other function blocks. IEC 61131-3 does not support multiple inheritances; as such, unlike languages such as C++ or Python, you cannot inherit from more than one class. IEC 61131-3 inheritance works similarly to C# or Java in that you can, at most, extend from exactly one function block.

Though you can only extend one function block, you can inherit properties from other function blocks through what is known as the **inheritance chain**. Essentially, the chain is like a flow-down system. If you extend a function block that extends another function block, you will be able to access the properties from both function blocks. For example, if we created a function block called `BabyLion` with a method called `Play` that extends `Lion`, we would be able to access both the `Lion` function block methods, the `BabyLion` methods, and the `Felion` methods. This is an important concept to remember and to be mindful of as a change to any of the parent classes will affect the behavior of all subsequent child classes.

If you notice the output message in *Figure 7.5*, you will see that each of the animals is saying `"meow"`; however, we know that only cats say `"meow"`. As such, we need to employ another OOP concept known as polymorphism.

## Polymorphism

**Polymorphism** is the concept of changing an attribute from one function block to another. There are many ways to implement polymorphism; however, the easiest and probably the most common way to implement the concept is by simply redefining the attribute in the child function block. For example, lions do not say "meow" they say "roar". To get this desired attribute, we need to add a new method to the child function blocks. So, to get the new functionality, you need to add a new method to the Lion function block and ensure it is named speak. When you are done, your variables block should look like the following:

```
METHOD PUBLIC speak : WSTRING
VAR_INPUT
END_VAR
```

Meanwhile, your main logic should look like the following:

```
speak := "roar";
```

When you run the code, you should see the following:

| Expression | Type | Value |
|---|---|---|
| cat | Cat | |
| lion | Lion | |
| tiger | Tiger | |
| catS | WSTRING | "meow" |
| lionS | WSTRING | "roar" |
| tigerS | WSTRING | "meow" |

Figure 7.6 – "roar" output

As can be seen, the lion is now saying "roar". This shows that when you declare an implementation of an attribute in a child function block, that attribute takes priority over the inherited attribute from the base function block.

The best way to think of polymorphism is as a means of chaining a behavior. In our cat example, we changed our speak method to display the proper animal sound. In other words, we morphed the behavior of the object.

Even though you implement a different variation of a method in the child function block, you can still use the parent class's implementation. You can accomplish this with the SUPER keyword. To demonstrate this, we're going to call the Felion, speak method. Essentially, we're going to keep all the code the same except for changing the Lion class. In the case of the Lion function block, we're going to change the main logic to look like the following code:

```
speak := SUPER^.speak();
```

When you run the code, you should be met with a screenshot similar to the following:

| Expression | Type | Value |
|---|---|---|
| + cat | Cat | |
| + lion | Lion | |
| + tiger | Tiger | |
| catS | WSTRING | "meow" |
| lionS | WSTRING | "meow" |
| tigerS | WSTRING | "meow" |

Figure 7.7 – Felion method call from Lion

As you can deduce, the `speak` method in the `lion` function block was ignored. There are many reasons why you would want to invoke the base function block's version of the method. It is quite common to want to do this when you need returned data from it, or when it is more appropriate to use it in a given situation than the child function block's version.

So far, we have touched on all four of the pillars of OOP. As you might guess, inheritance and polymorphism are very rich and complex topics and we have merely touched the surface. However, we know enough about the concepts to start playing with and exploring them more. For now, we are going to start exploring other concepts that are not necessarily a part of the four pillars but are still powerful OOP concepts. One very common area that needs to be explored is composition and how it contrasts with inheritance.

## Inheritance versus composition

Inheritance is a very important concept and is, without a doubt, a great way to recycle code under the right circumstances. However, many new or inexperienced programmers will often use inheritance as a means of importing code. This is a bad practice because instead of producing clean organized code, they produce jumbled-up code that has no true relationship between function blocks. When developing object-oriented code, it is very important to consider the relationships between function blocks. One very common way to implement object-oriented relationships is with a concept known as **composition**.

### When to use composition

For many inexperienced, traditional programmers, composition is often an ill-understood but, ironically, often-used concept. Composition is where you include object references from one function block in another. In other words, composition allows us to assemble things. Whereas inheritance consists of an *is a* relationship between functions blocks, composition is a *has a* relationship. In other words, we can summarize when to use composition or inheritance with the following:

- **Composition**: If something *has* something else
- **Inheritance**: If something *is* something else

Essentially, the best way to choose which method to use is to ask yourself whether the function block you're working on *is something* (such as *if a cat is a feline*) or *if something has something* (such as *does a car have an engine?*).

The core idea behind composition is that we are building function blocks from other function blocks. Think of a car. If we were programming a car, we would need an engine, wheels, and so on. If we were using inheritance, if we wanted to change the engine of our car, we would have to change what we are inheriting from. However, with composition, we can simply change the reference to the wheels we are using in our program and we are good to go.

Many developers will often favor composition over inheritance for many reasons. One of the biggest reasons developers usually opt to use the *has a* relationship and composition over the *is a* relationship and inheritance stems from the fact that the inheritance produces code that is more tightly coupled. This means that a change in the base function block can have a ripple effect that changes the behavior of the child function blocks. Since composition is assembling complex objects from other objects, a change to one of the component classes does not necessarily mean that the changes will have the same ripple effect.

As we saw with inheritance, there are limitations. For example, we can only inherit from one function block, and inheriting from another function block can cause a tight relationship between the blocks. Though subjective, it is my opinion that it is usually better to opt for a composition function block structure and build new function blocks from other function blocks. However, there is a time and place for everything, and what matters most is that you are using the correct methodology for the proper relationship. As such, whether to use composition or inheritance in practice stems from the function blocks you are creating and the way you architect them.

## Composition in practice

To demonstrate composition, we're going to create a `Car` function block. To do this, let's analyze some components of a car. A car is composed of many parts that have functionality, such as the following:

- Engine: `rev`
- Transmission: `shift`
- Brakes: `stop`

In this case, we are going to create a `Car` function block that is going to be composed of an `Engine`, `Transmission`, and `Brakes` function block.

To do this, we are going to create a `car_parts` folder in the application directory with the previously mentioned function blocks in it. When you are done, your project structure should look like this:

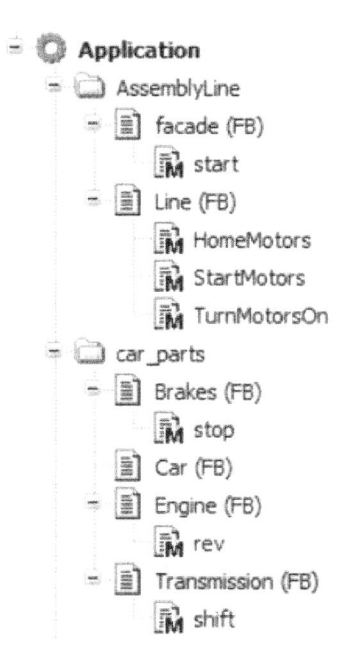

Figure 7.8 – Car project structure

The code for the function blocks is straightforward as the methods themselves only have one line of code and no variables, all have a return type of WSTRING, and they are set to *Public*. The code for each function block should look like the following:

- Brakes function block stop method:

    ```
    stop := "stop";
    ```

- Transmission function block stop method:

    ```
    shift := "shift";
    ```

- Engine function block rev method:

    ```
    rev := "rev";
    ```

The Car function block will be a bit different. This function block will not have any methods and the only code for the function block will be variables that reference the other function blocks, like so:

```
FUNCTION_BLOCK Car
VAR_INPUT
END_VAR
```

```
VAR_OUTPUT
END_VAR
VAR
        brakes              : Brakes;
        engine              : Engine;
        transmission        : Transmission;
END_VAR
```

The final piece that needs to be constructed is the `PLC_PRG` file, which will look like the following:

```
PROGRAM PLC_PRG
VAR
        car   : Car;
        drive : WSTRING;
        stop  : WSTRING;
        shift : WSTRING;
END_VAR
```

The logic will look like the following:

```
drive := car.engine.rev();
shift := car.transmission.shift();
stop  := car.brakes.stop();
```

When the code is run, you should be met with the output in the following screenshot:

| Expression | Type | Value |
| --- | --- | --- |
| car | Car | |
| drive | WSTRING | "rev" |
| stop | WSTRING | "stop" |
| shift | WSTRING | "shift" |

Figure 7.9 – Car output

As you can see, we accessed the methods from the three function blocks with the `Car` function block. Essentially, what we did is encapsulate three reference variables in the `Car` class. If you look at the lines of code, we referenced the `Car` function block variable, which allowed us to access the internal reference to the `Engine`, `Brakes`, and `Transmission` function blocks.

In this example, we essentially built a car. A car is not an engine, a transmission, or brakes, as such inheritance is not appropriate to use here. However, a car *has* an engine, a transmission, and brakes. This means that to create a car as we did, it is more appropriate to use composition. In this case, we are still able to recycle our function blocks without becoming totally dependent on any given function block. In essence, we can remove the old engine and replace it with a high-performance engine without completely overhauling the code as we would with inheritance, which would require us to model a specific car and commit to using that throughout the program's life cycle.

In short, this is probably one of the most important concepts in OOP. Though the composition technique is not a pillar of OOP, you will find yourself using it much more often than inheritance. At this point, to get a good grasp on composition, find a few items and try to represent them with composition.

In the real world, we often work from templates. For example, engineers will not overhaul the design of a car. In reality, whether or not they realize it, certain patterns are being followed when designing the car. For example, all cars have four wheels, brakes, an engine, a steering wheel, and so on. However, what will change from car to car is the way the parts work. In other words, the overall functionality is the same but the way in which the components operate will vary. In the programming world, we can model these parts with what are called **interfaces**.

## Examining interfaces

If you read a textbook on a traditional language such as Java or C#, you will see that interfaces are often referred to as contracts. When you opt to use an interface, you are telling CODESYS that you agree to, at the very least, implement all the methods prototyped in the interface. However, in my opinion, this is a little confusing.

Generally, when I describe an interface to a new programmer, I usually describe them as a model for something. For example, if we are building an airplane, we will need certain things such as wings, an engine, and a cockpit regardless of whether we are building a prop plane or an F-35 jet fighter. Obviously, for each type of plane, these parts are going to be different. As such, when we implement an interface, we are telling our function block that we are going to use the methods that are declared in the interface but those methods may have different implementations.

To demonstrate how an interface works, let's implement one. For this example, let's pretend we are making an airplane. As was stated before, regardless of the type of plane, each will have a cockpit, engine, and wings. The first thing we are going to do is create a folder called `Plane` and in the `Plane` folder, we will create an interface by right-clicking the `Plane` folder, clicking **Add Object**, and then selecting the interface.

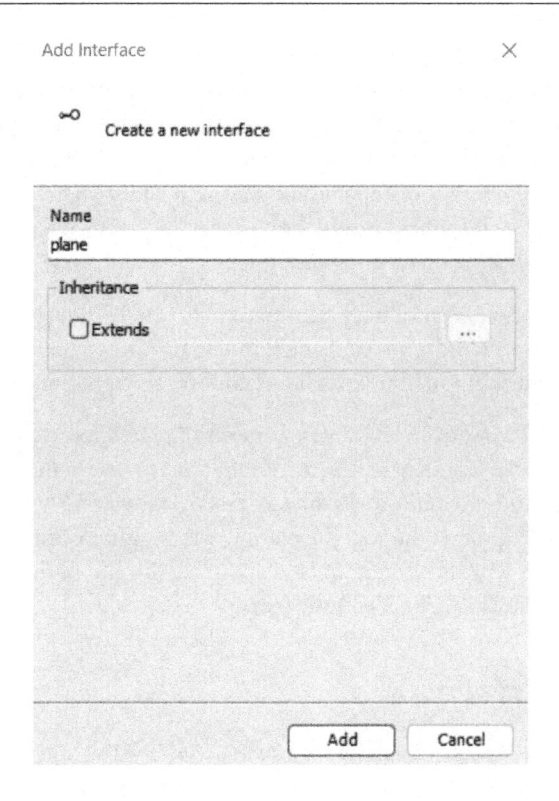

Figure 7.10 – Interface creation wizard

The preceding screenshot is the wizard window that you will see when you follow the steps correctly. Now, we have to add methods to it. Adding methods to an interface is as simple as adding methods to a function block. We will add an `engine`, `cockpit`, and `wings` method to the interface.

The methods are quite simple. Once you create them, you will only need to add an argument to the `engine` method so that it matches the following:

```
METHOD engine : INT
VAR_INPUT
    rpms : INT;
END_VAR
```

Next, we are going to implement the interface in two different function blocks called `F35` and `Prop`. After right-clicking on the `Plane` folder and adding a POU, click the **Implements** box and select the `Plane` interface. After you've completed building everything, you should have a structure similar to the following screenshot:

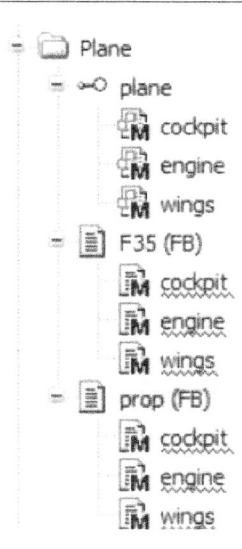

Figure 7.11 – Completed structure

The F35 block methods will consist of the following code:

- F35 cockpit method:

    ```
    cockpit := "1 seat";
    ```

- F35 engine method:

    ```
    engine := rpms * 1000;
    ```

- F35 wings method:

    ```
    wings := 2;
    ```

> **Important note**
> For the following and the preceding examples, write this code in the declared function block methods.

Once the F35 block is squared away, set the prop methods to the following:

- Prop cockpit method:

    ```
    cockpit := "2 seats";
    ```

- Prop engine method:

    ```
    engine := rpms * 100;
    ```

- Prop wings method:

  ```
  wings := 2;
  ```

As usual, once those are completed, we're going to make two reference variables and a series of variables to hold the output in the PLC_PRG file:

```
PROGRAM PLC_PRG
VAR
    f35          : F35;
    prop_plane  : prop;

    f35_cockpit : WSTRING;
    f35_engine  : INT;
    f35_wings   : INT;

    prop_cockpit : WSTRING;
    prop_engine  : INT;
    prop_wings   : INT;
END_VAR
```

This is the logic that will call the functions:

```
f35_cockpit := f35.cockpit();
f35_engine  := f35.engine(10);
f35_wings   := f35.wings();

prop_cockpit := prop_plane.cockpit();
prop_engine  := prop_plane.engine(5);
prop_wings   := prop_plane.wings();
```

When the code is run, you should see the following:

| | | |
|---|---|---|
| ⊞ ● f35 | F35 | |
| ⊞ ● prop_plane | prop | |
| ● f35_cockpit | WSTRING | "1 seat" |
| ● f35_engine | INT | 10000 |
| ● f35_wings | INT | 2 |
| ● prop_cockpit | WSTRING | "2 seats" |
| ● prop_engine | INT | 500 |
| ● prop_wings | INT | 2 |

Figure 7.12 – Interface program output

This example shows that by using an interface, we can automatically import methods and, more importantly, model something. As we can see in the example, the methods have the same name but have different implementations, which means that though we are building two different types of planes, we are using similar components, but the way the components work is different.

Another interesting aspect of the interfaces is since they are implemented and not extended, you can use multiple interfaces in a single function block. In short, there is a difference between implementing an interface and inheriting from another function block. Since you can implement multiple interfaces, you can combine them to model different things.

New or inexperienced programmers usually do not see the benefits of using interfaces. At first glance, what we did may simply seem like a roundabout way of declaring methods. However, a well-written program uses an interface and there is an old rule that says that you should code to an interface.

Coding to an interface will allow you to create more flexible code. So, if you need to add or remove more parts, you can do so without breaking implementation elsewhere. It is also a good way to ensure that projects written by a team are all consistent with method names and are implementing the correct functionality for their section. This is much better than coding to an implementation, in which case you would be creating things such as branch statements.

With this in mind, one area where interfaces shine is with **design patterns**. In short, design patterns are program design structures that allow us to solve a common task. As such, the next section is dedicated to understanding the idea of design patterns.

# Getting to know design patterns

Design patterns are solutions to common problems. You will mostly see design patterns in OOP to solve a wide variety of problems such as adding abstraction to method calls, creating single object references, and others. If you have a web development background, you may be familiar with the MVC design pattern, or the MVVM pattern if you've developed WPF applications in the past; however, there are many other patterns. Outside of the MVC and MVVM patterns, common patterns are as follows:

- Singleton

- Factory
- Builder
- Facade

Each one of these patterns has a different purpose and solves different problems. Whole books are dedicated to patterns; however, as an automation programmer, the one pattern that I used to gravitate to the most was the facade pattern.

The facade pattern is one of the most powerful patterns an automation programmer can use. In automation programming, doing things as simple as turning a machine on or off can be a very complex task. For example, multiple methods may have to be called to turn on a machine, such as turning on multiple motors, homing motors, and so on. In reality, calling each method can easily bloat the code, method calls can be missed, and any number of other issues could arise. The facade pattern provides a remedy for this.

The facade pattern is a class or function block with a series of methods that call all the necessary operations to perform a task. In other words, a facade block will have a series of methods that will condense multiple method calls into a single call. As we transition to the final example of the chapter, we are going to explore the facade pattern in action.

## Final project – creating a simulated assembly line

Our final project will consist of a production line. The line will consist of a function block for a facade and another function block that will have the following methods:

- Turn on the motors
- Home the motors
- Start the motors

The first thing that we will want to do is create the mentioned methods with an access specifier of *Public* and a return type of `BOOL`. Once that is done, create a GVL called `outputs` and set the following variables:

```
{attribute 'qualified_only'}
VAR_GLOBAL
    motorState   : WSTRING;
    startMotors  : BOOL;
    MotorsOn     : BOOL;
END_VAR
```

The `HomeMotor` method will consist of the following:

```
outputs.motorState := "motors homed";
```

The `StartMotors` method will consist of the following:

```
outputs.startMotors := TRUE;
```

Finally, the `TurnMotorsOn` method will consist of the following:

```
outputs.MotorsOn := TRUE;
```

The next thing we need to do is set up the facade function block. This function block will consist of only a single method called `start` and a reference variable:

```
METHOD PUBLIC start : BOOL
VAR_INPUT
END_VAR
VAR
    line : Line;
END_VAR
```

The `start` method's body will be like the following:

```
line.TurnMotorsOn();
line.HomeMotors();
line.StartMotors();
```

Finally, we will start the assembly line in the `PLC_PRG` file by using the following code:

```
PROGRAM PLC_PRG
VAR
    faca : facade;
END_VAR
```

The method call is achieved with the following code:

```
faca.start();
```

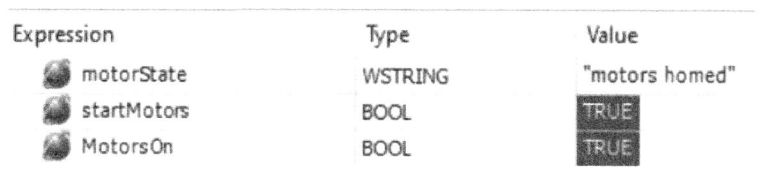

| Expression | Type | Value |
| --- | --- | --- |
| motorState | WSTRING | "motors homed" |
| startMotors | BOOL | TRUE |
| MotorsOn | BOOL | TRUE |

Figure 7.13 – Assembly line output

As can be seen in the preceding screenshot of the GVL outputs, we fired three methods with one method call. In reality, you would have multiple methods in the facade block that would stop the lines, pause the lines, and so on. It is recommended that you expand the function block using the principles to add extra functionality. In short, this is a real-world example of how you would use multiple OOP principles.

## Summary

This chapter has explored the more advanced features of OOP. OOP is a very powerful and new concept in the PLC world. When fully embraced and mastered, your complex project can become greatly simplified. As you become more familiar with OOP and the associated pillars, you will have no redundant code and will have a very maintainable codebase. As you master these concepts and learn how to integrate patterns into your code, you will be able to do more with less code.

Now, that we have a grasp of OOP, we can focus on creating portable code projects that can be used in many different projects.

## Questions

Answer the following questions based on what you've learned in this chapter. Cross-check your answers with those provided at the end of the book, under *Assessments*.

1. List the four pillars of OOP.
2. Is there a limit on the number of interfaces you can implement?
3. How many function blocks can you inherit from?
4. What is the difference between *Private* and *Public*?

## Further reading

Have a look at the following resources to further your knowledge:

- CODESYS interface documentation: `https://help.codesys.com/api-content/2/codesys/3.5.12.0/en/_cds_obj_interface/`
- CODESYS object methods: `https://help.codesys.com/api-content/2/codesys/3.5.13.0/en/_cds_obj_method/#e4507ebe4233ac0c0a8640e00a37b12-id-3375759d0dd23b38c0a864630d4cd159`

# Part 3 – Software Engineering for PLCs

This section will introduce you to software engineering. The goal of this section is to explore the **Software Development Life Cycle (SDLC)**, SOLID programming, and libraries. These concepts are often unknown to PLC programmers as many PLC developers are usually not formally educated in software engineering. The chapters in this section will approach PLC programming from a software engineer's mindset, which is a mindset that is often lost in the PLC programming world.

The following chapters are included in *Part 3*:

- *Chapter 8, Libraries — Write Once, Use Anywhere*
- *Chapter 9, The SDLC — Navigating the SDLC to Create Great Code*
- *Chapter 10, Advanced Coding — Using SOLID to Make Solid Code*

# 8
# Libraries — Write Once, Use Anywhere

Usually, when code is developed by a third party, it is shipped as a **library**. In CODESYS, a library is precompiled code that is designed to augment your code. In traditional programming, libraries are everywhere and are the backbone of most projects. In short, libraries offer a way to interface with unique hardware or add advanced functionality to a project without you having to worry about developing code for it.

Libraries are everywhere. If you ever wondered how Android and iPhone devices talk to hardware such as glucometers, household robots, and so on, the answer is libraries. You will not be able to function as a programmer without understanding libraries. Without libraries, you will not be able to talk to custom hardware, such as motor drives, encoders, and so on. In short, you will be very limited in what you can do as a programmer if you don't know how to use libraries.

In this chapter, we will explore the following:

- What a library is
- Guiding principles for developing a library
- Building a simple custom library

To round out the chapter, we are going to build a custom parts library that can compute the number of parts lost, and the total number of parts produced.

## Technical requirements

As per all previous chapters, this chapter will require nothing special other than a copy of CODESYS installed on your machine. If you have skipped ahead and have not read the past chapters, you will need to download and install a copy of CODESYS. The code for this project and all other examples can be found at the following URL: `https://github.com/PacktPublishing/Mastering-PLC-programming/tree/master/Chapter%208/Library`.

## Investigating libraries

We have touched on libraries a bit in the introduction. However, there is a lot to libraries, and an in-depth knowledge of how they work is needed before we can proceed.

A library is a prebuilt code that can augment your code by allowing you to easily talk to hardware, perform networking, and so on, without having to write code. In many cases, especially when it comes to proprietary systems such as hardware components, it would be difficult or impossible to effectively write code to interface with it. Many hardware manufacturers will simply provide a library to interface with the device.

### Why do we need libraries?

With that in mind, what is the purpose of a library? Libraries exist for multiple reasons, including the following:

- To avoid developing the same functionality multiple times
- To interface with custom or proprietary components
- To augment existing code with third-party libraries
- To distribute code to other developers

As stated before, libraries are the backbone of any modern programming language. For many programming languages such as C++, libraries are required for proper operations of the program you are trying to write. For example, for most C++ programs, the **Standard Template Library** (**STL**) must be imported to use many of the relative basic features of the languages. Libraries in CODESYS are a bit more optional compared to languages such as C++; however, they are nonetheless no less important. Libraries can be responsible for many different things, and some common examples include the following:

- Communicate with IoT devices
- Communicate with hardware interfaces
- Machine learning/artificial intelligence
- Communication protocols and conversions

### Libraries versus frameworks

In traditional programming, the terms *libraries* and *frameworks* are often used interchangeably. Even as an automation programmer, I would often hear inexperienced programmers and old-school PLC programmers use the terms interchangeably as well. However, there is a difference.

A framework calls your code, whereas your code calls a library. The way I like to demonstrate the differences between a framework and a library to my undergraduate students is as a skeleton.

A framework is like a skeletal system and the code you write is like the tissue that binds everything together and allows the skeleton to move. In traditional programming languages, frameworks are used as a template to build a specific application, such as the Python framework Django, which is used to create web applications.

On the other hand, a library is like the soft tissue in a skeleton. In the case of using libraries, your skeleton is your program. Where a framework is a temple for a program, a library is just a series of prebuilt commands that are used by your code to accomplish tasks. In other words, a library is like an assistant, whereas a framework is a roadmap.

Since automation is so unique, you will not use frameworks nearly as much, if at all, compared to libraries. In fact, during my time as an automation programmer, I don't recall using a framework. As an automation programmer, you are far more likely to use a library over a framework.

## Distribution

If you opt to use a third-party library, you need to pay attention to the licensing agreement. Many libraries and plugins are free to use; however, this may not mean that you can freely distribute them. In other words, just because you don't have to pay for a library, it doesn't mean you can use it in a product that you're going to ship. This is an issue that is more related to traditional programming languages but can still bite you if you're not careful.

As with many traditional programming languages, you can download third-party libraries from vendor websites, GitHub, or anywhere else. From many downloadable sources, the plugin is free; however, it will come with a license agreement that will tell you how you can use the library and distribute projects that utilize it. For some licenses, you can do whatever you want to with the library; for others, there are restrictions on modifications, while others are much more strict on what you can and cannot do. It is wise to remember that there are many different interpretations of *free*, and you would be well advised to understand the types of licenses the software you're employing has.

## Third-party libraries

We have been using the term *third-party library* quite a bit recently. Generally, the context for the term, at least for this book, is a library that is distributed by a person, organization, or so on. These libraries can usually be downloaded from sources such as GitHub, vendor websites, or any other download source.

In terms of PLC programming, many libraries come from a vendor. However, you can still get libraries for sources such as GitHub. If you opt to use a library that you pull off a source such as GitHub, you must import it. To demonstrate this, we are going to use a custom library that has one function block that consists of one method that adds two numbers. This is a custom library that was developed for this book. The library can be downloaded at the following link: `https://github.com/PacktPublishing/Mastering-PLC-programming/tree/master/Chapter%208/Library`.

156    Libraries — Write Once, Use Anywhere

The library is named `adderLib`. For this example, you will need to pull down the library to your development machine.

## Installing a library

The project in this chapter will have the library installed in it, so to follow along, use these steps:

1. Create a new project and click on the **Library Manager** icon.

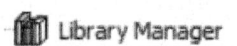

Figure 8.1 – Library Manager icon

When you click the icon, you will be met with a screen similar to this:

Figure 8.2 – Library Manager screen

2. Once you see this screen, you will want to click on the **Add Library** button in the top-left corner. When you do this, you will see something similar to what is in the following screenshot:

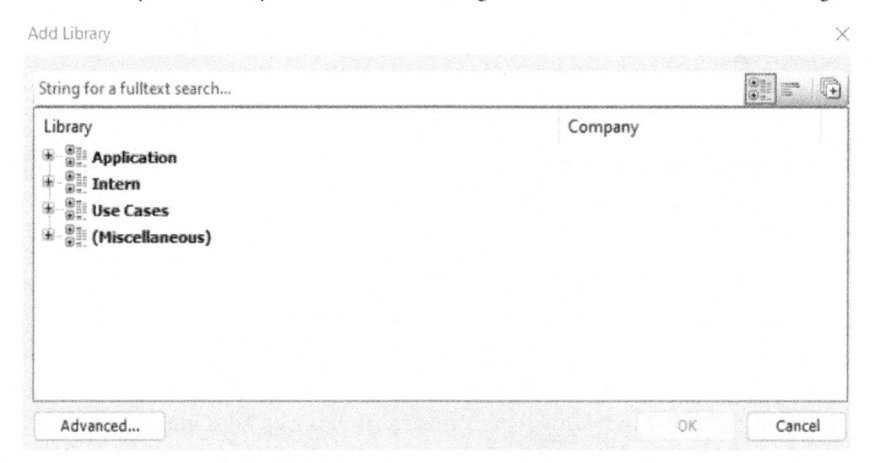

Figure 8.3 – Add Library screen

3. To import the library, you will need to press the **Advanced...** button in the lower left-hand corner. When you do this, you will be met with another screen. On this screen, you will need to click the **Library Repository...** button.

Figure 8.4 – Library Repository button

4. This will open up yet another screen and, on this screen, you will need to click the **Install** button and navigate to the directory where you stored the library project that you pulled down from GitHub. Once you find it, double-click the library and it should load.

5. If the library does not automatically import, you will need to click **Add Library** in the pop-up screen and click **(Miscellaneous)**.

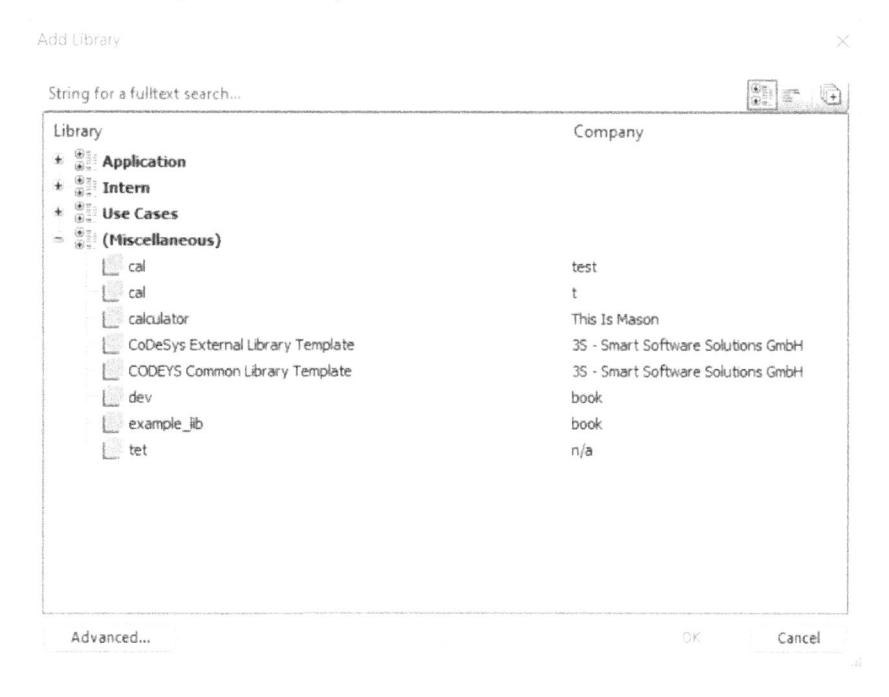

Figure 8.5 – Library selection

6. Double-click on `example_lib` and you should be good to go. To use the library, navigate to the `PLC_PRG` file and create the following variables:

```
PROGRAM PLC_PRG
VAR
lib_ref : addition;
```

```
        out : REAL;
END_VAR
```

You will also require the following logic:

`out := lib_ref.sum(33,33);`

When you run the application you should see an output similar to this:

| Expression | Type | Value |
|---|---|---|
| lib_ref | addition | |
| out | REAL | 66 |

Figure 8.6 – Library output

This is a general way to install a third-party library that you have pulled off the internet. It is a pretty straightforward process, and using the library is as simple as creating object references, as we have done before.

Now that we have a basic understanding of how to implement a library, we can look into creating a library. However, the first step to creating a library is understanding the architectural principles of creating one.

## Guiding principles for library development

Developing an effective library can be tricky. Where you have a clear-cut application in mind when developing a PLC program for a machine, developing a library will be a bit different. When you're developing a library, you have to think of everything at a very generic level. You will not know ahead of time who will use the library, how they will use the library, or what they will use the library for. Hence, creating a good library can be a very tricky and daunting task. There are no clear-cut ways to create a perfect library but there are a few rules that I came across that have helped me develop some decent libraries.

### Rule 1 – Keep it simple, stupid (KISS)

**KISS** is the golden rule for many programmers. When it comes to developing libraries, you must keep it as simple as possible. Generally, you need to have a very clear issue in mind that the library is designed to solve. Now, libraries can have many different functionalities and do many different things; however, you must have a clear issue in mind that it will solve. You don't want one library attempting to solve multiple problems.

For example, it is generally not a good idea to have a library that will provide support for databases, write CSV files, control motion, and so on. This library would become an overwhelming mess to maintain and use. Instead, it would be better to have one library that can control the motion of multiple different motors, another library that can establish a connection to different types of databases, and a third library to write CSV files. By having separate libraries, we have the following advantages:

- Not bloating the library with extra and potentially unwanted functionality
- Not overcomplicating the library with potentially useless functionality
- Not overwhelming the developer with potentially useless functions blocks
- Using logical names for our function blocks and methods

Now, with that being said, there is no golden rule as to what goes into a library and what does not. For example, you may be working on a motor drive system that can log the position of the motor. In this case, writing to a CSV file is optional, as it is a feature of the motor system. The developer can opt to write to a CSV file or not. In this case, you would want to have motor control capabilities and you will need the ability to write a CSV file.

In short, there is a fine line between what should be in a library and what is bloat. No developer, no matter how talented, will be able to fully anticipate the end user's needs. If you are deploying a library, you will want to keep a keen eye on how it is being used and what the developers are saying about it. You may have unnecessary functionality in the library or you may be coming up short on the functionality; either way, be prepared for your library to need to evolve.

## Rule 2 – Abstraction and encapsulation

Going with the theme of removing the possibility for your end user to shoot themselves in the foot, all of the function block attributes should be well encapsulated with a decent level of abstraction. In short, when developing a library, it is very important to show the consumer the absolute minimum they need to work with the library. My general rule of thumb for all attributes, especially ones that are in a library, is if I'm not planning on calling it, it gets an access specifier of `private`.

Due to the nature of the library and the wide variety of applications, it is important to hide as much of the inner workings as possible. Generally, I like to teach my students to write a program for the most inexperienced person in a room. This principle is even truer in library development. In other words, you cannot be sloppy with data hiding in library development. Expanding on my general rule, I usually tend to create what I like to think of as an **entry point** for the method. This entry point will have all the necessary arguments and return types; however, if there is dependent logic that breaks my one-sentence rule, I will break that out into other private methods. Sometimes this will be possible but other times it won't. However, in my experience, it is best to have a single method call that can accomplish the task than need to burden the end user with multiple other method calls to accomplish the same task. This principle kind of leads to the concept of design patterns.

## Rule 3 – Patterns make for perfection

In the last chapter, we briefly explored the concept of design patterns. Design patterns are a must for any well-written program, regardless of whether it is a library or not. However, a clean program architecture of a library is even more important. A clean library architecture will ultimately give the user a better experience using the library. As with any other programming project, many patterns can be used; however, there are two patterns that I have always leaned toward: the **facade** and **factory** patterns.

### *Facade pattern*

The facade is an excellent pattern to use for a library. As we saw in the previous chapter, it adds an easy level of abstraction for the end user. This pattern nails the principle of *keeping it simple*. This pattern is particularly handy if you're developing a library for hardware. When developing a hardware library, you need to keep the hardware calls as simple as possible. In other words, you want to try to execute a complete process with one call. As we know, this can be quite difficult at times since a simple process such as positioning a motor may require several different operations. In these cases, using a facade pattern or even a method that will provide this level of abstraction so that one method can call several. In these situations, it is generally advisable to declare all the operation methods (that is, the methods that will perform the embedded operations) as `private` and only the facade method or class as `public`. Generally, doing this will greatly reduce the complexity for the user. With that being said, there is another pattern, called the factory pattern, that can be used to help select different methods depending on the given inputs.

### *Factory pattern*

The factory pattern is another particularly useful pattern when working with libraries designed to operate different physical hardware. For example, suppose you work for a company that produces several different types of motor drives. Each drive may operate differently – for example, one motor drive may need to do a health check and then wait 5 seconds before it turns on, or one might have to be homed each time before it can operate. For this type of library, you will want to use something akin to the factory pattern.

Using patterns and keeping a library simple is a must. However, you can have a very well-developed library that is easy to use, but without the proper documents, it will be useless. So, the next area of library development we need to explore is documentation.

## Rule 4 – Documentation

You can develop the greatest library in the world; however, if it is not documented, it is about as good as useless. It is important to remember that a library is a compiled project and as such, ordinary comments will not be viewed by the consumer. You must use other means to communicate to other developers how to properly use the library. There are many ways to document the proper usage of a library, including custom documentation such as PDFs, websites, GitHub pages, and so on. You can also provide documentation in CODESYS itself.

There are a few things that must always be documented in a library, as follows:

- **Library information**: It is necessary to provide information on what the library is designed to accomplish.
- **Function blocks**: You will want to provide a simple synopsis of the function block.
- **Methods**: You will want to provide a synopsis of what the function does and provide information such as return types and arguments.
- **Variables**: You will also need to document the function block and method-level variables. You will want to provide information on what the variable is meant for.

All of these attributes can be easily documented in CODESYS with minimal effort. In terms of providing code documentation, there are many ways to document things; however, the syntax that I usually gravitate toward is the following:

- **Declaration header**: Denoted with ///
- **Member header**: Denoted with (**)

To demonstrate this, let's modify our example library. Open up the project and modify the code to match the following sum method code:

```
///this method returns sum
METHOD sum : REAL
VAR_INPUT
    x : REAL; (*input 1*)
    y : REAL; (*input 2*)
END_VAR
```

Next, modify the function block code to match the following function block code:

```
///This function block will add two numbers
FUNCTION_BLOCK addition
VAR_INPUT
END_VAR
VAR_OUTPUT
END_VAR
VAR
END_VAR
```

Libraries — Write Once, Use Anywhere

Once you are done with that, import the library into an example project or view it in the `Chapter 8` project and double-click on **Library Manager**. When you do this, you should see what is in the following screenshot:

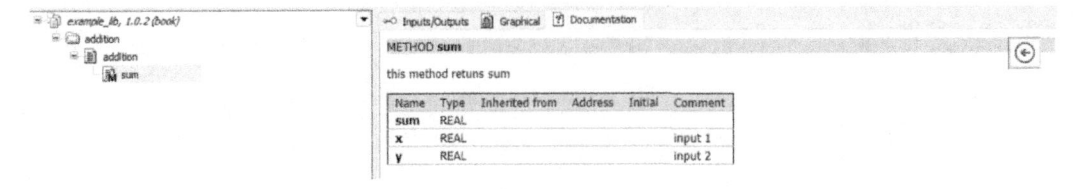

Figure 8.7 – Method documentation example

If you click on the `addition` function block, you should see what's in the following screenshot:

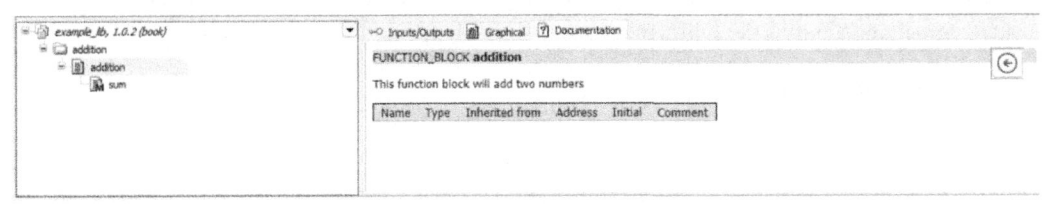

Figure 8.8 – Function block documentation

The takeaway from the screenshots is that the triple slash (///) will generate a general message on the top. In other words, the triple slash is more of a general attribute description while the parentheses are used more for the general description of variables.

There are other ways to add documentation, such as putting logically related function blocks into folders and documenting what the folder is for by right-clicking the folder and then clicking on **Properties**, then finally, selecting **Documentation**, which will render the following:

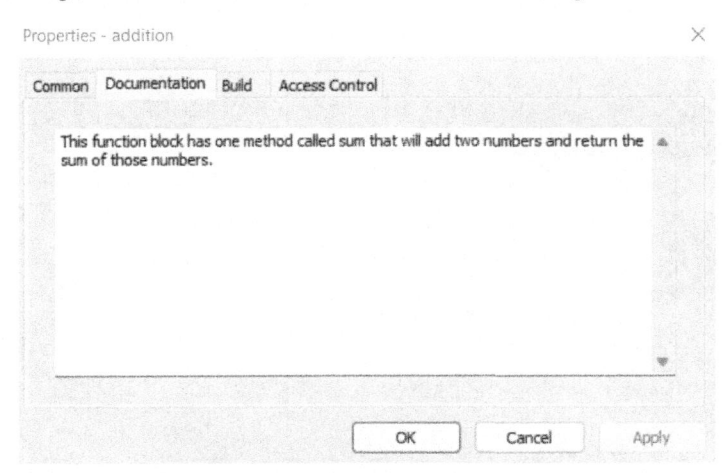

Figure 8.9 – Folder documentation window

When you click the **OK** button, the documentation will be generated. As with the other documentation, you will be able to view the documentation after the library has been imported. To view the documentation, click the **Library Manager** icon again and click on the folder. You should see something similar to the following screenshot:

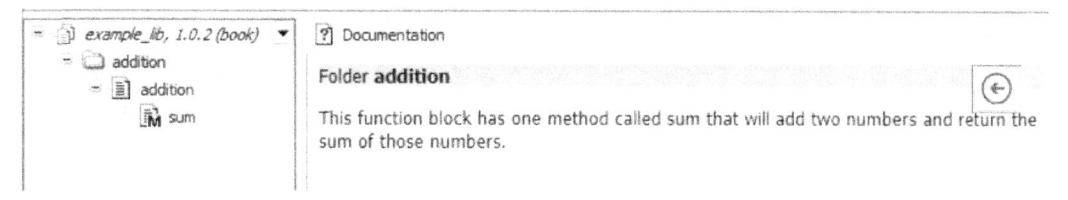

Figure 8.10 – Folder documentation

The final aspect that needs to be documented is the general information about the library. To do this, you will click the **Project Information** section in the library project area, which will generate a popup like the following:

Figure 8.11 – Library documentation

There are many other ways to document projects. What we have explored so far is just the tip of the iceberg. It is highly recommended that you view the following URL: https://help.codesys.com/webapp/docAreas;product=LibDevSummary;version=3.5.17.0.

Now, a lot of the button clicks may not make sense as we have not built a library yet. However, if you did pull down the library code and explored it, you would have been able to follow along. Before we move on, it is important that you understand the principles that were explored in this section. Developing a library for deployment is not like developing a normal program. Things must be named well, documented well, architected well, and easy to use. Once you understand these principles and have a grasp on them, we can attempt to create a simple library.

## Building custom libraries

By exploring third-party libraries and the guiding principles of developing libraries, we have touched on building custom libraries. So, before we move on and attempt to build a working math library, we are going to develop a simple library using some of the principles we have learned so far in the book.

### Requirements

For this project, we are going to build a simple library that can perform the following functions:

- Home the motor
- Turn the motor on
- Turn the motor off
- Stop the motor
- Position the motor

In short, this will be a very simple library and will not require complex architecture. For a library as simple as this, we don't have to worry too much about complexities such as design patterns; however, the facade pattern may help a little. Turning the motor off and on will be a bit more complex as we will need to automatically zero the motor out, which means that the homing function will also need to zero out. Let's break down the methods we will need:

- `Zero`: Will zero out the motor
- `Home`: Will return the motor to a zero position, that is, it's a facade method
- `Turn on the motor`: Will zero out the motor and put the motor in a standby state
- `Turn off the motor`: Will zero the motor and turn the motor off
- `Stop the motor`: Will halt the motor without zeroing it out
- `Position the motor`: Will move the motor to a position

Here, we have several methods; many of the methods require the motor to be positioned. `Home`, `Turn on the motor`, and `Turn off the motor` will all need to utilize this method. We probably don't want the end users to be able to use this function, so we're going to make that one private.

Building custom libraries    165

As for the rest of the methods, these will need to be used by the end user, for which we'll need to keep these methods public.

## Implementation

The first thing we need to do is to create a new project; however, unlike creating a normal project, we are going to do the following:

1. Select **Libraries** and **Empty library**, as in the following screenshot:

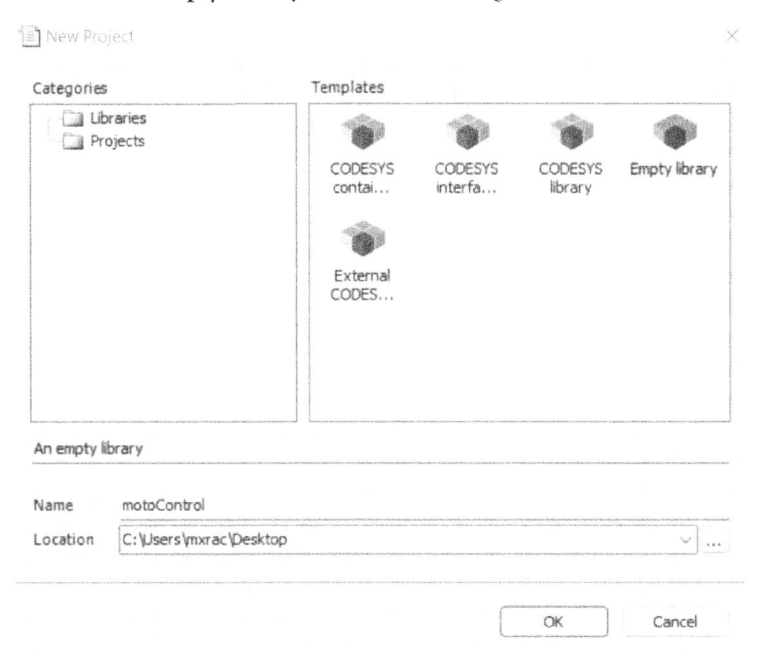

Figure 8.12 – Library creation

There are other ways to create a library with a full structure, such as by selecting **CODESYS library**. However, this option will give you a full project with potentially unnecessary files and structure. You can opt to use this if you would like, but for now, use an empty library to remove bloat.

2. After you complete that step, you should see a project tree like the following:

Figure 8.13 – Library project tree

3. Next, right-click `motoControl` and add a function block named `MotorControl` with the methods in *Figure 8.14*. Set the return type of all the methods to `BOOL` and the access specifier to *Public*.

```
motoControl
    MotorControl (FB)
        HomeMotor
        PositionMotor
        StopMotor
        TurnMotorOff
        TurnMotorOn
        ZeroMotor (private)
    Project Settings
```

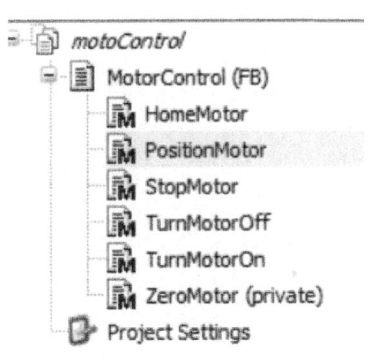

Figure 8.14 – Library project tree

4. Now that we have all the methods set up, we can start implementing the code. The first thing that we need to do is declare a variable that holds the motor's positions. For this, we will need to go into the `MotorControl` function block and set a variable, as in the following code:

```
FUNCTION_BLOCK MotorControl
VAR_INPUT
END_VAR
VAR_OUTPUT
END_VAR
VAR
    motorPosition : INT;
END_VAR
```

5. Next, we will implement the zero method with the following code:

```
motorPosition := 0;
```

6. This will be all that is required for this method as no internal variables will be set. Next, we will implement the `motorOn` method with the following:

```
ZeroMotor();
motorOn := TRUE;
```

7. After you have set up the `motorOn` function, we will now implement the `motorOff` method with the following:

Building custom libraries    167

```
ZeroMotor();
motorOn := FALSE;
```

8. `motorStop` will be similar to the `motorOff` function as it will consist of only the following:

   ```
   motorOn := FALSE;
   ```

9. The `HomeMotor` method is also quite simple. This method will turn the motor on if it is off and then call the `ZeroMotor` method. Once these operations are complete, the motor will be shut down. To do this, implement the following code:

   ```
   IF motorON = FALSE THEN
       motorON := TRUE;
   END_IF
   ZeroMotor();
   motorON := FALSE;
   ```

10. The next method to tackle is the `PositionMotor` method, which will take in an argument and set the `motorPosition` function block variable to it, as in the following code:

    ```
    IF motorON = FALSE THEN
        motorON := TRUE;
    END_IF
    ZeroMotor();
    motorON := FALSE;
    ```

    The logic for the method will be composed of the following:

    ```
    motorPosition := pos;
    ```

11. Now that all the variables are set up, we will need to save the project as a compiled library.

    Figure 8.15 – Saving the library

12. When you save the library, you will be met with the **Project Information** screen, as in *Figure 8.11*. You will want to input the following information from *Figure 8.16* to save the library:

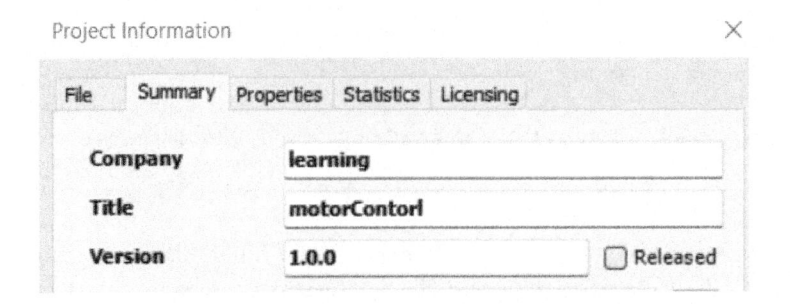

Figure 8.16 – Information fields for the library

This will create the `Project Information` file. This file will hold the metadata for the library. You can change the version number, name, or anything else by double-clicking the file once it is created.

13. At this point, your library is now saved and ready to be imported, similar to how we did in the *Third-party libraries* section. Once you import the library, you can modify the `PLC_PRG` file with the following to consume the code:

```
PROGRAM PLC_PRG
VAR
    motor1 : MotorControl;
END_VAR
```

At this point, we can now access all the methods in the function block. For example, we now have access to the following:

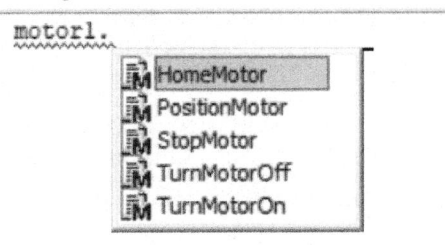

Figure 8.17 – Library methods

As can be seen, we can now access any of the public methods in the library. Now, as an exercise, try to use the library as it is without reviewing the code we compiled. Chances are it is pretty hard to figure out what is going on without any documentation. The reason why no examples were provided for this example was to demonstrate the necessity of code documentation. Chances are you're probably having trouble figuring out how the library should be consumed, how the methods are interacting with each other, and so on. The main takeaway from this lesson is how to create a library and why documentation is so important.

Now that we have built a sample library, we can move on to our final project and create a simple math library that can compute the total number of parts and the total number of parts lost.

# Final project – part computation library

In automation, it is common to have to program PLCs to keep track of many aspects of the job that is running. For example, it is common for a plant to want to know the number of parts in a job, and the total amount of parts that were rejected. For an operation like this, the calculations are never going to change. Writing the calculation for each different machine is rather pointless and redundant. Depending on what you work on, it may be best to just create a library and consume the library in multiple projects. Our first order of business is to figure out our requirements for the project.

## Requirements

The first step in developing a library is to gather the requirements. For this project, we need to create a library that will need to do the following:

- Compute the total parts created for a job
- Compute the number of parts lost for a job

With these requirements, we can deduce that we will need the following methods and variables:

- A method to calculate the total parts created
- A method to calculate the total number of parts lost
- A variable to hold the initial number of parts for the job

This will be a very simple library to implement. We will not need to worry about helper methods or design patterns for this project.

## Implementation

To create the project, we are going to follow the same steps we did in the *Building custom libraries* section to create this library:

1. This library will consist of one function block called `PartsCounter` and two methods, named `TotalParts` and `PartsLost`. Both methods will have a return type of `INT` and an access specifier of `Public`. When you are finished creating the library, your project tree should look like the one in the following screenshot:

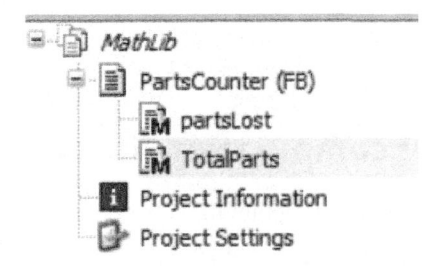

Figure 8.18 – Library tree

2. In the function block, we are going to add one variable, `partsOrder`, as in the following:

```
FUNCTION_BLOCK PartsCounter
VAR_INPUT
    partsOrder : INT;
END_VAR
VAR_OUTPUT
END_VAR
VAR
numPartsLost   : INT := 0;
END_VAR
```

3. The next method we are going to tackle is the `PartsLost` method. In short, this method will simply add 1 to `numPartsLost` each time the method is called. We can use the following code to implement this method:

```
numPartsLost := numPartsLost + 1;
partsLost := numPartsLost;
```

4. Now, with this implemented, we can move on to the `TotalParts` method. This method will subtract the number of `numPartsLost` from `partsOrder` every time the method is called. The code can be implemented as in the following:

```
TotalParts := partsOrder - partsLost;
```

5. Now, we can compile and save the library, as we did with the motor control library we made in the last section. When prompted to give the library information, simply fill out the wizard as before. When you're done, import the library into the project.

6. We can now move on to consuming what we have developed in the previous steps. In other words, we can actually use the library by implementing the following in the `PLC_PRG` file:

```
PROGRAM PLC_PRG
VAR
        tParts      : REAL;
    parts       : PartsCounter;
    lostPart    : BOOL;
END_VAR
```

7. The body of the `PLC_PRG` file will be composed of the following:

```
parts.partsOrder := 100;
IF lostPart = TRUE THEN
    parts.partsLost();
    tParts := parts.TotalParts();
END_IF
```

In this case, the Boolean variable will control whether there is a part lost. In the real world, this would be tied to a sensor. When the variable goes to TRUE, the calculation will get triggered. Play with the example; what output values do you get when the code is run? Notice that we can get a negative value for `tParts`; this shouldn't be. We should never have a negative number when we run the example code. Take a look at the library we made and the example `PLC_PRG` code. What and where can you modify the code to fix that bug?

In all, this was a crash course in libraries. There is much more to libraries, and their power cannot be fully appreciated until you have used them in the field or have developed them on your own. In all, you should have a basic idea now of what they are and how they work.

## Summary

In this chapter, we have explored libraries. We have learned what they are, how to use them, what third parties are, basic development principles, and so on. You should now be able to use libraries from external sources or create your own. What you will find is that by using libraries, you can now truly port code to different projects and cut down on your overall development time and effort.

There is a lot more to libraries, such as namespaces and so on, that was not explored in this chapter. A whole book could be dedicated to this subject. It is recommended that you explore libraries more on your own, as this chapter was just a crash course to get you familiar with the concept and consumption of libraries.

After covering many of the foundational elements of developing PLC software, we need to start exploring the software development life cycle so we can better manage the development of not only libraries but also entire codebases.

## Questions

Answer the following questions based on what you've learned in this chapter. Cross-check your answers with those provided at the end of the book, under *Assessments*.

1. What is a library?
2. Why is documentation important?
3. How is a library imported?
4. What are some good design patterns to use in a library?
5. What is the difference between (**) and /// in library documentation?
6. Go back and document the final project.

## Further reading

Have a look at the following resources to further your knowledge:

- CODESYS library documentation: `https://help.codesys.com/webapp/docAreas;product=LibDevSummary;version=3.5.17.0`
- CODESYS guidelines for creating libraries: `https://help.codesys.com/api-content/2/codesys/3.5.14.0/en/_cds_guidelines_for_creating_libraries/`

# 9

# The SDLC — Navigating the SDLC to Create Great Code

I didn't realize until I was about halfway through my graduate degree that software engineering is so much more than just writing code. Sure, I had several years of experience under my belt, but the full notion of what software engineering was didn't fully sink in until I took a class that focused on navigating the **Software Development Life Cycle** (**SDLC**). Luckily, I was in good company. While taking the class I realized I wasn't the only one who viewed the notion of the SDLC as exotic. The idea was reinforced after I graduated and progressed to a point in my career where I was working with less-experienced developers. In short, I found during the first part of my career that many developers were not aware of the SDLC, what it stood for, or more importantly, what it was.

Software engineering is much more than developing great code. Like any other engineering discipline, software engineering is a process with a series of pre-defined steps that must be completed in a strict sequence to produce viable code. Many experienced developers simply want to write code. They usually put very little effort into the planning or design of the code and, as a result, they will usually produce very fast code, but the code that will suffer from a poor design and quality issues, and may not be what the customer wanted in the first place. This is a common problem for inexperienced developers. Therefore, this chapter will cover the following topics to help you understand the SDLC:

- Understanding the SDLC
- The general steps of the SDLC

The chapter will end with a deployment of a working temperature conversion library similar to the ones we have developed in the past. However, this time, we will build one properly using the steps of the SDLC.

## Technical requirements

The source code for this chapter can be found in the GitHub repo for this book. You can use the following URL: `https://github.com/PacktPublishing/Mastering-PLC-programming/tree/master/Chapter%209`.

We will also need some type of rendering software for drawing UML diagrams. For this chapter, I'm going to use draw.io for the rendering. If you prefer another piece of software, you are free to use it. Alternatives that you can use can include things like Google Charts, Visio, or any other software. You can use draw.io for free at the following link: `https://app.diagrams.net/`.

## Understanding the SDLC

The SDLC is the steps in the software development process. In short, much like any other engineering process, the SDLC is the process that should be followed in some way to ensure you are correctly building the correct program. Depending on who you ask or what you read, the SDLC is usually broken down into the following:

1. Gathering requirements
2. Designing the software
3. Building the software
4. Testing the software
5. Deploying the software
6. Maintaining the software

Now, some models will only use five steps, and some will use more. However, no matter the model, the steps are the same, just broken out differently.

### Why care about the SDLC?

To many PLC programmers, the SDLC is as exotic a concept as alien life is to astrobiologists. Sadly, this comes from the mentality that software is an unimportant component of automation. However, to properly implement the concepts that we have covered thus far and to take our PLC software to the next level, we need a clear understanding of the SDLC.

In the introduction, we kind of touched on why the SDLC is important. In this section, we are going to go a little further. Many PLC programmers are usually engineers; chances are if you're reading this book and are not a student, you're probably something akin to a mechanical engineer, electrical engineer, electrician, technician, or whatever. Chances are also in favor of you having written PLC code in the past. If you are not an experienced software engineer, chances are that code is just a complement to your hardware. As we have touched on in this book, this is a very poor ideology to have as software must be treated like any other engineering project.

As with any other well-engineered project, you will need to know what the product is meant to do, and whether it has a competent design, construction, and test routine before it is deployed. After all that is done, you will need to make modifications and repairs to the product according to the customer's wishes. In traditional software development, it is common for developers to try to shape a problem to fit a solution or produce a solution that is so poorly designed it cannot be adapted to meet new challenges that the end users will encounter. As with any other process, there are proper ways to implement the SDLC.

## How is the SDLC implemented?

There are many ways that the SDLC can be implemented. For example, there is the mission-critical Waterfall method, the almighty Agile methodology, and many more. Out of all the methodologies, Waterfall and Agile are by far the most popular, with Agile becoming a major buzzword over the past few years. In all, I'm a firm believer that to understand the SDLC you must understand how it is implemented.

### *The Waterfall methodology*

Until recently, the waterfall methodology was used for mission-critical software projects and was heavily favored in realms such as the military, medicine, space exploration, financial markets, and so on. In short, anywhere that software had to be engineered to a precise standard, it was employed. Those times are changing but it is still a widely used development model.

Graphically, the Waterfall model resembles the following:

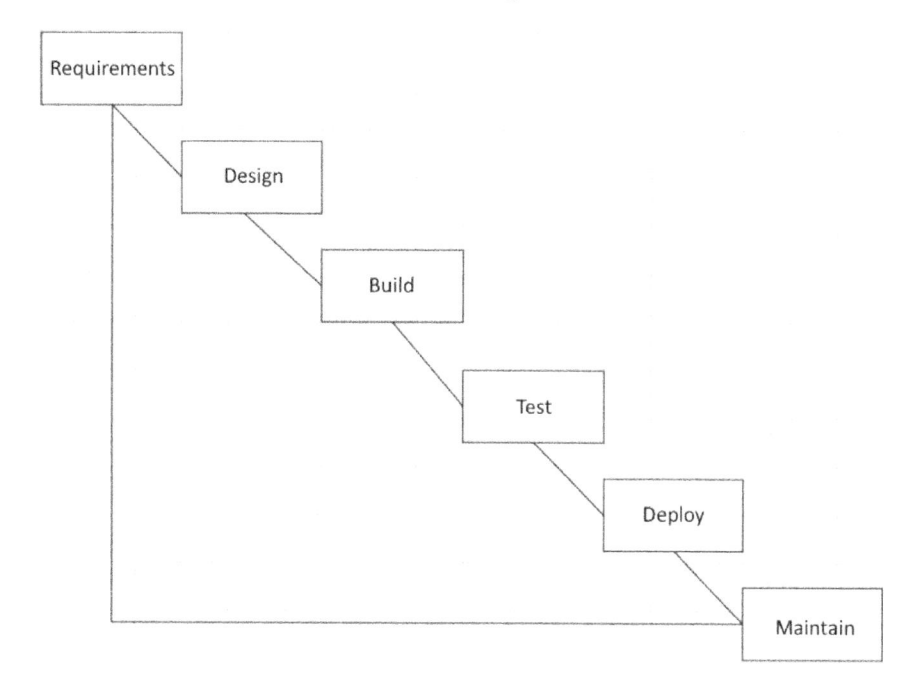

Figure 9.1 – The Waterfall model

As can be seen, the model resembles that of a waterfall that goes back to the top when a change needs to be made.

This model is very stringent to use and requires a lot of documentation to move from one phase of the SDLC to another. In short, if something goes wrong during deployment, you have to start at the top, in the requirements phase, to find and fix the problem. If you are in the testing phase of the project and an error occurred in the design phase, in keeping with the model, you would have to start at the design phase and work yourself back down to the deployment phase. Depending on what you're working on or trying to accomplish, this can be a very daunting task. For example, if you have a whole factory of machines that are utilizing a program that is hundreds of thousands of lines long, you would need a very hefty overhaul to fix the problem.

However, times are changing and the process by which software is developed is also changing as well. Recently, a newer methodology has risen to challenge the Waterfall model.

### *The Agile methodology*

Agile is the new kid on the block that is slowly overtaking the Waterfall model. Whereas in the Waterfall method, every step is done sequentially, with one phase needing to be fully completed before the development team can move on to the next phase of the project, Agile offers a less stringent approach to this. In Agile, the flow of the project is determined by what are called *sprints*. Sprints are were developers work on small, meaningful sections of the job.

By using the Agile method, you can accelerate the flow of the project, deliver code faster, and if a serious error does occur, you only have to fix a small portion of the project as opposed to the whole project. In short, the Agile methodology is much more forgiving and fast. There are several ways to implement an Agile project by using frameworks such as Scrum, **extreme programming** (**XP**), Kanban, and so on. There are also offshoots of these frameworks that can be adopted as well.

Many organizations are transitioning to an Agile framework, with Scrum being the most popular. However, organizations that try to implement Agile often either get stuck in a weird hybrid phase between Agile and Waterfall or try to switch to Agile too fast. It is important to remember that Agile is a methodology to govern a process. It is important to understand that adopting an Agile framework, the Waterfall methodology, or any other process is a cultural issue. This means that if you do opt to choose one methodology or framework over another, you must understand that it is important to adopt the features that work best for your team and organization. If your team is used to a certain framework or methodology, it is important to take it very slow and not get caught up in minor details when implementing a form of whatever other methodology you are adopting. Whole books are dedicated to the implementation of the Agile method. For now, however, we are going to focus on understanding the elements of the SDLC.

## Investigating the general steps of the SDLC

We have briefly touched on the steps of the SDLC. However, we have not dived into what they are or what they consist of. This section is dedicated to exploring the steps in the SDLC so we can implement them.

## Requirements/planning

If you ask a person off the street or an inexperienced software developer what the most important aspect of developing software is, chances are, they will answer coding. It makes sense since software development is about developing software after all. However, two steps must be completed before you or another developer even thinks about touching a keyboard. The first of which is the *requirements/planning* phase. Some break this phase into two distinct phases while others simply call this an analysis phase. Regardless of whether you consider this a phase or phases, it is the backbone of the project. In short, without this step, you simply do not have a project. If the SDLC is a building, the requirements/planning phase is the very foundation and like any other building, if you don't have a good foundation, your building will eventually crumble.

In short, requirements can be thought of as a punch list of functionality that is required by the program. In other words, the requirements are the functionalities that are necessary for the program to solve the problem. Without properly gathering the correct requirements, there is little chance that the program will solve the problem at hand and you'll end up having to fit the problem into the solution, which is quite literally the worst thing you can do.

During this phase of the SDLC, you will plan out the rest of the SDLC, choose the technologies you are going to use, and so on. This phase is the most pivotal step in the development of your software. If this phase is not completed properly, regardless of what methodology you're using, your project is as good as sunk. During this phase, you will want to think about the following deliverables:

- Technology stack
- A plan of how the SDLC will be executed
- Who will work on what components
- The requirements of the project
- User acceptance criteria

Software development is so much more than coding. Coding isn't even among the top two most important skills to have as a software developer when you think about it. This may seem oxymoronic, but consider this: you just got done building the most awesome, well-written, fast, PLC program to ever grace the earth, but there's one tiny problem: it doesn't do what the customer wants. Guess what, your super awesome program is now ready for the cyber trash heap. Efficient code is not the key to a quality project; having the code meet the customer's needs is. It is all too common for developers to try and fit a problem into a solution. In short, young, edgy programmers are usually more worried about showing off their chops than getting the job done. A good manager and programmer can put that aside and implement a strict set of requirements to develop with.

*Tips for collecting requirements*

During this phase of the development life cycle, it is important to develop a good picture of what you're trying to accomplish and who your end user or users will be. If you can, it is a good idea to try to communicate with your end users during this stage. For example, I would generally like to speak with the operators in a one-on-one conversation just to get a feel for who they are and what they know. Writing a program is a lot like writing a paper – you want to gear the program toward your audience.

In the Agile methodology, there is a concept that is known as the *user story*. Essentially, the user story is a single sentence that describes the role of the user, the action, and the added value of the action. A general user story is as follows:

*As a <role> I want to perform <action> to get < value>.*

For example, we can write a user story like the following:

*As a technician, I want to set the sensitivity of the sensor to get better readings.*

User stories are an Agile technique that is often not used in a PLC programming environment; however, in my opinion, this is one of the most powerful requirement-gathering techniques, regardless of which methodology you choose to develop your software with. Writing down user stories is an excellent way of conceptualizing functionality, especially in an environment that may consist of user-restricted operations. They also help prioritize functionality while allowing for progress metrics to be used with proper Agile. Just from a pure organizational point of view, I would strongly recommend trying user stories in at least one project.

There is a lot to the art of gathering requirements. Learning how to gather requirements is an art and it takes a skill that cannot be taught. In short, to gather requirements, you have to learn what questions to ask, which means you have to at the very least understand your end user to a degree. There is a lot to the art – so much so that whole books have been written about this. What we have covered is just the very tip of the iceberg. So, that being said, now that we have a general idea about the requirements, it is time to look at turning these requirements into a coherent design that will allow for expandability.

## Design

Another area that gets more attention than requirements gathering but is still generally ignored by inexperienced developers is the actual design of the software. Before you even think about touching a keyboard, your program should have a flushed-out design. Due to the nature of automation programming, this means you will need a finalized design of the necessary hardware before you can architect the software. After you iron out your hardware design, you're going to need to iron out your software design.

Generally, your software will have an overarching design. This will need to be a high-level design that will flesh out how all the software components will interact. At this level, you won't need a detailed design of how each component will work as the goal is to just flesh out how all the software modules will interact. In short, this is the program's architecture. Think of this level of design as a wiring diagram.

On a lower level, you will need to design each of the software components. This is like designing a device at the component level. This is basically the nitty-gritty of the software design. At this level, you are going to be concerned with things such as function block relations and function block designs. Essentially, at this level, you're one step above actually writing code.

When both of these levels are properly designed, your code should almost write itself. With a good program architecture and good modular designs, the coding phase should just be typing out the logic. Now, similar to how you will draw out component-level diagrams and electrical schematics, you will want to generate diagrams for the software. There are many types of ways to diagram software, with the most common being **Unified Modeling Language** (**UML**).

### Getting to know UML diagrams

Object-oriented software design is synonymous with UML. UML depicts the relationship between classes or, in the case of automation programming, function blocks. UML is an excellent tool for designing software at the function block or class level. Essentially, UML is like designing a component-level electrical schematic. UML diagrams will depict, at a minimum, the following:

- Function blocks
- Function block relations
- Private and public function block variables
- Private and public function block methods

This is a very important step for developing software modules. In short, when you UML out your software before you start coding, you will save yourself time, money, and ultimately, make your life as a developer much, much easier. The rule of thumb in software development goes that a little extra time up front will save you a lot more time later on.

Now that we have an idea of what UML is and what it is for, we need to explore what a UML diagram looks like.

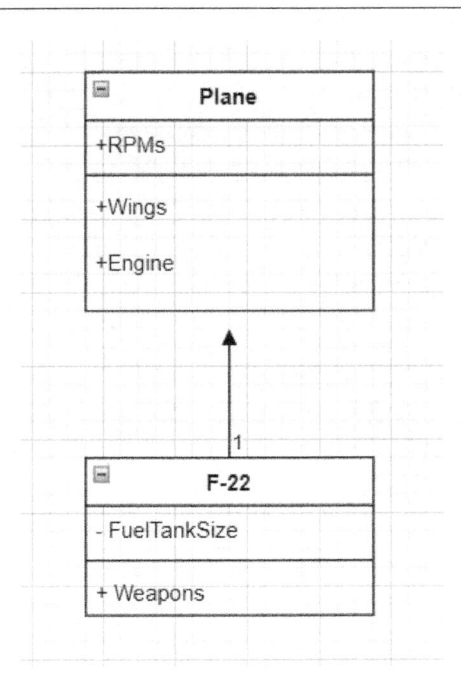

Figure 9.2 – Simple UML diagram

In this simple UML diagram, we have modeled an `F-22` fighter jet. We have a base function block called `Plane` and a child function block called `F-22`. With this diagram, a developer will know what function blocks to implement, what function block level variables to implement, and what methods to implement.

Okay, in theory, this UML model sounds great; however, much like electrical schematics or wiring diagrams, this means nothing if you don't know how to read it. So, with that being said, let's dissect this diagram and figure out what's what.

### Reading a UML diagram

Much like with CAD diagrams, the representation of the general outline will vary from program to program. However, general rules of thumb will apply. For example, let's only examine the `F-22` block for now.

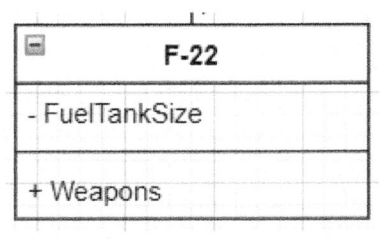

Figure 9.3 – F-22 function block

As we can deduce, a block in a UML usually represents a class or, in the case of IEC 61131-3, a function block. At the very top of the block, we have the function block named `F-22`. This will tell the developer that is tasked with implementing the function block what to call it.

Under the name, we have a minus symbol next to a name (`FuelTankSize`). The attributes under the name are usually variables. In this particular block, it is hard to tell what is a method and what is a function block level variable; however, this will become more clear as we progress through this project. So, as a general rule, you write all your variables under the function block name.

Now, next to any attribute, you will usually put an access specifier. The three access specifiers are as follows:

- +: Public
- -: Private
- #: Protected

So, from our block, the tasked developer can deduce that we will have a private, function block level variable named `FuelTankSize`.

The final attribute that we need to implement are the methods. The methods appear under the variables section. Again for this example, it is hard to distinguish because there are only two attributes. However, from the diagram, we are telling the developer to implement a method in the `F-22` function block called `Weapons`, which is public. For simplicity, we did not do it here, but it is usually a good idea to supply the method arguments with the arguments' data type if any exist. If you have a method with many arguments, it can be hard to do, and in cases like those, you may just want to make a note on the side about it.

So, with this in mind, let's dissect the base function block.

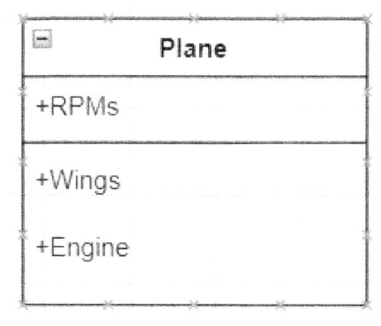

Figure 9.4 – Plane (base) function block

Working off the same principles, we can deduce from Figure 9.4 that we have one function block level variable called `RPMs` that is public and two public methods called `Wings` and `Engine`.

The next thing we need to look at is the relationship arrows. In short, this is the heart and soul of UML as they show how the function blocks will interact with each other. To do this, we use certain arrows to show how the blocks are related. In short, the arrows are as follows:

- Inheritance arrow

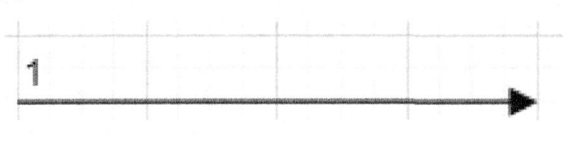

Figure 9.5

- Composition arrow

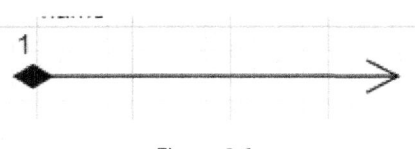

Figure 9.6

From this, if we examine *Figure 9.2* again, we will see that the `F-22` function block inherits from the `Plane` function block.

It is important to understand that the exact depiction of the arrow may change from rendering program to rendering program. Even the arrows in draw.io will sometimes vary. However, the general meaning and design will always be the same. You will also notice that many programs support arrows such as aggregation or association. These arrows are symbolic of other class or function block relations that are not covered in this book. So, for this book, they can be ignored.

In all, UML is a great tool for showing how things interact and what they are composed of. However, as you can deduce, there isn't much in the way of context, meaning what the block is supposed to do. Much like wiring diagrams or electrical schematics, this is a major drawback of UML diagrams. UML will allow you to communicate the guts of a function block and how function blocks interact, but it is a very poor tool for determining what function blocks and methods do as well as what variables are used for. Usually, it is a good idea to put notes to the side to add context to what function blocks, variables, and methods do. UML is a bit of an art and you, along with your team, will have to come up with a way of properly implementing it for your projects.

## Build

After a quality design, you can move on to what most developers live for, coding. Since everyone knows what programming is, and due to the amount of time we've already spent on programming in this book, this section will be relatively short. In short, coding is not the most important aspect of developing an excellent program. The heart of a program resides in the requirements and design. When those two steps are done correctly, the code should, for the most part, write itself. Now, this does not mean that writing code should be considered unimportant because, obviously, it's important. No code ultimately means no product. The point of this section and chapter, in general, is to merely hammer home the point that code is not everything when it comes to software development. Coding is very important but it is not the most important aspect of the development process.

With that in mind, how do we know whether our well-designed, well-crafted program meets the requirements and is working the way it should? The answer to that is testing. Testing was briefly covered in *Chapter 3*. In the next section, we are going to cover what testing is, and some types of testing.

## Test

After we build our code, we need to ensure that it is working to the specs laid out in the requirements. Testing is very important and there is a lot to it, so much so that there are countless books dedicated to the subject. This section will be a bit longer than the other sections; however, there is so much to testing software, it will be by no means fully comprehensive.

Depending on what you're trying to accomplish, there are many different types of testing. Some types are more relevant to automation programming than others. In traditional programming, most testing is accomplished using automation tools such as Selenium Python, JUnit, or other frameworks, depending on what you're testing for. Sometimes, developers will have the luxury of testing with simulators. For example, software engineers that are working with robots can use a system such as RoboDK to build virtual *worlds* and see how the code will behave without risking either the robot or injury to a person. Though some frameworks do exist for systems such as CODESYS or Beckhoff, it is my experience that most software testing is usually done manually for automation projects. However, regardless of whether you're testing by hand or with some type of automation framework, the principles are all the same.

The goal of testing is to ensure that the program is not buggy and to ensure that the program is performing in such a way that it solves the problem it was intended to solve. The first step in learning about testing is understanding what verification and validation are.

Verification and validation are two testing concepts that often come up in testing circles and interviews. When it comes to the testing phase of the SDLC, verification and validation are two of the most important concepts there are.

What validation and verification boil down to is ensuring you're building the right system correctly. It all goes back to not molding a problem to fit a solution. Verification and validation ensure that the code solves the problem in such a way that the code is reliable. To ensure the code is reliable and to ensure that the code is performing as it should, you have to test it.

## Verification testing

**Verification** ensures that the software is high quality and works. There is no silver bullet for this type of testing as there are no set standards. This means verification is rather subjective. You're ensuring that the program is bug-free and works as expected with no issues.

Unit testing and integration testing are two forms of verification testing. These types of testing are what you, as a developer, will carry out. Essentially, unit testing, integration, and the metric ton of other types of testing will ensure that you are building the system correctly, or in other words, verify the system.

### Unit testing

One of the most common types of testing is unit testing. Unit testing is testing out code blocks that provide some meaningful value. This can be testing out functions, methods, or whole function blocks. Generally, you are testing the smallest block(s) of code that can return a meaningful result.

When possible, I like to test at the method or function level, especially when these blocks are public. If possible, you will want to find and use a framework that will allow you to write automated unit tests; however, as I said earlier, in automation programming it is not always possible to do this due to the nature of the industry. Therefore, you will often need to test manually. Usually, what I like to do, and this could be argued as a bad practice, is to modify my code with a series of output statements or return values to track the code flow and output respectively. For example, if I'm working on a math library that consists of add, subtract, multiply, and divide methods as we have seen previously in the book, I would normally feed in dummy values and watch the outputs in the variable window as we have done.

When unit testing, it is a good idea to either find a unit test form online or create a spreadsheet that will keep track of the following:

- The test date
- Who performed the test
- The code blocks that were tested (if possible – not necessary)
- The input values
- The expected output
- The actual, recorded output
- Whether the test passed or failed

In short, you want to keep a paper trail of what you tested and the results, just to ensure that you did test the modules and that they passed. This basic outline is what is known as a test case. Technically, you don't have to keep any paperwork on what you tested and the results if you have a good memory, but from experience, it does help to keep things organized.

When it comes to unit testing, you want to shoot for at least 80% coverage. This means that at a minimum, you want to ensure that at least 80% of the code is tested with your unit tests. This does not mean that one test has to cover 80% of your code. What it means is that you have a series of tests that, when combined, have run at least 80% of the code in the codebase. To calculate your code coverage, you would use the following equation:

$$Code\ Coverage = \frac{number\ of\ lines\ executed\ by\ unit\ tests}{total\ number\ of\ lines\ in\ pogram} \times 100$$

Depending on what you're working on, 80% may not be enough. For example, if you're working on a medical device or a device for the military, the code coverage may need to be increased depending on industry standards. You will also want to run the same unit test multiple times with different values, including values that may accidentally be entered.

Now, sometimes what is called *dead code* can end up in a codebase and if you're counting the number of lines tested, it can interfere with code coverage. Dead code is code that is executed but doesn't serve any purpose. That is, it does not contribute to the success of the program's execution. If you do have dead code in your program, it is best to simply remove it and retest your code. Dead code can be tricky to find at times, especially if it is in a file that is no longer needed. However, if you are using an IDE such as Visual Studio, you'll usually be notified of it. It is generally best to remove useless code as quickly as possible, so a general rule of thumb is "if you don't use it, lose it!"

Another thing to watch out for is *unreachable code*. Unreachable code is code that can never run. Unreachable code might be caused by a control statement that has an unreachable branch, code after a RETURN statement, code in a file that does not execute, or any number of other situations. Similar to dead code, a good IDE will catch it. However, less sophisticated programming systems may not. So, you must pay attention to your code branches and remove code that will not run or you could end up skewing your unit test coverage. This is a task that can oftentimes be easier said than done.

Unit testing is meant to test the quality of code; in other words, this is to ensure the build process went or is going correctly and the code modules are performing as they should. Unit testing is not the only form of testing; you will need to carry out more types of testing to ensure the code is working as intended; as such, there are many other types of testing, such as performance testing, latency testing, and so on. At a minimum, you will want to perform some type of unit testing, whether it be very formal with an automation tool and a paper trail or informally by testing manually and not keeping a paper trail. Either way, when your code modules are working, you can move on to another form of testing called integration testing.

## Integration testing

Integration testing is pretty straightforward in concept. In short, your machine will have at least two software components. You will more than likely have an HMI and the PLC code. In some cases, you may have other software components such as databases, logging software, or any number of other software components. In any case, you will need to make sure the components are seamlessly working together. This is where integration testing comes in. Much like unit testing, there exist many frameworks for integration testing; however, much like unit testing, in my experience, integration testing is often done by hand in automation.

Integration testing is testing modules is test how modules interact. Though integration testing is traditionally concerned with software components, if your software is being integrated into a hardware system, such as a PLC-based system, you must factor in hardware related issues. This can add an extra layer of complexity to integration testing as a failed test case may be caused by either a faulty software component or a faulty hardware component. You will need to have an understanding of both the overall software components and hardware components to be able to troubleshoot problems that integration testing uncovers.

To conduct integration testing, it is common to write test cases that collect similar information that is collected in the unit test cases. However, you want to make sure that your tests will encompass all the modules you're trying to test. In other words, if you're testing the integration between your HMI and PLC code, you will need to write test cases that will encompass the HMI clicks and the PLC output. Consider the following steps:

1. Navigate to the **Homing** screen.
2. Click the **Home All** button.
3. Verify that the motor positions in the PLC variables area are all 0.

These steps are for a homing test. In short, all we are doing is testing the integration of the HMI's homing feature and the PLC code. In this case, we are only testing that the HMI and PLC are interacting properly.

Now, unit and integration testing are just two examples of verification. There are many other types of verification testing out there and it is highly recommended that you become familiar with a few more. However, with an idea of what verification looks like, we can move on to what validation testing looks like.

## *Validation testing*

**Validation** is the process of ensuring that the program solves the original problem. In short, validation is the process of making sure that the program satisfies the requirements. In other words, where verification determines whether you're developing the system right, validation ensures that you are developing the right system. There are several different types of tests to ensure you are validating, AKA developing, the correct system.

### Functional testing

Probably the most common type of validation testing is functional testing. In short, functional testing validates your software and hardware (in the case of automation), against the requirements that were set out during the requirements phase. For this type of testing, you want to use someone that might be familiar with the overall gist of the machine but probably not someone who spent a lot of time on the code. This isn't always possible but it is a luxury if you can afford it. Generally, functional testing is what is known as black box testing. During this phase of testing, you are not concerned with the code anymore; you are concerned with the behavior of the code and how it relates to the overall objectives of the project. This means you are testing the machine in the manner that the operator would use it.

For this type of testing, you will also need test cases, but these test cases will be overarching and system related. For example, instead of worrying about how fast, secure, and so on your code is, you are going to be concerned with the outcome of running what would be considered a real-world process. Take, for example, you're programming a packaging machine. Suppose your machine does the following:

1. Opens empty bags
2. Fills the bags with cement
3. Seals the bags
4. Weighs the bags
5. Sends bags with the correct weight to a holding area
6. Sends bags of the wrong weight to a recycle bin

To initiate this process, you will need to input the following information into the HMI:

- Input the number of bags into the HMI
- Select cement products to bag
- Input the weight

The actual test would consist of ensuring that the correct number of bags were produced, the proper cement was bagged, and the bags were of the correct weight.

Now, all of these testing methods that we have explored thus far are used when the product is being developed. In other words, pre-deployment. However, what about when an application has been changed after deployment? In that case, you need to consider regression testing.

## Regression testing

The final type of testing that you want to know about is regression testing. Regression testing is for the most part testing your system after the program has been changed in some way. This change may be an upgrade to the software, a bug fix, or any other time a developer changes a line of code in the system. This is very important in the automation world as software usually changes as the customer's processes change. Usually, if a plant or other machine owner decides they are going to change their process, they will update the PLC software to accommodate the new process. Requested changes will be either a bug fix or software modification such as changing a feature, adding a feature, or modifying a feature. In short, the moment you change or add any line of code, you will need to re-test your program again.

Luckily, when it comes to regression testing, you already have most of the test cases, especially if you're just fixing bugs that are found in the program. The goal of regression testing is to simply ensure that the modified code does not negatively impact the code that was not changed and that the new changes are working as intended.. If you're just fixing bugs, you can use the same test cases you used to validate the system before you deployed it, and if you are adding new functionality, all you have to do is create new test cases for the newly added functionality to use in conjunction with the original test cases.

Now that we have testing covered, we can move on to the next phase of the SDLC, which is the deployment phase. In short, it is now time to understand what software deployment is and what it means for automation engineers and programmers.

## Deployment

For some, deployment is the final phase of the SDLC. After you have designed, built, and tested your project, it is time to deliver the final product to the user. In the case of automation programming, this can mean that either the machine has been built and is ready to be installed or that the software, software modification, or whatever it may be is ready. Much like the requirements/planning phase of the SDLC, this is also an extremely important phase. During this phase, the end user will finally take ownership of the project. This will be a busy phase that will normally take some time to complete.

For many inexperienced developers, the deployment of the system is just dropping off the software, the end users happily taking ownership, and everyone going their separate ways. However, this phase consists of a few steps and, at least in terms of automation engineering, requires the customer and usually the operators that will be responsible for operating the machine.

During this phase of the life cycle, you can at the minimum expect to do the following:

- **Delivery of the system**: The delivery of the system can mean simply installing the new software on the machine or physically installing the machine at the operation plant.
- **Training**: After the machine is installed, you will need to train the end users on how to properly use the software. This can be accomplished with documentation, but it is usually a good idea to allocate at least a few days as the customer will usually want live demonstrations and to ask questions.
- **User acceptance testing**: The end users will need to verify that the software is solving the task that the machine was commissioned for. This can be very formal with well-defined test cases or it can be very informal with general operators running a series of test runs on the machine to ensure that the machine is behaving the way it should. Either way, expect to have follow-up calls from the end users and visits to the site of the machine.
- **Modifications**: It is not uncommon to have to perform on-the-fly modifications to the software before the customer signs off on it. These modifications can range from simply adding or removing graphical components on the HMI to adjusting or adding to the machine's behavior.

In all, the deployment phase of the SDLC can easily become a major effort. During this phase, it is important to allocate time to teach, answer questions, and modify the software or hardware on a whim. Now, modifications can occur during this phase. If you recall, when we explored the sections of the SDLC, we mentioned that maintenance is a step of the SDLC. In terms of the SDLC, maintenance can be considered a modification phase. Therefore, with a basic understanding of deployment, we can discuss maintenance in terms of the SDLC.

## Maintenance

Maintenance is synonymous with modifications. Depending on the methodology that you are using and how strictly you are adhering to that particular process, you will either need to start over at the requirements/planning phase for the entire project or for that particular sprint.

Though you should practice great diligence with any modification, you do not always have to be stringent with how closely you follow the full SDLC when working on a patch. Obviously, if your project as a whole is way off, you may need to start over from scratch and follow the SDLC very closely. On the other hand, if you are just adding something minor such as a button or changing the text or color of an element, you probably won't have to stringently follow the SDLC. However, if you are changing any behavior or functionality, you should ensure that you are documenting everything and running your original test cases. In short, you should be regression testing the machine.

At this point, we have covered the SDLC. As was stated previously, the SDLC can be broken out more depending on how detailed you want the phases to be. Regardless, we can now move forward and use the principles to build a simple library.

# Final project – creating a simple library

Now that we have explored the SDLC, we are going to apply what we learned and build a full project with those principles. The following section will be dedicated to building a temperature conversion library.

## Gathering requirements for the library

As we have discussed in this chapter, the first thing we need to do is determine the requirements for the project. Our goal is to create a temperature converter similar to the one we built before. However, our library will need to be portable without the possibility of anyone modifying it. It will also need to convert between all temperature units. We can say our requirements are the following:

- Should be a compiled library
- Should convert Fahrenheit to Celsius and Celsius to Fahrenheit
- Should convert Celsius to Kelvin and Kelvin to Celsius
- Should convert Fahrenheit to Kelvin and Kelvin to Fahrenheit

We can write some use cases, such as the following:

- As a developer, I want to convert Fahrenheit to Celsius and vice versa, so I can reuse that code in many projects
- As a developer, I want the library to convert Celsius to Kelvin and vice versa, so I can reuse that code in many projects
- As a developer, I want the library to convert Fahrenheit to Kelvin and vice versa, so I can reuse that code in many projects

From these user stories, we can deduce that we don't need anything fancy such as a user interface or high-precision code. Since it is a library, we will need to create a library project and will only need a Structured Text program. With the requirements established we can now move on to our design phase.

## Designing the library

There are different ways to accomplish this project; for example, we could use functions. However, to get practice with function blocks and methods, we are going to take an OOP approach. Once the project is complete, it is recommended that you go back and try a different design approach using functions. With that being said, the picture the requirements paint is very straightforward. From the sounds of it, we can create a simple single-function block library with many methods. In short, we can get away with the following:

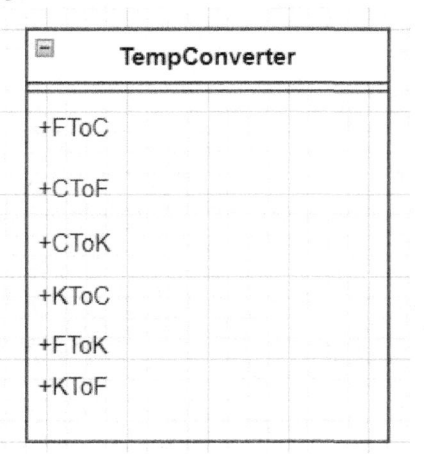

| TempConverter |
|---|
| +FToC |
| +CToF |
| +CToK |
| +KToC |
| +FToK |
| +KToF |

Figure 9.7 – Library design

As can be seen in *Figure 9.7*, this will be a very simple library to build. It will consist of a function block and `TempConverter`, which is composed of six simple, public methods. With this diagram and the requirements, we can deduce that each method will need to take exactly one argument since the requirements called on us to convert one temperature at a time. Also, since their temperatures often come in a decimal format, the arguments will need to be of type REAL, and since the returned value will often be a decimal value, we need to return a REAL as well.

Now, we inferred a few things that were not on the diagram with the return types and the argument types. Assumptions like this, though not necessarily a good thing, are common. Depending on the detail of the UML diagram, some design engineers will leave it up to the developer to infer certain things if the diagram is coming from someone else. This may stem from a lack of detailed knowledge on the designer's end, an idea that it is trivial, or, as in the case with our diagram, common logic Generally, this is a bad practice, but a practice that happens more times than not. So, if you are a developer, you should be prepared for instances like this. If you cannot infer certain aspects of the UML diagram and you did not produce the diagram, you should follow up with the designer for clarification. With a suitable design in place, we can now move on to implementing the `design`.

## Building the library

Now that we have finalized the requirements and design, we can move on to doing what developers love to do – write code. Since we have a decent design in place, we can now easily start implementing the code. Based on the UML diagram, our library structure will look akin to the following:

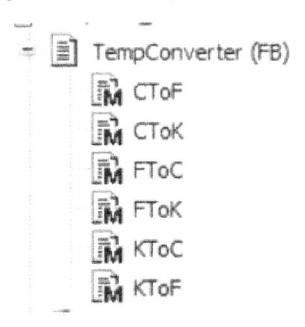

Figure 9.8 – TempConverter project structure

Each of these methods will have a single input argument that we'll call `temp`. The `temp` variables will be of the type REAL; hence, be sure to include that in each of the methods' variables blocks. To implement these methods, we are going to use the following code:

CToF method:

```
CToF := ((temp * 9) /5) + 32;
```

CToK method:

```
CToK := temp + 273.15;
```

FToC method:

```
FToC := ((temp - 32) * 5) / 9;
```

FToK method:

```
FToK := (((temp - 32) * 5 ) / 9) + 273.15;
```

KToC method:

```
KToC := temp - 273.15;
```

KToF method:

```
KToF := (((temp - 273.15) * 9)/5) + 32;
```

Now, let's save the project as a compiled library. Once you have saved it as a compiled library, you should create a new PLC project and add the library to it.

Proper code documentation is most important. As you may have noticed, we did not document the functionality of the library. We will skip it for now to avoid getting bogged down with too much detail. But with that complete, we can now move on to the next phase of the SDLC – testing!

## Testing the library

This library is incredibly simple. For a library of this size and complexity, unless specifically told otherwise, it is probably easier to unit test the library manually. Also, for a library of this size, we should be able to reach 100% code coverage. To do this, we need to create a series of simple method calls, record the output of the program and compare them to the expected values. We could devise a simple series of test cases like the following:

| Method | Input | Expected Value | Actual Value | Date | Pass(y/n) |
|---|---|---|---|---|---|
| FToC | 33 | 0.556 | | 7/31/2022 | |
| FToC | -100 | -73.33 | | 7/31/2022 | |
| FToC | 500 | 260 | | 7/31/2022 | |

Figure 9.9 – Test cases

Notice that we are testing multiple different values for a single method. This is because we want to ensure that it can handle a wide range of values. Hence we have a small number, a negative number, and a large number. In the real world, you may have many more test cases that test a wider range of input values. For our purposes, we are only going to test three values.

To create the unit test, we are going to prepare a program that was created to consume the library. Once you add the library to the project, modify the variable block to match the following:

```
PROGRAM PLC_PRG
VAR
temp : TempConverter;
    unit1 : REAL;
    unit2 : REAL;
    unit3 : REAL;
END_VAR
```

Next, modify the body of the program to match the following:

```
unit1 := temp.FToC(33);
unit2 := temp.FToC(-100);
unit3 := temp.FToC(500);
```

When the code is run, you should get what's shown in the following screenshot:

| Expression | Type | Value |
|---|---|---|
| temp | TempConverter | |
| unit1 | REAL | 0.5555556 |
| unit2 | REAL | -73.3333359 |
| unit3 | REAL | 260 |

Figure 9.10 – Unit test output

From this data, we can finish filling out our test case spreadsheet.

| Method | Input | Expected Value | Actual Value | Date | Pass(y/n) |
|---|---|---|---|---|---|
| FToC | 33 | 0.556 | 0.05556 | 7/31/2022 | y |
| FToC | -100 | -73.33 | -73.333 | 7/31/2022 | y |
| FToC | 500 | 260 | 260 | 7/31/2022 | y |

Figure 9.11 – Completed test cases

In short, we have successfully tested the `FToC` method. Keep in mind that this was only one method tested with three test cases.

With these simple test cases completed, as an exercise, create your own test cases for the other methods. As stated before, you should be able to reach 100% code coverage with this library.

### Deploying the library

At this phase, we have tested the library and we are ready to send it out into the world. For automation engineers, this usually means shipping your machine or installing your patches. However, for software such as libraries, this means making it publicly available to either the development community, your customers, or other developers on your team. For this book, we are only going to make the software available on GitHub, as we do in every other chapter.

If you're developing for the community, you may want to host the software on a platform such as GitHub. On GitHub, you can host software, allow other developers to contribute to the project, or simply allow people to download it. GitHub also allows you to create a web page of sorts about the project

. This is a great place to host your library for others to use. Now, if you make your software available in this manner, you don't have to directly interact with the end users; however, you will usually want to have a way to receive feedback and maybe answer questions about the product.

If you opt to keep your software component internal, you will simply store it on a local server or repository for others to pull and use. Even if you opt to go this route, you will still probably have to interact with other developers to teach them how to use the software. In all, no matter how you're deploying the software, you will have to take user feedback, fix bugs, and make modifications, which leads us on to the next phase of the SDLC.

### Maintaining the library

Now that you have a working product that you have deployed in some way, even if it is just deploying it to another project, you can look at how you want to modify it or fix any bugs that may arise in the software. This is a prime opportunity for you to take the library we built and expand on it while keeping true to the SDLC. Ideas you can use to expand on the library would be to include some of the following:

- Temperature input limits
- Add other unit conversion function blocks
- Improve the general usability of the software

At this point, you should try to think of something to do to modify the software. Whatever you do, start at the first phase and work your way down to re-deployment. I would recommend trying to add a few different features using this methodology just to get the hang of everything.

## Summary

It is pivotal to understand the SDLC for anyone who wishes to write code. The SDLC should be thought of as a guide to properly develop software. No matter what you're doing, you should always follow the SDLC as closely as you can so that your software will be easy to build, fix, and expand upon in the future.

This chapter has been a crash course in the SDLC, the methodologies that govern it, and the steps that it encompasses. Of all the chapters, I would argue that this is the most important. Too often, developers get caught up in what I like to call the *code culture* of just blindly building things with no roadmap of where they are, where they've been, or where they're going. Being able to navigate the SDLC will set you apart from those developers as in-depth knowledge of the SDLC is what separates an engineer from a programmer. With these principles under your belt, you can build software that will be extraordinary.

At this point, you should have a basic idea of how to navigate the SDLC and execute the phases at a basic level. You should have an idea of how to gather requirements, design a program, build a program, test a program, deploy a system, and finally, maintain the system. After you practice these steps a few times, you will be ready to move on to the next chapter and create SOLID software using SOLID programming principles.

## Questions

Answer the following questions based on what you've learned in this chapter. Cross-check your answers with those provided at the end of the book, under *Assessments*.

1. What is code coverage?
2. How much code coverage should you be shooting for?
3. If you have 100 lines of code and you test 50 lines, what is your code coverage? Is it enough?
4. Define the SDLC.
5. How many steps are in the SDLC?
6. What is UML?
7. What is the difference between regression testing and unit testing?
8. What is a test case?
9. What is validation and what is verification?

## Further reading

Have a look at the following resources to further your knowledge:

- Wikipedia, Waterfall model: https://en.wikipedia.org/wiki/Waterfall_model
- TechTarget, *Unit Testing*: https://www.techtarget.com/searchsoftwarequality/definition/unit-testing#:~:text=Unit%20testing%20is%20a%20software,developers%20and%20sometimes%20QA%20staff
- Agile Alliance, *The 12 Principles Behind the Agile Manifesto*: https://www.agilealliance.org/agile101/12-principles-behind-the-agile-manifesto/

# 10
# Advanced Coding — Using SOLID to Make Solid Code

The major downside to OOP is that many think that if they organize their code and stick to using class or function blocks, they're going to produce excellent software. However, more often than not, this is false. As the old saying goes, *"with great power comes great responsibility,"* and in terms of OOP, this is true. When writing OOP code, you can often shoot yourself in the foot just as easily as you can produce the next technological wonder. For many, OOP is just programming with classes or, in the case of PLC programming, function blocks. Though concepts such as class/function block relationships do help clean up code, more often than not, inexperienced programmers are just as likely to make a mess out of object-oriented code as they are to develop quality code.

Up until this point, we have explored the power of OOP and how it can allow us to reduce the amount of code that we have to write. However, OOP does not necessarily translate into code that is easy to maintain, expand upon, or, for that matter, understand. Even when using proper relationships between classes/function blocks, proper design patterns, and the like, code can still easily come out as a mess. This mess and rigidness can be a major issue for industrial automation where systems are constantly changing to adapt to an ever-changing world. So, if OOP is the future for industrial automation programming, how can we ensure that our OOP code will be of sufficient quality to last?

Enter the world of SOLID programming. SOLID programming is a set of rules that will help you drastically improve your code. In this chapter, we are going to learn about SOLID programming by exploring the following concepts:

- A soft introduction to SOLID
- How SOLID benefits code
- The principles of SOLID

Lastly, we are going to use SOLID principles to build a simulated industrial painter.

## Technical requirements

For this chapter, all you will need is a copy of CODESYS installed and working on your machine. The examples for this chapter can be found at the following URL:

https://github.com/PacktPublishing/Mastering-PLC-programming/tree/master/Chapter%2010.

As is standard for this book, you should download the code. Some of the code for this chapter will be a bit different from that of the other chapters. Much of the code in this chapter will be more akin to pseudocode that follows the IEC 61131-3 Structure Text syntax, as much of the code in this chapter will be used to merely demonstrate concepts. However, working examples will be provided, and these examples will either be noted or have screen outputs.

## Introducing SOLID programming

When first introduced to SOLID programming, I was incredibly confused about its purpose. My young, inexperienced self simply could not fathom that OOP did not ensure quality code. After all, as long as you're following proper OOP principles, you should be producing quality code, correct? Well, the answer to that is "Wrong." Quality code stems from well-architected code. Essentially, a quality program is a program where things can be easily added or removed, bugs can be easily found, and code can be easily changed without the risk of breaking other code. This is where SOLID comes into play. SOLID programming is a set of general rules that, when followed, will drastically improve the quality of your program's architecture.

So, what is SOLID programming? SOLID programming is a set of five **object-oriented design** (**OOD**) principles. SOLID is the brainchild of Rober C. Martin, a.k.a Uncle Bob. In short, Uncle Bob devised five principles that, when implemented properly, can allow your program to become flexible enough to be maintained so it can stand the test of time. With the introduction of OOP into the industrial automation world, having flexible code is a necessity.

Now that we have a brief overview of what SOLID is, why should we care about it?

## Benefits of SOLID programming

As every automation engineer knows, automation systems can stay in production for decades on end. Also as every automation engineer also knows, during that time, the process will change, which will require new software, hardware will become obsolete and have to be replaced with new components, and so on, which, as you can guess, will require software modifications. As someone who has spent countless hours sifting through thousands of lines of code on customers' sites for hours on end for multiple different employers, I can say that when it comes to architecture, it is worth it to put in the extra effort. Even when you're working on well-organized and architected codebases, you will find that tracking down a single error can be quite daunting. The task can become Herculean when the codebase is poorly designed.

With this in mind, it is best to think of SOLID as a series of general rules as opposed to hard standards that will produce code that is easy to maintain, expand, and, if necessary, modify. In other words, most think of SOLID as a set of best practices. When implemented correctly, these principles will allow you to build very robust code that is easy to maintain in the future.

Since SOLID is more of a best practice, what you need to understand is that the context in which it is thought of and how it is implemented will vary, similar to OOP. However, much like how the general rules don't change from language to language, neither will the rules of SOLID. With that in mind, let's explore the principles that govern SOLID programming.

## The governing principles of SOLID programming

As we stated in the last section, there are five underlining principles that govern SOLID programming. They are as follows:

- **S**: **Single-responsibility principle (SRP)**
- **O**: Open-closed principle (**OCP**)
- **L**: **Liskov substitution principle (LSP)**
- **I**: **Interface segregation principle (ISP)**
- **D**: **Dependency inversion principle (DIP)**

The first, and in my opinion, the easiest and most important principle to explore is the **SRP**.

### The single-responsibility principle

The SRP is, in my opinion, the most important of the five principles to implement. In short, the SRP states that a code module should do one thing and one thing alone, kind of like the way we defined what a function should do in *Chapter 5*. In short, this goes back to the one-sentence rule. If you have to use the word *and* to describe your module, you have violated the SRP and you should break the service out. Generally, this is a trick that many experienced developers use to ensure that code components are properly broken out.

With that being said, what qualifies as a code module? Well, that is kind of an open-ended question. However, as an everyday developer, the code modules that you will interact with the most will generally consist of the following:

- Functions/methods
- Classes/function blocks
- Microservices

As a developer, any time I'm developing a code module like the ones just mentioned, I will usually try to summarize its responsibility in a complete sentence without the word *and*. If the word *and* appears in the sentence, it will usually signal to me that I have a module that is doing more than one thing. This is especially true for methods and functions. Even in the realm of traditional software development, it is not uncommon to see functions/methods that do multiple things. This will usually lead to serious trouble when one of the module's responsibilities has to be changed. In short, the moment you change a line of code in a module, that module should be treated as new, untested, and potentially defective code. So, if you have a single function that can, for example, play the games blackjack and solitaire and you change the rules of one of the games, you've essentially broken both games. As such, you will have to spend time testing to ensure that both games work as opposed to only testing the game you modified. In terms of the profit-driven automation industry, this can result in delayed machine deployment, downtime/loss of money for the customer, and a loss of money and productivity for your organization. As such, it can be said that it pays to follow the SRP.

## *Implementing the SRP*

One key module that I have found that usually gets the slip when it comes to the SRP is functions/methods. For some reason, developers (both experienced and non-experienced) love to bunch several different responsibilities into a single module. So, when things inevitably have to be changed, guess what? You end up having to retest several different behaviors. To demonstrate this with code, consider the following function with the following variables:

```
FUNCTION Motors : BOOL
VAR_INPUT
     pos            : INT;
     motorPos       : INT;
END_VAR
VAR
     turnOnMotor : BOOL;
END_VAR
```

The body of the function is composed of the following:

```
//turn on motor
IF turnOnMotor = FALSE THEN
     turnOnMotor := TRUE;
END_IF

//home motor
IF motorPos <> 0 THEN
motorPos := 0;
```

```
END_IF

//set the new pos
MotorPos := pos;
```

The body of the function shows that this function is responsible for turning the motor on if it is off *and* homing the motor if it is not homed *and* finally, setting the position of the new motor.

Now, suppose that our boss wants to modify this so that the function toggles the motor state as opposed to just enabling the motor. This means that there is a possibility for new bugs to be introduced and we can't reuse any of this code. As such, in the best of cases, we will get unneeded and redundant code that may or may not have defects in it. Though redundancy in automation engineering isn't necessarily bad, redundant code in software development of any kind is. In any case, we now have to run regression testing to ensure that the motor still homes and the motor position is still set properly, as well as test to ensure that the new code is performing its intended functionality. Now, as opposed to only testing the new feature (in this case, the toggle function), we have to test all three responsibilities.

If we were to summarize this function in a sentence, we would get something like the following:

*This function turns on the motor and homes the motor and positions the motor.*

As can be read in the sentence, the word *and* appears twice. As such, it can be said that this function is performing at least three responsibilities. In these situations, it is usually a good idea to break out each responsibility that comes after the word *and* into a function of its own and call those functions from another. As such, a better solution would be to create three functions like the following:

homeMotor (FUN)
motorOn (FUN)
Motors (FUN)

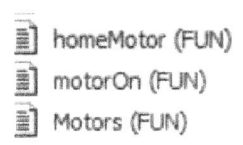

Figure 10.1 – Functions

The `homeMotor` function will consist of the following code:

```
FUNCTION homeMotor : BOOL
VAR_INPUT
END_VAR
VAR
    motorPos    : INT;
END_VAR
```

The body will consist of the following:

```
//home motor
IF motorPos <> 0 THEN
     MotorPos := 0;
END_IF
```

Next, the `motorOn` function will consist of the following:

```
FUNCTION motorOn : BOOL
VAR_INPUT
END_VAR
VAR
     turnOnMotor : BOOL;
END_VAR
```

The body of the function will be as follows:

```
//turn on motor
IF turnOnMotor = FALSE THEN
     turnOnMotor := TRUE;
END_IF
```

Finally, what consisted of the motor function will now be reduced to the following:

```
FUNCTION Motors : BOOL
VAR_INPUT
     pos          : INT;
     motorPos     : INT;
END_VAR
VAR
END_VAR
```

The body of the function will now consist of the following:

```
motorOn();
homeMotor();
//set the new pos
MotorPos := pos;
```

In this case, all we did was break out the logic into separate functions and call those functions within the original function. By doing this, we can now modify them without directly breaking the other functions and we can call them individually, which means no redundant code.

Now, this is just a general example with code that is little more than pseudocode. Though you are not directly changing or at risk of breaking outside code, it is still generally a good idea to test the dependent functionality. However, the main takeaway here is that we freed up code to be reused.

Now that we have a basic understanding of the SRP, we need to move on to the next principle of SOLID, the **OCP**.

## The open-closed principle

In software development, you generally don't want to modify code. Usually, the general rule of thumb is that once a code module is implemented, you don't want to ever touch it. This kind of leaves us in a pickle as, at some point, we are going to need to add new features; in short, we are going to have to eventually scale the program.

The OCP is mostly seen in OOP. If you perform a Google search on the OCP, you will usually find a definition that states the following:

*Objects or entities should be open to extension but closed to modifications.*

This essentially means that instead of modifying a class/function block, it is better to extend it using inheritance. This means the OCP is ultimately a design principle. Since the principle is a design principle, it is something that needs to be baked in at the design phase and it will usually take practice to properly implement the principle.

This principle also walks hand-in-hand with the SRP in my opinion; however, the OCP will be more concerned with modules such as classes/function blocks as opposed to modules such as functions. To properly implement this principle, you must have a clear understanding of the inheritance as an *is a* relationship and the SRP. In short, properly implementing the principle starts in the design phase of the SDLC.

### *Implementing the OCP*

Implementing the OCP takes practice to master, and like many other things in programming, it is often subjective what constitutes a properly open-closed architecture. However, consider that we have a function block that drives a motor, as in the following example.

For example, we are going to examine a function block that consists of two methods. The function block will be called `MotorControl` and it will have a `motorOff` and `motorON` method. Both methods will use a series of function block-level variables that will look like the following:

```
FUNCTION_BLOCK MotorControl
VAR_INPUT
```

```
    END_VAR
    VAR_OUTPUT
    END_VAR
    VAR
        motor : INT;
        wait  : WSTRING;
        state : BOOL;
    END_VAR
```

The `motorON` method will look like the following:

```
CASE motor OF
1:
    //company 1
      wait := "10 ms";
      state := TRUE;
2:
    // company 2
    wait := "2 ms";
    state := TRUE;
END_CASE;
```

The `motorOff` method will consist of the following:

```
CASE motor OF
1:
    //company 1
    wait := "10 ms";
    state := FALSE;
2:
    // company 2
    wait := "2 ms";
    state := FALSE;
END_CASE;
```

If you look at the code, you will see that both of the motor functions each has a case for different types of motors. The code shows that each type of motor will have a unique pause before the motor is either turned on or shut down. In short, this code will work for the two motors. However, suppose we want to add a third motor with a waiting period of 5 milliseconds. To implement this, we would have to modify two different methods. As such, we would be breaking the golden rule and changing the existing code. A change like this would include an extra case with a state change and a wait time. This isn't that huge of a deal, but a simple change like this can lead to broken code for motors that are not relevant to the upgrade. At the very least, to be thorough, you will have to retest each of the motors. As we established before, in the productivity- and profit-driven world of industrial automation, this can lead to extra downtime, which, in turn, will lead to a loss of money.

To summarize the code thus far, we created a program that does not follow the OCP, and, as such, we created code that is not scalable. Our program cannot be modified without manipulating old code. As such, our program is not architected well, and sooner or later, we will end up with issues such as breaking code. In all, if this code is deployed to a customer site, it will eventually end up costing downtime and money.

With that in mind, how can we alter the code to keep it in line and create a more flexible and SOLID architecture? The answer to that will reside in the `motorControl` function block. Conceptually, the `motorControl` block does follow the SRP because if we were to describe it with a sentence, it would read like the following:

*The function block controls the state of the motors.*

It can be argued that the motors being plural could mean that it is not fully in compliance with the SRP. This type of semantics is where the one-sentence rule can get gray. As such, as we design the software, this is one area that needs special attention as we are kind of in compliance with the SRP, but we're still experiencing issues with flexible code.

To fix these scalability issues with this program, let's redesign the program using **Unified Modeling Language** (**UML**).

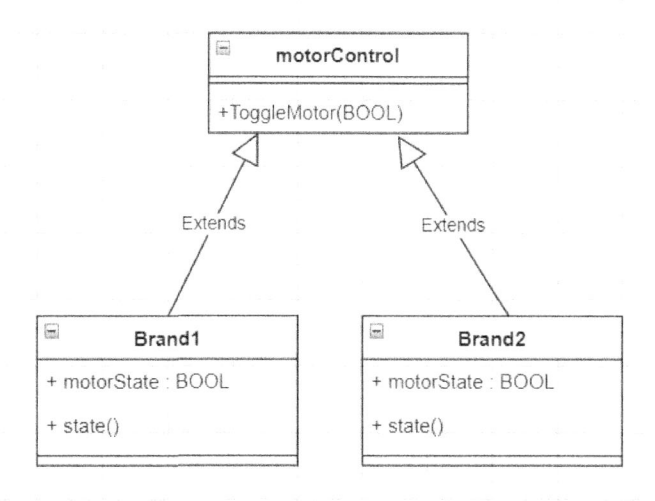

Figure 10.2 – Open-closed motorControl function block

In real life, there will probably be custom methods, overriding attributes, and so on. However, with this design, the motor brand classes will inherit from the `motorControl` function block. In this design, the only thing the `Brand` function blocks will keep track of is the state of the motor. The way this program is designed, if a new brand has to be added, all we would have to do is create a new function block that extends `motorControl` to control the new motor. This is in contrast to the old design where, if we wanted to add a new motor, we would have to modify at least two different methods.

If we were to turn UML into code, it would have the following structure:

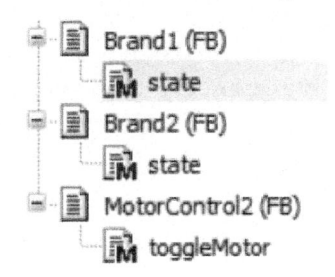

Figure 10.3 – Open-closed project tree

The code for the `toggleMotor` method in the `MotorControl2` function block should match the following:

```
METHOD PUBLIC toggleMotor : BOOL
VAR_INPUT
    state : BOOL;
END_VAR
```

The body of the method should look like this:

```
IF state = TRUE THEN
    toggleMotor := FALSE;
ELSE
    toggleMotor := TRUE;
END_IF
```

The variable logic for the `Brand1` and `Brand2` function blocks will be simple, like the following:

```
FUNCTION_BLOCK Brand2 EXTENDS MotorControl2
VAR_INPUT
END_VAR
VAR_OUTPUT
END_VAR
```

```
VAR
     motorState : BOOL;
END_VAR
```

The logic for the `Brand1` method will resemble the following:

```
//turn motor off
motorState := toggleMotor(TRUE);
```

By the same extension, the `Brand2` method will resemble this:

```
//Turn motor on
motorState := toggleMotor(FALSE);
```

As can be seen in the two methods for this example, both methods are very similar and only pass an argument to the `toggleMotor` method.

With this design, if another motor from `Brand3` were to be added to the software, all you would have to do is extend a third function block. As such, with this design, we can easily add a new motor from a different brand. All we have to do is create a new function block that extends the `MotorControl` function block.

In short, this design is much more flexible and scalable. With this design, scaling the functionality of the program is as simple as adding a new function block. This is a major improvement over the original design, which required many different code changes to existing code to simply add support for another motor.

With all that being said, it is unrealistic to think that you are never going to modify old code. Old code will eventually have to be changed – there is simply no way around that. You will eventually have to update a function block or method to add a new base feature, such as a speed control method or whatever. There is no way of knowing the future and, as such, there is no way of working everything into a design. Now, what the OCP is getting at is that you don't want to have to modify existing code to add support for new things such as extra motors or the like. Generally, if it is a functionality that is reflected across all child function blocks or targets a specific function block, it is, in my opinion, okay to modify. However, if you are constantly having to modify code to add support for new components, you probably need to revisit your design with the OCP in mind.

In summary, the OCP is a design principle. This is not something that you can bake into your code on the fly, as it will need to be addressed during the design phase. Mastering this rule will take practice to implement properly, and people may disagree with the overall best course of action. However, with practice, you will learn how to implement this rule, which, in turn, will increase the quality of your code. With all that being said, we can now move on to the *L* in SOLID, which stands for the **LSP**.

## The Liskov substitution principle

The LSP is without a doubt one of the hardest SOLID concepts to understand and implement. The concept is usually defined with something akin to the following:

*Objects of a parent class/function block should be replaceable with objects of a child class/function block without affecting the behavior.*

In short, this idea means that the child function blocks should not restrict or change the behavior of the parent function blocks. This is a simple idea but it can be hard to understand or implement.

In short, what the LSP boils down to is that you should be able to swap a parent function block reference with a child function block reference and get the same result. To properly implement this principle, you must have developers that are knowledgeable on the topic. This isn't a principle that can normally be enforced by outside software; instead, you usually have to find violations by testing and, in the case of automation programming, using code reviews. As such, it is my experience that when an organization tries to implement SOLID, this principle can be either overlooked or warped since enforcing it is usually a matter of style.

### *Implementing the LSP*

To best understand the LSP, it should be demonstrated in an example. To demonstrate the concept, it is common to use a square and rectangle as an example. This is a common example and it is a good example to research and grasp the LSP. In practice, both a square and a rectangle have an area equal to the following equation:

$$area = length * width$$

In mathematics, a square can be defined as a rectangle; therefore, we have an *is a* relationship so a square will use the same area equation and can inherit from the rectangle function block. As such, we will structure our program to match the following code:

```
FUNCTION_BLOCK rectangle
VAR_INPUT
        length : LREAL;
width : LREAL;
END_VAR
VAR_OUTPUT
END_VAR
VAR
        area : LREAL;
END_VAR
```

The body of the function block will simply be composed of the following:

```
area := length * width;
```

For this function block, we will add a single `getArea` method that will consist of the following logic in the body:

```
getArea := area;
```

Once that is complete, we will construct the `square` function block.

The `square` function will not have any variables, as all the variables will come from the `rectangle` class. In short, after inheritance, we will have a `length` and `height` variable.

The body of the function block will comprise the following:

```
width := length;
```

We will also add a `getArea` method that will be congruent to the one in the `rectangle` function block.

To run this code, we will need to configure the `PLC_PRG` file to match the following:

```
PROGRAM PLC_PRG
VAR
    rec : rectangle;
    sqr : square;
    recArea1 : LREAL;
    sqrArea1 : LREAL;

END_VAR
```

The body of the `PLC_PRG` file will need to match this:

```
rec(length:=5, width:=10);
sqr(length:= 5, width:=10);

recArea1 := rec.getArea();
```

Notice that on the first two lines, we have `rec` and `sqr`. When these lines are run, we should get an area of 50 and 25. Next, we replaced `rec1` with `sqr`. Technically, the arguments are not describing a square but, according to the LSP, we should be able to replace `rec1` with `sqr`, and if all complies, we should get the same answer. So, to test this code, run it, and you should be met with what is in *Figure 10.4*:

| Expression | Type | Value |
|---|---|---|
| ⊞ ◆ rec | rectangle | |
| ⊞ ◆ sqr | square | |
| ◆ recArea1 | LREAL | 50 |
| ◆ sqrArea1 | LREAL | 25 |

Figure 10.4 – Non-compliant substitution

We replaced the base function block with the child function block and, in the case of *Figure 10.4*, the areas for the square and the rectangle are not the same. Though a square logically cannot have a length and height that are not equal, the code should still have the same answer. In short, the LSP is not concerned with the validity of the answer, only the ability to swap out a base function block reference for a child reference and still have the same answer.

A logical question at this point is why is this and what does this mean? Well, in short, what it means is that even though a square is a rectangle, they are not compatible enough for an inheritance to be properly applied and for everything to still make sense. In other words, a square is a rectangle but a square is not enough of a rectangle for the square to inherit from.

So, how can we fix this? How can we make this program compatible with the LSP? Well, to answer the question, we need to add another function block. Our current program is configured like *Figure 10.5*:

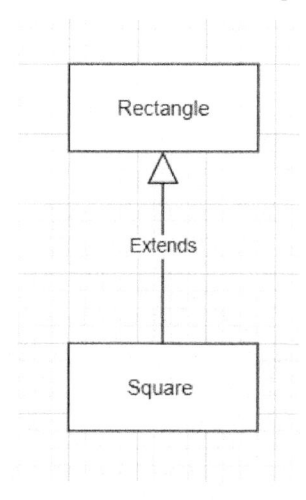

Figure 10.5 – Non-compliant program

One thing that we can do to fix this is to create a more appropriate base function block called `shape` and have both the `rectangle` function block and the `square` function block inherit from it, as in the following diagram:

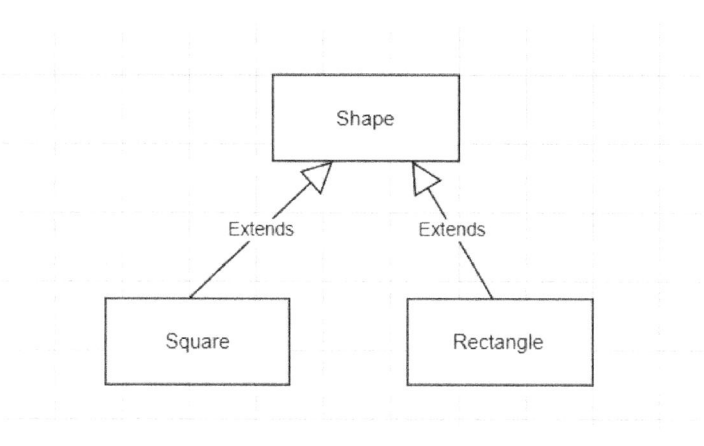

Figure 10.6 – Liskov solution

To better understand the LSP, let's look at a more practical example. For this example, suppose that we have a `brand1` function block that extends a `Motor` function block, as in *Figure 10.7*:

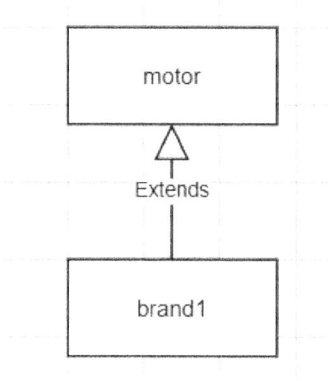

Figure 10.7 – Liskov-compliant motor program

The variables for the `Motor` function block are as follows:

```
FUNCTION_BLOCK Motor
VAR_INPUT
    speed : LREAL;
END_VAR
VAR_OUTPUT
END_VAR
VAR
```

```
        spd : LREAL;
END_VAR
```

Next, we will set the body of the function block with the following:

```
spd := speed;
```

With this logic set, we will create a getSpeed method in the Motor function block with the following:

```
getSpeed := spd;
```

This should do it for this function block. The next thing will need to do is set the Brand function block.

These will be the variables for the Brand1 function block:

```
FUNCTION_BLOCK Brand1 EXTENDS Motor
VAR_INPUT
        bspeed : LREAL;
END_VAR
VAR_OUTPUT
END_VAR
VAR
END_VAR
```

The logic for the function block will utilize the following:

```
spd := bspeed;
```

This should be all the preparation for the function blocks. The final file to set up will be the PLC_PRG file, which will utilize the following code. These are the variables that will be needed for the example:

```
PROGRAM PLC_PRG
VAR
     brand       : Brand1;
     mot         : motor;
     motSpeed    : LREAL;
     brandSpeed  : LREAL;
END_VAR
```

The body of PLC_PRG will comprise the following:

```
brand(bspeed:=10);
mot(speed:=10);
```

```
motSpeed    := mot.getSpeed();
brandSpeed  := brand.getSpeed();
```

Overall, when the program is run, you should be met with what is in *Figure 10.8*. If you study the code and the output, you will see that no matter what values are passed in, we can swap out the child and base function blocks. This means that this program is compliant with Liskov.

| Expression | Type | Value |
|---|---|---|
| brand | Brand1 | |
| mot | motor | |
| motSpeed | LREAL | 10 |
| brandSpeed | LREAL | 10 |

Figure 10.8 – Liskov-compliant program

Now, due to the nature of inheritance, the child function block can be (and should be) expanded upon. In short, you can and will have attributes that are not in the base function. This means, if you try to call an attribute that only exists in the child function block, the call will not work and the program will crash. As such, this means you can't swap out every single child object with its parent. Ultimately, this is not what the LSP is. The LSP is merely a principle to ensure that you are not limiting the base function block and, as such, you should only concern yourself with swapping out objects where base function block attributes are called.

The LSP is an in-depth principle, and it can be very tricky to fully utilize or understand, for that matter. To master this principle, you will need practice and a decent set of eyes watching over your shoulder to ensure that you are doing it right. These examples are very oversimplified examples so you can get the gist of the idea. Many experienced developers have difficulty understanding this concept, so it is recommended you spend some time researching the concept in detail and working through some more examples. However, once you master the skill, you will be able to take your design to the next level, and your function blocks will reach a very pure level of inheritance.

With all that being said, we can now move on to our next principle, which is known as the **ISP**.

## The interface segregation principle

As we have explored, anytime we implement an interface, we have to implement all the methods that come with it. This can be good and bad in a way. In a sense, when we implement an interface, we never have to worry about accidentally missing a method. On the other hand, if we are not careful and our interfaces are not designed well, we can end up with methods that are not used in the function block, which is a bad practice. In programming, we don't want unused code in our programs as it can clutter and bloat the codebase, and in terms of the interface, there may be major issues with the implementation if it is deleted by accident.

This is where the ISP comes into play. Essentially, the meaning of the principle can be summed up with the following:

*A function block should not have to implement an interface it does not use nor should it depend on methods that it does not use.*

In other words, the best way to think of this principle is to *use it or lose it*! It is very sloppy but common practice to leave unimplemented code in your program due to interfaces. Though common, it should be avoided whenever possible, and if you see your function block is dependent on methods that it does not use, it is a definite time to redesign your interfaces.

So, how can this principle be accomplished? The solution is simple; you can usually implement multiple interfaces in a function block. It is important to remember that interfaces are models. If your function block is a hybrid of multiple things, you can use multiple models to craft it. For example, consider a checking and savings account. Both accounts will allow you to withdraw money and deposit money; however, a savings account will build up interest over time. Though this type of program would not normally be utilized in a PLC, for the sake of example, let's look at some code to explore this concept.

### *Implementing the ISP*

An inexperienced programmer may do something like the following to implement `accounts`:

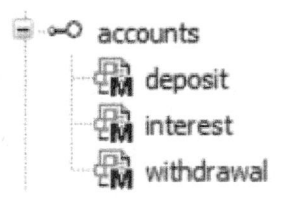

Figure 10.9 – Accounts interface

In this case, we have three methods: `deposit`, `interest`, and `withdrawal`. When we implement this interface, we get the following:

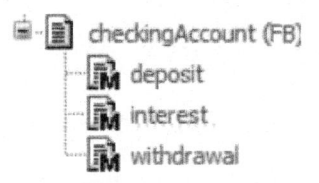

Figure 10.10 – checkingAccount function block

As can be seen in this case, the `interest` method was automatically implemented when we implemented the `accounts` interface. This means that we have unused code and, as we stated before, this is not good. We violated the ISP and, as such, we are now dependent on unused code. Essentially, the ISP boils down to having client-specific interfaces over general interfaces. So, with that in mind, how can we fix this?

Well, the cleanest way would be to create a savings accounts interface that extends a `checkingAccount` interface. To do this, we are going to create a saving and checking account function block as well as a saving and checking account interface, similar to the following:

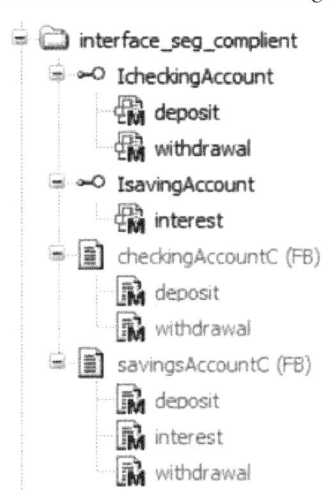

Figure 10.11 – Interface segregation compliant

In this example, the `savingAccount` function block implements the `IsavingAccount` interface. The `IsavingAccount` interface extends the `IcheckingAccount` interface and, as such, the `savingAccount` interface will consist of the `deposit`, `withdrawal`, and `interest` methods. With this implementation, the `savingAccount` function block will have all three necessary methods.

On the other hand, the `checkingAccount` function block will only implement the `checkingAccount` function block. This means that it will not require the `interest` method, which, in turn, means that the function block will not have unnecessary code. As such, the program is now in compliance with the ISP. Now that we have the ISP under our belt, we can look at the final principle, the **DIP**.

## The Dependency inversion principle

When developing modern-day software, you want your code to be as loosely coupled as possible. It is common, especially in automation, to have to work with various types of libraries and APIs. The kicker to this is that this software will change as will the parts in the machine. For example, it is not uncommon for the customer to opt to put a different brand of hardware in the machine. This means that if you have something such as a motor drive that requires a certain library, your software will need to be changed to accommodate. As has been the whole point of this chapter, you don't want to change existing software or change it as minimally as possible.

This is where loosely coupled architecture comes into play. When it comes to consuming low-level software components such as APIs or the like, you don't want your software closely tied to it. To use something like a drive library, you want to create a middleman component to act as a go-between. In other words, you don't want to talk directly to the API; you want to talk to a function block that will talk to the API for you.

The middleman API is like a facade function block. The middleman will always have a set of methods that will never change. For example, you could have an on, off, and ready method that your high-level code will call. The middleman function block will be responsible for determining which API method(s) to call to accomplish the task. In terms of a block diagram, the DIP can be viewed as the following:

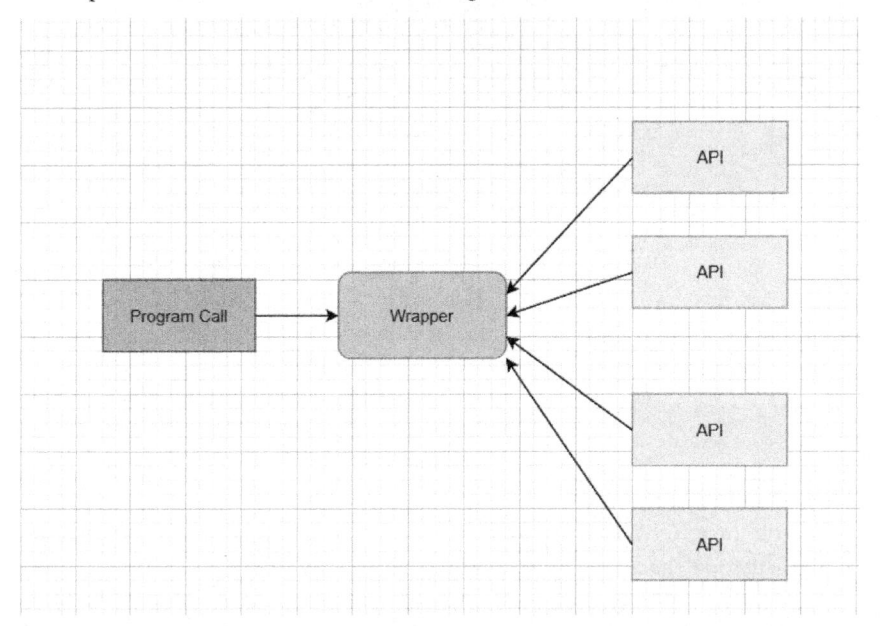

Figure 10.12 – Dependency inversion diagram

### *Implementing the DIP*

Suppose we have two APIs, called `api1` and `api2`. These two APIs will represent a simple control API that turns a device on or off. To promote a loose design, we are going to create a `drive` function block that will serve as a middleman or facade for the `cart` class, which will be the high-level code that the user is manipulating with calls from a control panel.

For this example, we are going to create a GVL file and the following function blocks:

- `cart`
- `drive`

- `api`
- `api2`

When we are done, the structure should look like the following:

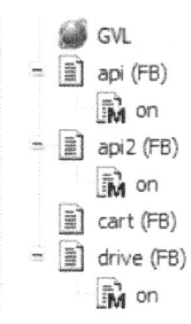

Figure 10.13 – Dependency inversion project

Once you have this in place we can start to set up the code.

The `api` function block will only have a single line of code in the `on` method, which will be the following:

```
GVL.msg := "api1";
```

The same can be said for the `api2` function block, which will consist of the following:

```
GVL.msg := "api2";
```

The `drive` method will consist of the following:

```
FUNCTION_BLOCK drive
VAR_INPUT
END_VAR
VAR_OUTPUT
END_VAR
VAR
    a1 : api;
    a2 : api2;
END_VAR
```

The method will consist of the following variables:

```
METHOD on : BOOL
VAR_INPUT
     api : INT;
END_VAR
```

The body will consist of the following:

```
IF api = 1 THEN
     a1.on();
ELSE
     a2.on();
END_IF
```

The next function block to set up is the `cart` block, which will comprise the following:

```
END_VAR
VAR_OUTPUT
END_VAR
VAR
     d : drive;
END_VAR
```

The body of the `cart` method will comprise the following:

```
d.on(in);
```

The last file to set up is the `PLC_PRG` file, which will consist of the following variable:

```
PROGRAM PLC_PRG
VAR
     c : cart;
END_VAR
```

To call the `drive` function block and utilize our mock APIs, we will use the following line:

```
c(in:=2);
```

When the program is run, you should see the following output in the GVL:

| Expression | Type | Value |
|---|---|---|
| msg | WSTRING | "api2" |

Figure 10.14 – GVL output for the program

If you replace 2 in the final line of the `PLC_PRG` file with 1, you should see the other `api` method getting called. In short, what we have done is created a loosely coupled program that can support minimal modification to the program. Now, this code was merely for demonstration. The `if` statements are not the most effective or robust way of implementing this program. They were used as a means to easily demonstrate the concept of using a middleman. A much better solution would be to either create a factory function in the drive class or simply pass around a reference to the objects; either of those would be a better and more SOLID solution to the situation.

Regardless of how you do it, the goal is to demonstrate what loosely coupled software would look like. The goal here is to show that we should be designing our software in a way that our high-level code is not reliant on the low-level code, such as API code, to function. In short, by doing this, you will need fewer code changes to adapt and have an overall more flexible design.

By this point, we have explored all of the SOLID principles. SOLID is not an approach that can be easily mastered, as it is more of a mindset that must be practiced to be understood and integrated into your future code. Now, that we have a basic understanding of SOLID, let's try to build a simulated painting machine.

# Final project – a painting machine

Painting machines are often complex devices that have many moving parts. For our final project, we are going to build a simulated device that can move the part on a conveyor belt and paint a sentence on the part. For this project, we are going to set the following requirements:

- Drive the conveyor belt (belt on/off)
- Select between two paint APIs
- Paint a message on a part

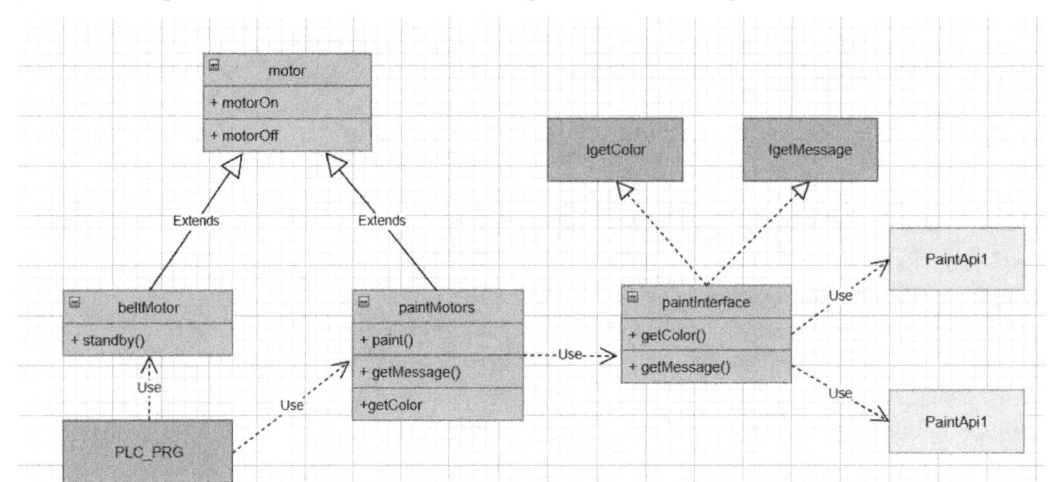

With these requirements, we can formulate a design like the following:

Figure 10.15 – Painter design

Compared to other programs that have been presented in the book so far, this one has many more components and lines of code. As such, no code for this example will be displayed in the book. However, a working example can be found at the URL that was provided in the *Technical requirements* section. For this chapter example, the code can be found in the final project directory.

In this case, the `PLC_PRG` file is going to act as our orchestrator. The file will control when the belt is running and when the machine is painting. We have a `beltMotor` function block that will inherit the attributes from the base motor function block, as will the `paintMotors` block. The `paintMotors` block will use the `paintInterface` function block as a middleman. The `paintInterface` function block will use two different paint APIs, and it will also implement two interfaces, called `IgetColor` and `IgetMessage`.

In short, this example is SOLID. All of our methods are describable with one sentence and our `motor` function blocks are open-expansion but closed to change, which is why we are extending `beltMotor` and `paintMotors`. There is one flaw in this design: this configuration does not distinguish between the painter motors and the belt motors. However, this is merely for a demonstration and, as such, we are not going to worry about it as our goal is to utilize SOLID programming. Moreover, the motor blocks also enforce the LSP, as the `paintMotors` and `beltMotor` function blocks can be swapped out with the parent function block, `motor`, and everything will still work in terms of turning the motors on and off.

The `paintInterface` function block will enforce the ISP as the two interfaces only provide the relevant methods for the function block to use. As such, if at some point we decided that we no longer need a `getColor` method, we can remove this interface and the method with no leftover code. The `paintInterface` function block is also a facade wrapper that provides a buffer between the APIs and the `paintMotors` function, which tells `paintInterface` to start. Essentially, this function block is just there to ensure that the API function blocks are loosely decoupled.

The two APIs are just function blocks that exist to simulate a real API. In short, all these APIs do is ensure that they return the color and message. In real life, these APIs would be much more complex; however, we kept them simple for this example. Since we have a middleman function block between the APIs and the `paintMotors` function, we are compliant with the DIP.

As such, this example has incorporated each principle of SOLID programming. Designing a program with SOLID in mind will be complex and may have more components, as was seen in *Figure 10.15*. By this point, you should have a decent understanding of SOLID programming.

## Summary

As was seen in this chapter, SOLID can take some extra effort during the design phase. However, when done correctly, these five principles can ensure that your code is easy to fix and expand upon. In the fast-paced world of industrial automation, this is a must. You need to be in and out of a customer site as quickly as possible. As such, when implemented correctly, SOLID will support this.

Each of these principles will take some time to master but, once you do, the payback will be well worth it. At this point, you should have a good enough background to start expanding your knowledge and experience of these concepts. Hence, with all this in mind, we can now move on to another very important aspect of automation programming: HMI design.

## Questions

Answer the following questions based on what you've learned in this chapter. Cross-check your answers with those provided at the end of the book, under *Assessments*.

1. What are common code modules?
2. What should you do if the word *and* appears in the summary of your module?
3. What is the Liskov substitution principle?
4. What is the interface segregation principle?
5. Name the five principles of SOLID.

## Further reading

Have a look at the following resources to further your knowledge:

- *SOLID: The First 5 principles of Object-Oriented Design*: `https://www.digitalocean.com/community/conceptual_articles/s-o-l-i-d-the-first-five-principles-of-object-oriented-design`
- *SOLID*: `https://en.wikipedia.org/wiki/SOLID`
- *Exploring the Liskov Substitution Principle*: `https://www.infoworld.com/article/2971271/exploring-the-liskov-substitution-principle.html`
- *LSP*: `https://medium.com/@gabriellamedas/lsp-the-liskov-substitution-principle-e43910b638bc`
- *SOLID Design Principles – Liskov Substitution Principle (LSP)*: `https://medium.com/@gabriellamedas/lsp-the-liskov-substitution-principle-e43910b638bc`

# Part 4 – HMIs and Alarms

This section explores HMI design. More often than not, PLC programmers are not well versed in the art of HMI design. Generally, they do not understand how to properly lay out HMIs so operators can easily use them. Normally, HMI development is not touched on in books about PLCs, nor do PLC developers usually have a good grasp of HMI development.

The section also explores alarms. Alarms are used to avoid catastrophic issues and to report potential issues to the operator. Many PLC programmers do not understand how to effectively implement or setup alarms. As such, this section will give you an in-depth look at HMIs, HMI controls, HMI Layouts, alarms, and alarm programming.

This section includes the following chapters:

- *Chapter 11, HMIs — UIs for PLCs*
- *Chapter 12, Industrial Controls — User Inputs and Outputs*
- *Chapter 13, Layouts — Making HMIs User-Friendly*
- *Chapter 14, Alarms — Avoiding Catastrophic Issues with Alarms*

# 11
# HMIs — UIs for PLCs

Everything has a **User Interface** (**UI**) of some type nowadays. The website you used this morning, your car's radio, and even the app you're reading this book on if you're reading a digital copy, your device has a UI of some type. Automation programming is no different. Everything uses a UI of some kind to either interact with the hardware or with other software.

There are two ways that your end users will interact with your PLC. They can either use some type of control panel that is built using physical hardware (for many applications, this is no longer a viable option), or they can use a **Human Machine Interface** (**HMI**). With the drop in the cost of computing power over the past 20 or so years, HMIs are now the primary way for end users to interact with a PLC program. In short, no matter what you're doing, chances are you're going to have an HMI for the operator to control the machine.

HMI development, in my opinion, is as much an art as it is a science and the best HMIs are usually designed by experienced programmers. HMIs are easy to develop and as such, they usually get built by the least experienced programmers, which can lead to issues. However, I usually like to classify this as an advanced skill because the HMI is usually the focal point of the machine when it is deployed. As such, if you don't have a quality, well-laid-out HMI the end users are probably not going to be happy with the machine no matter how well-constructed the PLC side of the software is. In short, I have seen many very well-engineered machines fall by the wayside due to a poor HMI. Therefore, this chapter is going to be an introduction to HMI development and will cover the following topics:

- What an HMI is
- Why create and use an HMI?
- How are HMIs developed?
- What should HMIs do?
- HMIs versus SCADA
- How the SDLC applies to HMI development
- Wireframing

Finally, we will round out the chapter by attaching an HMI to a PLC project. By the end of this chapter, you should have a good overview of HMI development.

## Technical requirements

For this chapter, the example can be found at the following URL: `https://github.com/PacktPublishing/Mastering-PLC-programming/tree/master/Chapter%2011`.

There won't be any development in this chapter as this is an introduction to HMIs. However, our final project can be pulled from the same URL.

## Understanding HMIs

When I have to explain what an HMI is to a developer that does not have a background in automation, I usually describe it as an industrial UI for machines. In short, an HMI serves as a digital control panel. An HMI is a program that will have digital buttons, switches, readouts, and so on that will run on some type of touchscreen computer. The HMI serves as the control panel that the operator will use to send commands to the PLC.

In my experience, I have seen HMIs make or break projects. HMIs are presentation layers for your project. As such, a poorly designed HMI can make your machine difficult, if not impossible, to use. It is important to keep in mind that these are soft control panels and, as such, a poor layout of controls will make the machine hard to operate. The customer will also pay particular attention to the HMI. For people not familiar with software engineering, the only thing they have to gauge quality on is how the controls are organized. In other words, you as a programmer should be more concerned with the internal functions of the PLC software to ensure that you can easily adjust the code to accommodate new needs. However, the customer or end user is going to care more about operating the machine on a day-to-day basis, and the way they do that is through the HMI. In short, I have seen machines that have terrible codebases be praised as super successful due to the HMI. On the inverse of that, I have seen excellent codebases be called utter failures due to a poor HMI.

HMIs are very tricky to implement. They require a good balance of engineering and attention to design, as well as scalability. As such, depending on what technology you use to build your HMI, you can have your work cut out for you. With that in mind, why should we use HMIs, and are there any benefits to using an HMI over a physical control panel?

### Why create and use an HMI?

Why should we use an HMI? That is a question I asked myself when I first learned what HMIs are so many years ago. What benefit could a digital control panel offer over a physical control panel? To explain this answer, let's look at what it takes to build a control panel. To build a working control panel, you need the following:

- Physical components such as buttons and switches, which cost money
- The time it takes to lay out the panel
- The time it takes to fabricate the panel to accommodate the components
- Wiring and troubleshooting of the panel
- Control redesign when something needs to change
- Time and money to build a new control panel when a change is made

This also does not take into account the time and cost it takes to replace a broken component when the part does go out, which it eventually will.

In contrast, you can expect the following when you build an HMI:

- The only physical component is a touch screen or computer to run the HMI
- The only panel fabrication is for the display mount
- There is no component cost since they are digital
- There is no worrying about electrical troubleshooting or failing physical components except for the machine the HMI is running on
- If a change has to be made, it will be as simple as adding in the digital control and hooking up the logic to the PLC
- You can pack different control panels in a single HMI program

In short, HMIs offer a much better user experience and are more cost-effective than a physical control panel. You will be able to more effectively change the behavior of the HMI and the layouts without ordering expensive hardware.

HMIs are, for the most part, the way to go for modern control systems. However, in terms of safety, you never want to use an HMI to control emergency shutdown functions. No matter how good the computer that runs the HMI is, the computer can crash, freeze, or do any number of other things that prevent an **emergency stop** (**E-Stop**) from engaging. Emergency shutdowns should always be done with a physical switch, no matter whether you have a redundant software control or not.

Now, what do HMIs look like? If you were to walk into a plant with an HMI that is meant to control two valves, how might it be laid out? Well, *Figure 11.1* shows a possible layout.

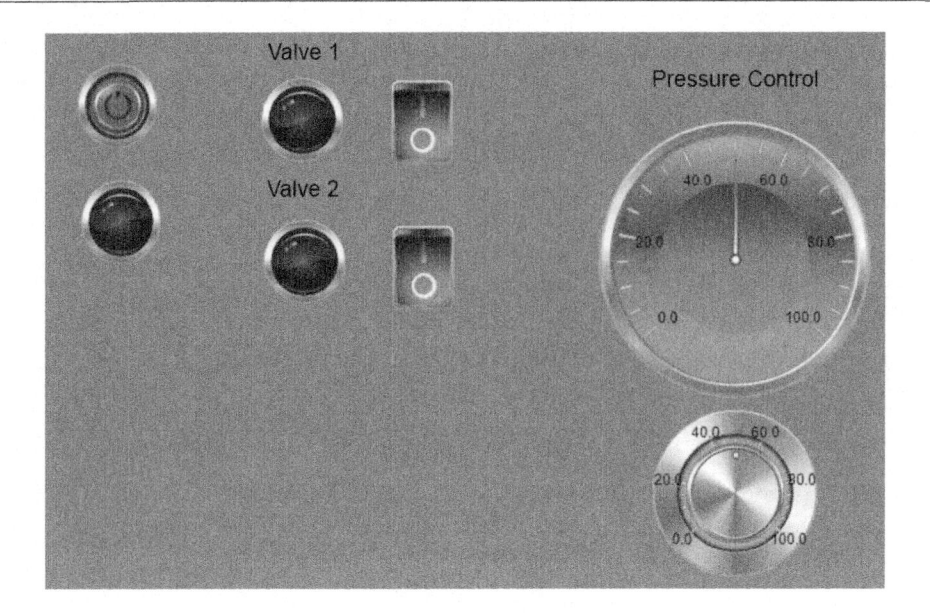

Figure 11.1 – A mock HMI layout

In this case, we have the main power button with an LED to signify whether the power is on. Next to that, we have our valve controls and each has a switch for the operator to turn them on or off, and an LED to show the valve's state. Finally, we have a pressure display and a potentiometer to control the pressure in the valves. So, with all that in mind, how are HMIs developed? How can we build a digital control panel like the one in *Figure 11.1*?

## How are HMIs created?

HMIs can be developed in any number of ways. You can use custom software packages or you can use traditional programming languages. Some common development tools are as follows:

- VB/C#.NET with either TCP/IP, TwinCat API, or another communication API
- Java with proper communication APIs
- C-More
- RedLion
- ClearSCADA
- Movicon
- Ignition
- The CODESYS HMI package

Many of the technologies that are not traditional programming languages will usually have drag-and-drop control layouts and have a version of BASIC, C, Python, or some other scripting language built into them to enhance HMI functionality. However, some systems, such as RedLion and C-More, will require you to use a custom touch screen panel that is specifically designed to run their HMI software. This can be problematic sometimes as replacements for these custom screens can be hard to obtain after they are no longer manufactured. In any case, these systems are excellent choices for developing simple HMIs.

On the other hand, systems such as Movicon, Ignition, and so on are true SCADA packages. These are usually more powerful and can interface with things such as databases and so on. Much like the aforementioned counterparts, they are also drag-and-drop with the ability to write scripts. However, unlike the raw HMI systems, these can usually be run on a regular computer. These systems can be more complex to use but offer more power.

## Programming languages to develop HMIs

In terms of raw power, using a general-purpose programming language such as C# or Java will provide the most bang for your buck. When developing an HMI using these technologies, you will use a framework such as .NET's WinForms or **Windows Presentation Foundation** (**WPF**). You will also need to communicate with the PLC using an API (a programming interface that acts like a driver) such as TwinCat, or some other means, such as TCP/IP, UDP, and so on. If you are using a PLC that has an available API for a traditional programming language and you choose to develop using a traditional programming language, it is recommended that you use that API over raw TCP/IP, UDP, or the like as that'll probably be the simplest and safest route.

Using a general-purpose programming language is an excellent means of developing HMIs. However, there are drawbacks. When you are developing using a general-purpose programming language, you must realize that you are developing a second program from scratch. As such, you will need to be proficient in the programming language of your choice and you will need to take the time to craft a design for the HMI. A general-purpose language such as C# or Java will not have the same features that a system such as Movicon or RedLion will have. Therefore, more care must be given to these HMIs compared to their traditional counterparts as more can go wrong and it will be easier to introduce more – and larger – bugs into the system. However, if you do choose this route, you will be able to produce much more powerful and sophisticated HMIs. It is similar to the old saying, "*greater risk, greater reward!*"

If you opt to use a general-purpose programming language and some form of communication API to develop your HMI with, you're probably going to want to use a framework with a designer built into it. A designer is a tool that gives you the ability to drag-and-drop controls similar to what CODESYS allows you to do. However, if you choose this route, you will have to write code for the HMI to behave the same way a canned HMI or SCADA system will automatically behave. If you do find yourself using one of these tools, it is best to use a framework such as the following:

- .NET -> WPF or WinForms

- Java -> WindowBuilder/JavaFX
- Python -> QT
- C++ -> QT

These are just a few designers that are out there. Choosing the right designer will boil down to what you need to do. With all this in mind, what should an HMI be responsible for in the grand scheme of the system?

## What should an HMI do?

Even as I became more experienced with HMI development, I could never quite get over the habit of wanting to put as much of the computing as possible in the HMI, especially when I used a general-purpose programming language such as C# or Java. It was too tempting to use all the cool, built-in .NET functions to try to perform calculations and so on. However, this is a really bad thing to do. Generally, all your heavy computing should be done on the backend by the PLC. If you have a background in web development, this principle may seem familiar. Your HMI (or as web developers would say, your frontend) only serves to send and receive data from the backend or PLC. As I grew as an automation engineer, I found myself only including things such as the following into the HMI:

- Sending signals such as button clicks or data such as numbers to the PLC
- Displaying data from the PLC such as temperatures, part counts, and so on
- Loading data such as XML or JSON data from other files so the data can be sent to the PLC
- Logging data that was received from the PLC in a file or database format

In short, if you ask an HMI developer what an HMI should do, they'll usually respond with, *"the HMI should be as dumb as possible"*. In other words, if the PLC can perform the task, allow it to. This is especially true if you're using a canned HMI system. A good HMI will do the bare minimum necessary to properly send and receive data from the PLC. A good rule of thumb that I eventually learned to live by was *the less the HMI had to do, the better it was designed*.

Many of the features that are listed are usually only accomplished using a general-purpose programming language such as C# or with canned SCADA packages. Though often confused with HMIs, SCADA packages are powerhouses compared to mere HMIs. Now that we have explored what an HMI should be responsible for, let's take a look at the difference between an HMI and a SCADA package.

## HMIs versus SCADA

It is very common for people, even experienced automation engineers, to confuse HMIs with SCADA systems. **SCADA** stands for **Supervisory Control And Data Acquisition**. When a person correctly refers to SCADA, they are referring to systems that include sensors, PLCs, RTUs, control software such as HMIs, and so on. SCADA systems are more for larger systems, for example, systems that will supervise whole plants.

In contrast, an HMI is designed to control a single machine. A machine's HMI will usually be placed near the machine, and it exists to operate that machine or a very limited group of related machines. Depending on the system you used to develop the HMI, you can network HMIs to a SCADA system. To do this, you will need to have a SCADA system that can support this type of functionality.

As such, the best way to think of the differences between HMIs and SCADA systems are as follows:

- **HMIs**: HMIs control a machine or small groups of machines. The HMI may or may not be a part of a larger SCADA system. Also, HMIs are usually positioned close to the machine they are meant to control. HMIs are programs, with the exception of the screen or computer they are running on. Their job is to input and display data from the PLC. As such, they are not composed of hardware such as sensors or PLCs. It is important to understand that they talk to a PLC, but the PLC is a separate part of the system.
- **SCADA**: SCADA systems are large systems that are designed to control and monitor whole processes. A SCADA system will usually control a whole plant and allow remote access for persons that are not on site. A SCADA system is composed of many different modules, such as PLCs, HMIs, RTUs, sensors, and so on. In other words, SCADA systems are high-level supervisory systems that tell other modules what to do. They will also perform actions such as logging data into databases. Whereas an HMI is just a UI, SCADA can best be thought of as a system that includes HMIs, PLCs, sensors, and so on.

In short, a SCADA system is a remote monitoring system that is composed of many different hardware and software components, while an HMI is a software component that is local to a machine or set of machines. Now that we have an understanding of HMI and SCADA systems, we can move on to how the SDLC applies to HMI development.

## How the SDLC applies to HMIs

Sadly, HMI development is often less rigorous than PLC software, which, as we have established, gets little attention compared to physical hardware. However, as we have stated in the past, the HMI will be the focal point of the software for the end user. If the HMI is not a quality HMI, the machine will run the risk of being pushed to the wayside. As such, it is important to follow the SDLC even with HMI development.

In my opinion, the design phase is of the most importance. You will usually design the layout of the HMI during the design phase. The design phase is usually the make-or-break phase of the SDLC and in terms of HMI development, even more so. As we will see in the next section, a very important part of the HMI design is wireframing, which is a technique used to create a layout of the HMI before you try to build it. Due to the nature of HMIs, you will want to be in contact with the customer/end user during the design phase.

If you opt to choose a development HMI tool such as the one in CODESYS, RedLion, Movicon, and the like, you will find that the actual build phase is straightforward and relatively easy. However, what you will want to pay attention to in the SDLC is the testing phase.

After the HMI and PLC software is completed, you're going to want to pay extra attention to integration testing. In short, HMI and PLC integration is critical. If an HMI does not send the correct signal to the PLC or vice versa, the system is useless. As such, to ensure that the HMI and PLC are behaving in unison and as expected, you will need to develop quality test cases and have a very clear understanding of the requirements to ensure that the system works as expected.

Bugs are bound to happen – nothing will ever go as expected. So, if you do end up in a situation where a bug appears to be present, you will need to track it down. This means that instead of troubleshooting one piece of software, you will need to troubleshoot two. Now, debugging multiple pieces of software can be a daunting task. In terms of debugging an HMI and PLC program, a good strategy is to start with the HMI and work toward the PLC code and try to isolate the bugs that way. Generally, you will want to keep an eye on your PLC code and look for signals from the HMI to ensure they are coming in. If the signals are being properly received by the PLC from the HMI, you can assume your PLC code is defective or you're operating the machine wrong.

In all, even with the simplistic nature of HMI development, the HMI should be considered a true software project and the SDLC should be followed as such. As mentioned before, wireframing is a very important concept and as such, we are now going to explore it.

## Exploring wireframing

Wireframing is a simple yet vital concept for any type of UI development. Generally, this is a practice that isn't carried out much in the automation world, but when it is, it can greatly benefit the overall HMI design. So, what is wireframing?

In short, wireframing is a design practice where you lay out UI/HMI designs on either paper or some type of rendering program before you start the development process. For example, a wireframe may look like *Figure 11.2*.

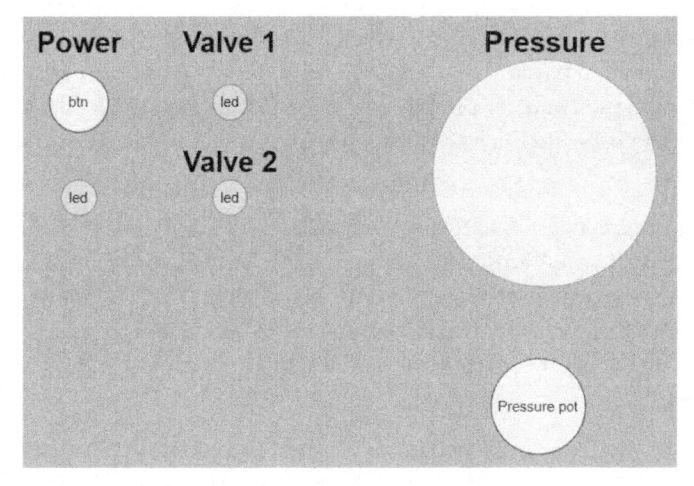

Figure 11.2 – Wireframe for mock HMI

As can be deduced, this is a simple diagram of the mock HMI in *Figure 11.1*. However, in the diagram, there are labeled components such as the LEDs and so on.

Generally, you don't have to wireframe in any particular software package as you can simply draw them out on paper or a whiteboard. You usually want to wireframe on a whiteboard or paper during things such as brainstorming sessions. If you're using an HMI development package, a SCADA development package, WPF, WinForms, or anything that has drag-and-drop control functionality, you only need to worry about wireframing to work out ideas, or present ideas during a brainstorming session, . Technologies such as the aforementioned that do provide drag-and-drop controls can make wireframing redundant as you are essentially doing the same thing. Usually, when I use a system such as the CODESYS HMI tool, I would just submit a screenshot in place of a wireframe. However, it is important to know that not every programming system supports drag-and-drop functionality, in which case you should wireframe. Now that we have all that down, we can move on to creating an HMI project.

## Final project – creating an HMI

The easiest way to create an HMI project is to simply create a standard project as we have done throughout the book. Once you create a project, you will want to right-click **Application**, navigate to **Add Object**, and select **Visualization** as in *Figure 11.3*.

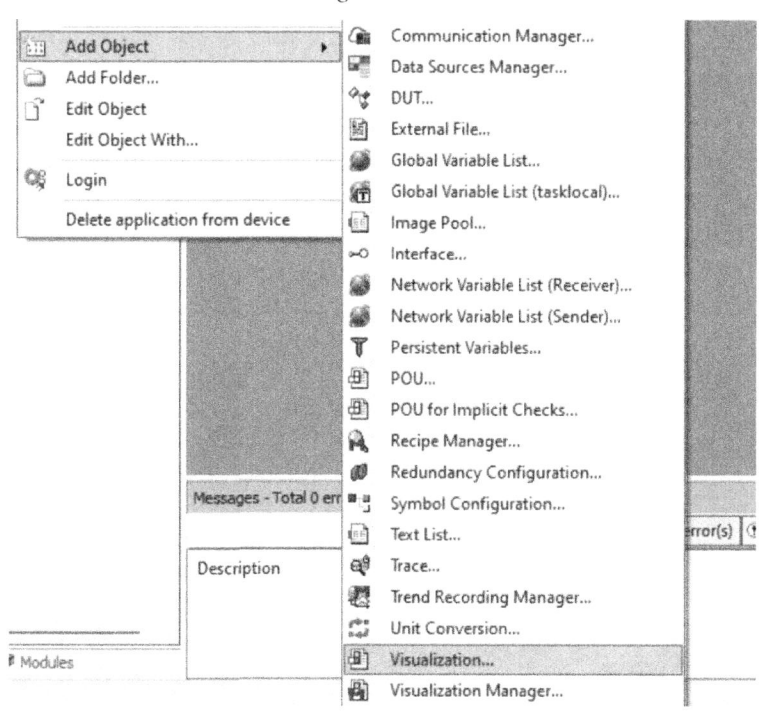

Figure 11.3 – Adding an HMI to a project

When you do this, you should be met with *Figure 11.4*.

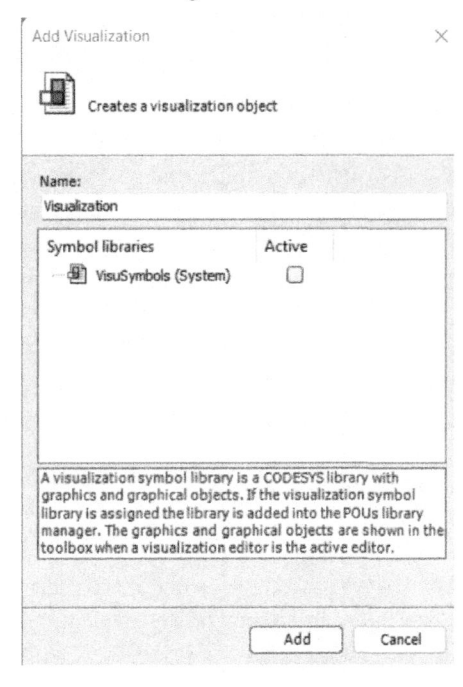

Figure 11.4 – Add screen

Click **Add** and wait a few minutes for the controls to render. Once you are done, you will be met with a new area to the right of the screen with HMI controls in it, similar to *Figure 11.5*. As can be seen in the figure, there are many different controls, such as LEDs and switches to choose from.

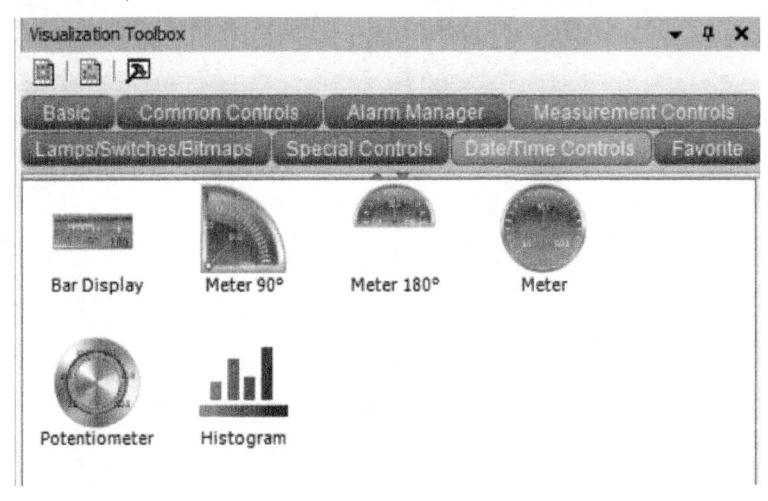

Figure 11.5 – HMI controls

There are also many different tabs that each have different types of controls grouped by functionality. To add a control to the screen, you simply click on it and drag it over to the screen area, which will add it to the layout. This is a very handy HMI development tool and it is the tool that we will use for the rest of this book to create our HMIs. The remainder of our projects will have HMIs attached to them so it is recommended that you become familiar with the layout.

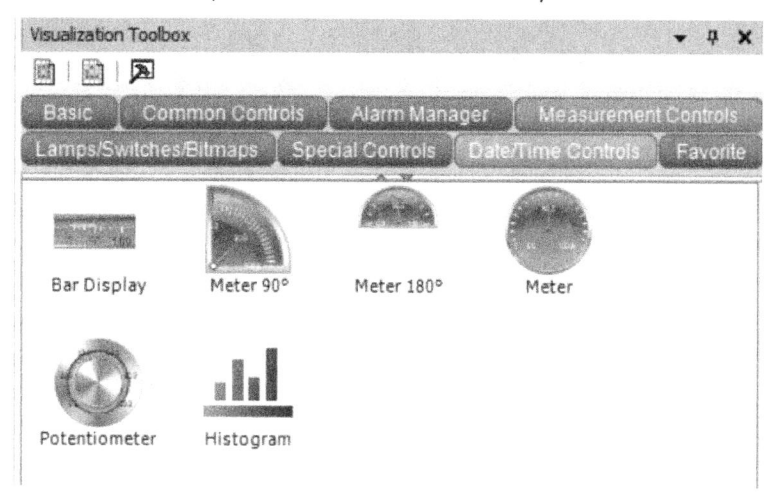

Figure 11.6 – Measurement Controls tab

The tab you are currently in will be highlighted in green. I would also strongly recommend becoming familiar with all the controls in each tab as we will be using many of them in the coming chapters. For now, click **Lamps/Switches/Bitmaps** and drop a light and a rocker switch onto the visualization screen. When you are done, it should look like *Figure 11.7*.

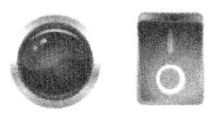

Figure 11.7 – Final HMI project

Next, modify the PLC_PRG file to match the following:

```
PROGRAM PLC_PRG
VAR
    light : BOOL;
END_VAR
```

Click on each element, click on the **Variable** field, and select the light variable for both. Run the program and click the rocker switch. Observe how the light will turn on and off.

## Summary

This chapter has explored HMIs. HMIs are pivotal pieces of software for any automation project. With the rise in power and drop in the cost of computing, HMIs are now permanently ingrained in the automation world. In short, you cannot be an automation programmer without knowing what an HMI is and how to develop one.

To the end user, the HMI will make or break the machine, they will gauge the quality of your machine based on the HMI. If the HMI is well designed and laid out, they will see the machine as better quality as opposed to a machine with a poor layout. In short, I always tell young automation programmers that the customer doesn't care about the hardware or PLC software; what they care about is how easy the machines are to operate and the key to that is a quality HMI.

As such, we have explored what HMIs are and how they are made. We have also explored different ways to develop HMIs and the difference between HMIs and SCADA packages. Lastly, we learned how to add an HMI to a project. In all, we can say that we have started our journey in HMI development. In the next chapter, we will start exploring the HMI controls and creating our first working HMI!

## Questions

Answer the following questions based on what you've learned in this chapter. Cross-check your answers with those provided at the end of the book, under *Assessments*.

1. What is an HMI?
2. How would you describe an HMI to a person that is not an automation engineer?
3. What is a SCADA system composed of?
4. What is wireframing?
5. Name three HMI development packages
6. Can you use C# or Java to build HMIs?

## Further reading

Have a look at the following resources to further your knowledge:

- PLC, DCS, SCADA, HMI – What are the differences?: http://dcs-news.com/plc-dcs-scada-hmi-differences/#:~:text=HMI%20%E2%80%93%20Human%20Machine%20Interfaces&text=The%20main%20difference%20is%20that,away%20from%20the%20machine%20itself
- Differences Between SCADA and HMI: http://www.differencebetween.net/technology/industrial/differences-between-scada-and-hmi/
- What is HMI?: https://www.inductiveautomation.com/resources/article/what-is-hmi#:~:text=A%20Human%2DMachine%20Interface%20(HMI,context%20of%20an%20industrial%20process

# 12
# Industrial Controls — User Inputs and Outputs

HMIs are industrial UIs that are designed to talk to hardware. As such they offer us ways of entering data into a PLC and displaying data that was received from the PLC. HMI development packages can make this process very simple. In short, most HMI development packages will fall into a low-code or no-code category. Even if you use an advanced SCADA package, you'll find that the actual coding will be minimal. The only time you'll write copious amounts of code for your HMI is when you use a traditional programming language such as C# or Java.

With that in mind, most HMI development packages, such as the one in CODESYS, will give you a wide variety of input and output controls to choose from. Also, since there is no coding with the CODESYS HMI development tool, attaching the controls to the PLC code is very straightforward.

Now, with that being said, HMIs are very important pieces of software that can make or break a project. Due to the simplistic nature of HMI development with low- or no-code solutions, they are often delegated to inexperienced engineers with minimal training in the subject. I have seen inexperienced automation engineers make some very questionable design choices when developing HMIs. Some of the poor choices are due to a lack of knowledge of common controls and how they work in an HMI. As such, this chapter will be concerned with the following:

- Switches
- Buttons
- LEDs
- Potentiometers
- Sliders
- Spinners
- Measurement controls

- Control properties

After we explore all of the common controls, we're going to build a sample HMI panel along with some PLC code for the controls to talk to.

## Technical requirements

The HMI tool that CODESYS offers is built into the system. As such, when you downloaded and installed CODESYS, the HMI tool was also downloaded and installed. The project that will be developed in this chapter can be downloaded at the following URL: https://github.com/PacktPublishing/Mastering-PLC-programming/tree/master/Chapter%2012.

HMI development is as much an artistic endeavor as it is an engineering practice. I strongly recommend you pull down the code and modify it using the principles that have been covered so far to get a good feel for everything.

## Exploring common HMI controls

All systems need some way for the operator to send input signals and receive feedback. For purely physical systems, these input devices are things such as switches, buttons, and so on for the inputs while things such as LEDs and gauges are used for the outputs. However, this can be costly and, in the modern computer-driven world, unnecessary as we can simply program in our controls. As such, the remainder of this section will explore software-based controls.

### Flip switches

As we all learned in high school, a **switch** causes a break in a circuit that will essentially cause the flow of electricity to stop when it reaches the switch. In other words, with the switch closed, the electricity is free to flow in the circuit, which will cause the equivalent of a TRUE condition. If the switch is open, the electricity will not be allowed to flow throughout the circuit, which will cause a FALSE condition. In terms of HMIs, a switch can be thought of in a similar sense. A digital HMI switch will behave the same way as a physical switch. When the switch is *on*, the variable that it is attached to will be set to TRUE, and when the switch is *off*, the variable will be set to FALSE. Two common types of HMI switches are the rocker and dip switches, which can be seen in *Figure 12.1*.

Figure 12.1 – Switches

These switches are flip switches that behave similarly to a common wall switch. When you flip the switch up, it will produce a TRUE condition, and when it is flipped down, it will cause a FALSE condition. The only way to toggle the state is to toggle the switch itself. Switches like these are commonly used to put things into a given state until the operator decides to toggle that particular state. For example, the valves in the mock HMI in *Chapter 11* used rocker switches to put the valves in an on or off state. Now that we have established an understanding of flip switches, we can move on to push switches.

## Push switches

Another type of switch is the push switch. A push switch will behave the same way as a flip switch but instead of flipping it, you will push it. In their non-pressed state, the push switches will produce a FALSE state; while pressed, they will produce a TRUE state. The CODESYS HMI builder offers a few different types of push switches. Two common push switches can be seen in *Figure 12.2*.

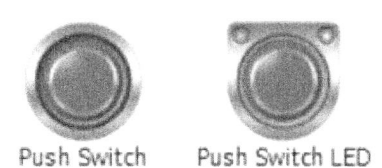

Figure 12.2 – Push switches

With switches now explored, we can move on to **buttons**, which are derivatives of switches.

## Buttons

As with flip switches, buttons behave the same way as their physical counterparts do. An HMI button will only change states while the button is pressed. In general, when the button is released, the state will change back to what it was before. However, many HMI development systems will allow you to configure the button to latch a bit or pulse for a single scan. In general, you will want to use a button to perform operations such as jogging an axis into position, starting a process, inputting data, changing/opening an HMI screen, and so on.

For most HMI systems, buttons are much more customizable in terms of appearance. You can customize the color, the text on the button, whether it will be normally on or normally off, and so on. A common button is the one in *Figure 12.3*, which is a raw button. In other words, when you add a button to the screen, it will look similar to the following:

Figure 12.3 – Button

However, after customizing a button, you could make it look like the button in *Figure 12.4*.

Figure 12.4 – Customized button

Buttons are very important components. In my experience, I have always found myself using buttons more than switches. There is no set rule of when to use one over the other as long as you get your desired results. However, as a general rule, I would use buttons in the following situations:

- Changing an HMI screen
- Starting a process that will end and needs to be restarted
- Entering data
- Jogging components into place

Switches and buttons are very common, powerful HMI components. At their core, they are inputs for Boolean variables. With that being said, if switches and buttons are inputs, LEDs are their outputs. With buttons explored, we can now move on to **LEDs**.

## LEDs

One of my favorite HMI components is LEDs. Since I was a child, I have loved playing with LEDs and as an adult, not much has changed. I have to constantly keep my love of adding lights to my HMI control panels in check.

In CODESYS, LEDs are referred to as lamps; however, in everyday speech, any light indicator is referred to as an LED. So, for this book, we will use the term LED to refer to lamps. LEDs are, in my opinion, one of the simplest and most powerful status indicators a HMI developer can use. An LED or lamp in CODESYS will look like what is shown in *Figure 12.5*.

Figure 12.5 – CODESYS LED/lamp

LEDs can serve multiple purposes in an HMI. LEDs can determine the status of a device: whether a component is on or off, whether there is an issue, whether there is about to be an issue, and so on. In CODESYS, LEDs can be set to five different colors, which are red, yellow, green, blue, and gray.

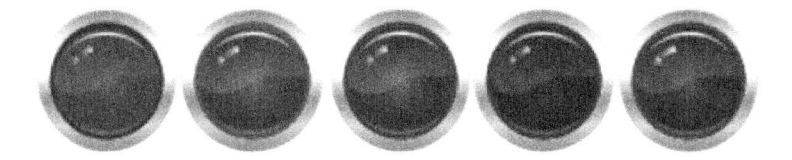

Figure 12.6 – The five LED colors

The primary way LEDs are used to relay status information is with their color. Colors such as red, yellow, and green all have meanings that can signify a certain status. What these colors mean will be explored in the next chapter. However, for now, we are going to switch gears and talk about **potentiometers** (**pots**).

## Potentiometers

Outside of switches, arguably the other most common HMI control is the potentiometer or, as they are most commonly referred to, pots. In a nutshell, pots allow you to input a numerical value in a range. In the same way a physical pot will allow you to adjust the resistance in an electrical circuit, an HMI pot will allow you to adjust a value in the software.

Pots have many different uses and are very common in HMI development. They are commonly used in applications that require temperature control, such as ovens, speed control input, and even as inputs for things such as part counters. A pot is depicted in *Figure 12.7*.

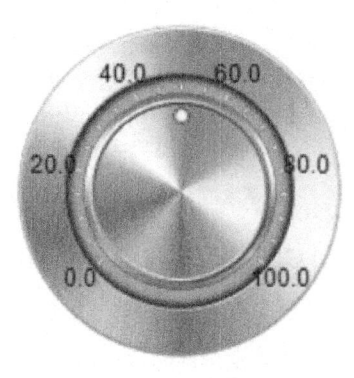

Figure 12.7 – Potentiometer

By default, a pot will have a range from 0 to 100; however, this can be adjusted. You can adjust the range of the pot, the value of the dial, and so on. When working with pots, it is very important to remember to adjust your range. This is a common mistake that even the most experienced HMI developers make. So, when you're working with pots, it is worth keeping that in mind. Similar to switches and buttons, pots also have a cousin in HMI development that is known as a **slider**. As such, with a grasp on pots, we are going to switch gears again and take a quick look at sliders.

## Sliders

Sliders can best be thought of as pots that are depicted as straight lines. Sliders work similarly to pots. For the most part, anywhere you can use a pot, you can use a slider, and vice versa. As such, the differences between pots and sliders are mostly aesthetic. A sliders in CODESYS is depicted in *Figure 12.8*:

Figure 12.8 – Slider

Sliders can be customized in much the same way pots can. You can customize attributes such as the range in the same way you can with pots.

In my experience, both sliders and pots can be difficult for operators to work with. A large part of this is due to operators usually wearing work gloves while interacting with the screen. In my opinion, sliders are a bit harder to work with than pots. Sliders are usually smaller and they usually don't have a range on them, similar to *Figure 12.8*. However, this is just my opinion and you may disagree based on your and your customers' experience. Either way, pots and sliders are excellent input controls and any HMI developer should have a basic understanding of both.

In terms of the data type of the variable, it is common to use an integer type for both sliders and pots. You can use a REAL type but having control over the decimal value will be difficult to obtain. For example, it will be hard to hit a specific number after a point. If you need a floating number, you may want to opt for a numerical input such as a keypad. Generally, I never like using pots or sliders for any inputs that need a value of less than 1. In other words, if you need a high level of precision, you probably don't want to use either of these types of controls. With all that in mind, we can now explore **spinners**.

## Spinners

Spinners are another input control. As opposed to pots or sliders, spinners have an up-and-down button that allows you to adjust the value. Essentially, spinners are very simple and very handy. Usually, for many applications where a slider or pot may be inefficient, a spinner can be an excellent alternative. A simple spinner can be viewed in *Figure 12.9*.

Figure 12.9 – Spinner

If you use a spinner, you should set a maximum and minimum value. These values will essentially act as the range for the spinner. The buttons that set the value can be seen in *Figure 12.9*. When the *up* button is clicked, the value in the spinner will increase, and when the *down* button is clicked, the value in the spinner will decrease. The one downside of spinners is that the buttons on the spinner can be hard to click when they are small. As such, when you do use spinners, it is important to consider the button size for the operator's ease of use.

In all, spinners are very simple but very powerful controls. Similar to pots and sliders, spinners are excellent for operator inputs.

Now that we have explored spinners, we can focus our attention on **measurement controls**, which display data from the PLC or input controls.

## Measurement controls

Whereas an LED is a common readout for a control such as a switch or a button, measurement controls such as gauges and bar graphs are used as readouts for controls such as pots and sliders. Readout controls usually vary the most between HMI development systems. In CODESYS, the readouts are mostly analog. As such, they are mostly for displaying things such as temperature or pressure. CODESYS offers several different styles of readouts to choose from.

Let's look at a few graph readouts that can be used for many different applications.

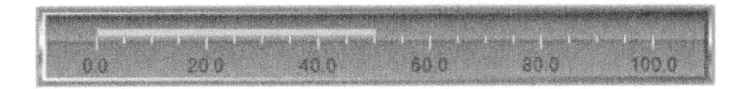

Figure 12.10 – Bar graph

The bar graph is simply a straight line with a green bar inside that points to a value. As the value of the variables increases, the green line will as well. A bar graph is great for things such as showing the percentage of the job that is complete and so on.

Figure 12.11 – 90° gauge

The 90° gauge works the same way as the bar graph does but simply has a more compact and different style.

Figure 12.12 – 180° gauge

The preceding figure depicts a 180° gauge, while the following figure depicts a 360° gauge. Both function similarly to the other gauges, with the only difference being the style.

Figure 12.13 – 360° gauge

For the most part, the only real differences between the different displays are their shapes. Much as with pots and sliders, you will need to set the range on these as well. As you can see in all of the gauge images, they range from 0 to 100 by default. This means that if you're planning to have input values that are more than 100 and you don't set these values, you're going to peg out the gauge before you reach the upper limit of the control. So, the important thing to remember is whenever you use pots, sliders, or gauges, you need to remember to change your ranges at the very least.

## Histogram

Though it is not a gauge, another measurement control that can be used is a **histogram**. Histograms show the current values in an array. As such, they are excellent for applications such as reading temperatures from multiple thermal couples. An example of the output for a histogram can be viewed in *Figure 12.14*.

Figure 12.14 – Histogram

Now, as stated before, histograms work off of an array. Unlike gauges or LEDs that read a single variable, histograms read an array of variables. In short, the code that generates the graph data in *Figure 12.14* can be seen in the code snippet:

```
PROGRAM PLC_PRG
VAR
     hist : ARRAY [1..4] OF INT;
END_VAR
```

Setting the variables is accomplished with the following code:

```
hist[1] := 10;
hist[2] := 30;
hist[3] := 25;
hist[4] := 42;
```

As can be seen from the histogram and the code, the value of each element in the array will correspond to a bar in the graph.

## Text field

The final control that we will explore is the text field. A text field is a simple control that allows an operator to input text via a built-in keypad or display data on the HMI screen. Similar to other controls, you can drop a text field onto the screen; however, setting one up requires a bit more work. To configure the text field for an input, you will first need to add one to the screen and add the following to the `Texts` property:

```
Input: %s
```

Once you do that, you should have a text field that is similar to *Figure 12.15*.

Figure 12.15 – Partially configured text field

Next, you will want to add a variable to hold the input, similar to what we did with the other controls. Once that is done, navigate to **Inputs** and then **OnMouseClick**. Once you do that, you will want to click the field with the three dots and configure the popup to match *Figure 12.16*.

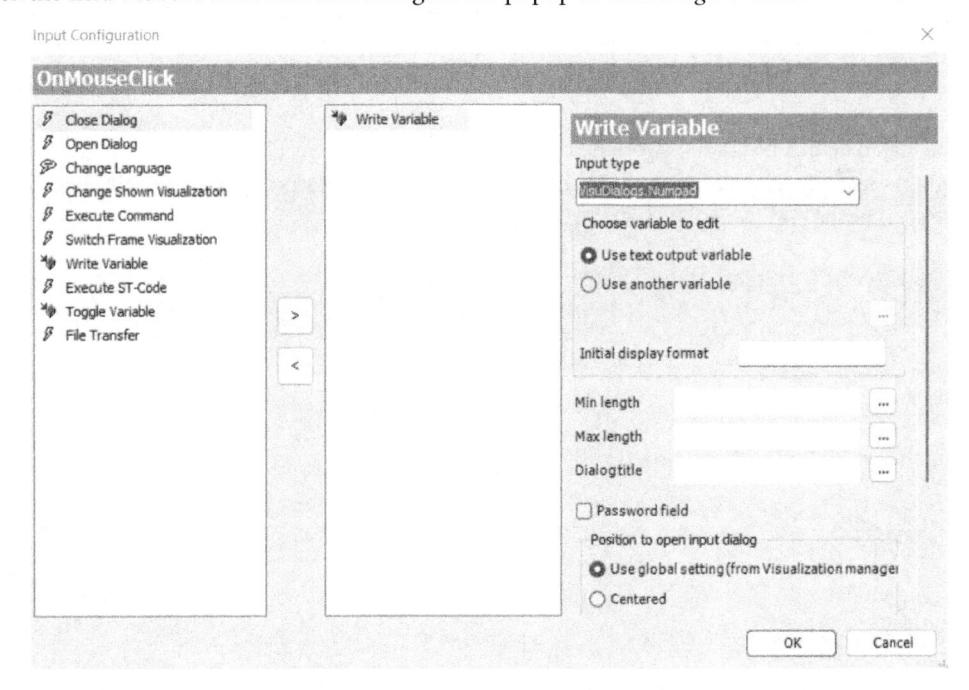

Figure 12.16 – Text field configuration

Once finished, click **OK** and run the program. When you click the text field, you should see a keypad appear, as in *Figure 12.17*.

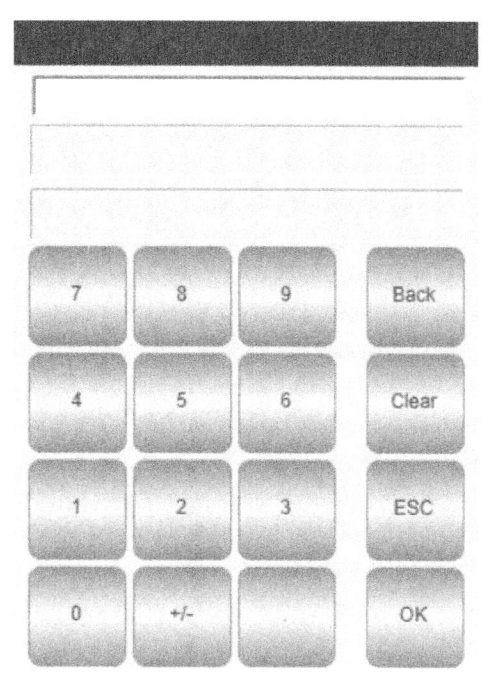

Figure 12.17 – Input pad

Now, enter a number and click the **OK** button on the pad. Go back and check the variable that you assigned to it; notice how the variable matches what you typed in.

Text fields are very common, but keypads can be a little annoying to use. In a real-world setting, you're going to use these more times than not, whether it is a built-in keyboard like the one we just saw or one you created. It is also common to use these to display data. However, for the remainder of this book, we are only going to use spinners as they will be easier to use in our examples.

At this point, we have explored many of the controls that CODESYS offers. CODESYS offers many more tools but, for the most part, these are the basic controls that you'll need to know about to get you through a project. Now that we know what the controls do, we need to know how to customize the controls via the **Properties** menu.

## Control properties

The core of any control is setting up the control's properties. The **Properties** screen will vary for each control. An example of a screen can be seen in *Figure 12.18*.

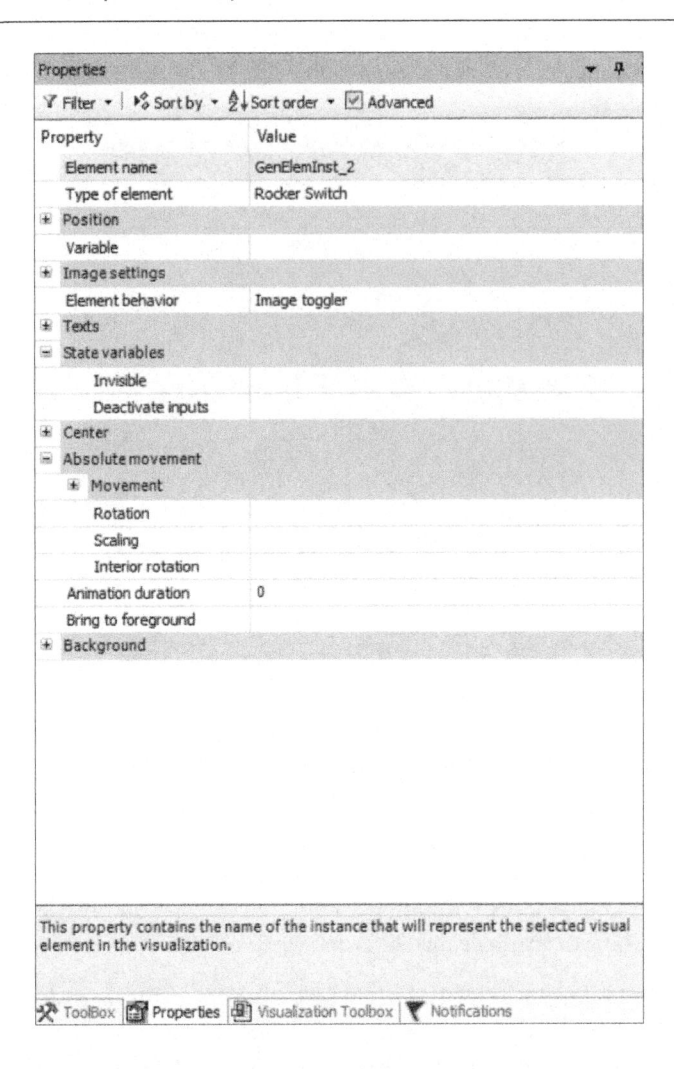

Figure 12.18 – Rocker switch Properties menu

*Figure 12.18* shows the **Properties** screen for a rocker switch. You will need to set the properties for each component you have on the HMI, even if the controls are the same. For example, if you have 20 rocker switches, you will need to set the properties for each one to ensure they behave properly.

Now, as stated before, it is very important to remember that each control type will have a different set of properties that need to be configured. There will be common fields for each control but each control will ultimately have different fields to set. As such, it is important to get familiar with the controls that we have studied as they will vary. At this point, we have explored many common controls and explored the basics of how to configure the behavior. Therefore, we can now try to create a simple HMI.

# Final project – creating a simple HMI

For this project, we are going to create a simple HMI that can control a histogram. The HMI we are going to create is going to be straightforward. When a switch is flipped, an LED is going to turn on and a pot will become visible. When the pot appears, we will be able to turn the pot to adjust one of the lines on the histogram. With that in mind, let's set up some basic requirements.

## Requirements for the HMI

The HMI will need the following:

- Four rocker switches that will control the visibility of four different pots
- Four LEDs that indicate when the rocker switch is on
- Four pots that will only be visible when the rocker switch is on
- Each pot will control exactly one bar on the histogram
- Both the pots and the histogram will have a range of 0 to 100 (default range)

With these requirements, minimal code is required to make the HMI function as intended. The requirements also dictate that there will be the following controls:

- Four rocker switches
- Four LEDs
- Four pots
- One histogram

## Design of the HMI

From these requirements, we can move into the design phase and layout of our HMI. Much like coding, there is no set way of laying out an HMI. The next chapter will explore some best practices, but for now, we're going to use a layout like the following:

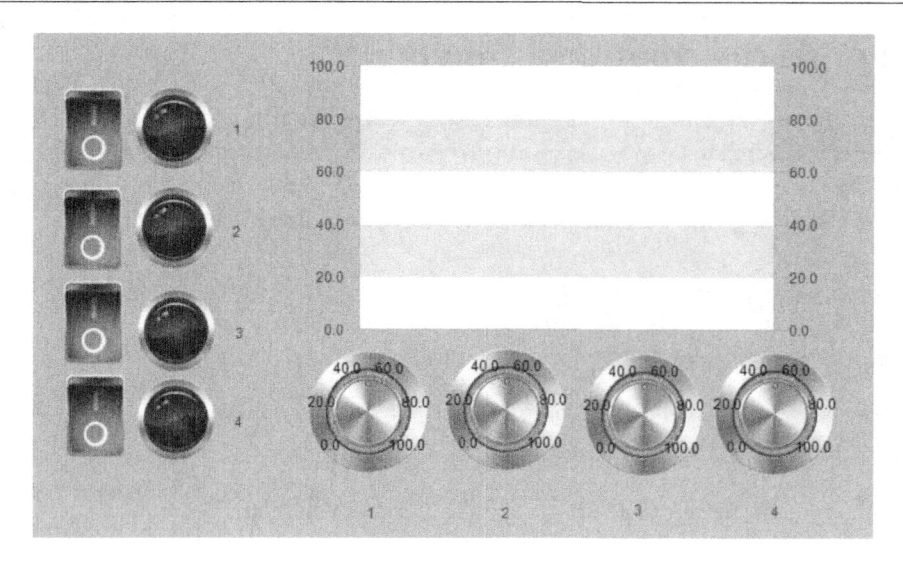

Figure 12.19 – HMI layout

In this layout, we have four rocker switches next to their corresponding LEDs. The focal point of the HMI is the histogram in the middle. Underneath the histogram resides the four pots that will control the bars in the histogram. In short, this is a condensed design that will get the job done. However, this isn't the only design that can accomplish the job. At this point, I would recommend that you play around and see whether you can improve the design of the HMI.

Now that we have a design in place, we can move on to building the HMI.

## Building the HMI

The code of the HMI will consist mostly of variables. There should be four switch variables that are tied to LEDs and the **Invisible** field. However, to make the pot visible when the switches are on, we will have to add four additional variables that will be the inverse of the state of the switch. This may seem to be counter-intuitive but a True variable in the **Invisible** field for the pot will cause the pot to be invisible. This will create a counter-intuitive situation. As such, we will remedy the problem with PLC logic. For this example, we will use PLC logic for learning purposes, but there is a better method of accomplishing the same task, which we will explore later on.

The variables for the HMI will look like the following:

```
PROGRAM PLC_PRG
VAR
    hist : ARRAY [1..4] OF INT;
    sw1  : BOOL;
    sw2  : BOOL;
```

```
        sw3  : BOOL;
        sw4  : BOOL;
        pot1 : BOOL;
        pot2 : BOOL;
        pot3 : BOOL;
        pot4 : BOOL;
END_VAR
```

The logic will look like the following:

```
pot1 := NOT sw1;
pot2 := NOT sw2;
pot3 := NOT sw3;
pot4 := NOT sw4;
```

This logic will simply invert the Boolean state of the rocker switch. These inverted variables will be responsible for causing consistent visible/invisible behavior with the current state of the switch.

If you observe the variables, you will notice we have an array declared. That array will be assigned to the histogram. Each pot will be assigned an element of the array. As such, the graph will dynamically update when the pots are turned.

After the variables are in place, you can start to add the controls to the screen. This is a simple task as all you have to do is select the control and, with your mouse button, down-drag the element to the screen. Once you have the element on the screen, you can move it around as needed and resize it. Lay out the elements in a similar fashion to what can be seen in *Figure 12.19*.

With all this set up, we need to start assigning variables to the HMI components. The first controls we are going to hook up are the switches and the LED:.

1. All we have to do is select the control and find the **Variable** field, as in *Figure 12.20*.

Figure 12.20 – Empty Variable field

2. To assign a variable, click on the button with the three dots on the right and you should see a popup like the following:

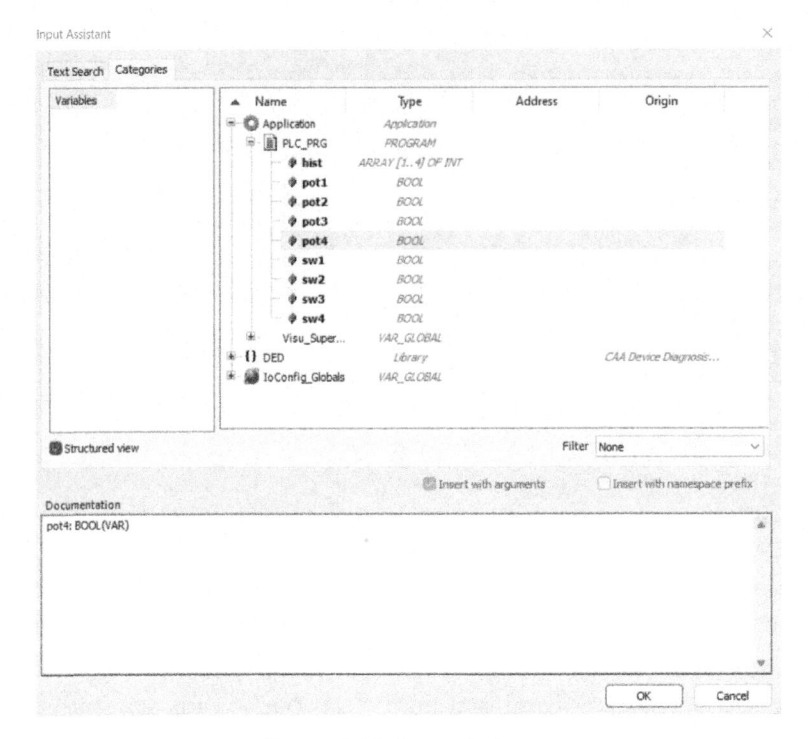

Figure 12.21 – Input Assistant

3. Select the corresponding variable for each switch and LED. In short, you will assign a switch variable to both the rocker switch and the LED that is next to it.

Once you are done with that, you can move on to setting up the pots. The pots will require a little more setup than the switches and LED:.

1. For the pots, you will need to set up the **Variable** field and the **Invisible** field.

Figure 12.22 – Pot Invisible field

2. Assign the corresponding pot variable to this field using the same process that we used with the LEDs and switches. When you are finished, your field should look like the following:

Figure 12.23 – Pot Invisible field with variable

3. Repeat this process with each pot.
4. Once you have completed that, you will need to assign the `hist` array element to the pot's **Variable** field.

| Variable | PLC_PRG.hist[1] |

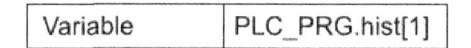

Figure 12.24 – Pot variable

5. Now, you will have to add the square brackets with the element number to the end of the variable.
6. Repeat this process for each pot variable.
7. Once you have done this, you will need to set up the histogram. The histogram will be very simple. All you will have to do is assign the array to the **Data array** field in the histogram's **Property** menu. Adding the array will be the same as adding a normal variable. As such, when you are done, the field should look like the following:

| Data array | PLC_PRG.hist |

Figure 12.25 – Data array field for histogram

This will be the final component to set up and you can now start the HMI by performing the same set of steps that we used to run the PLC code.

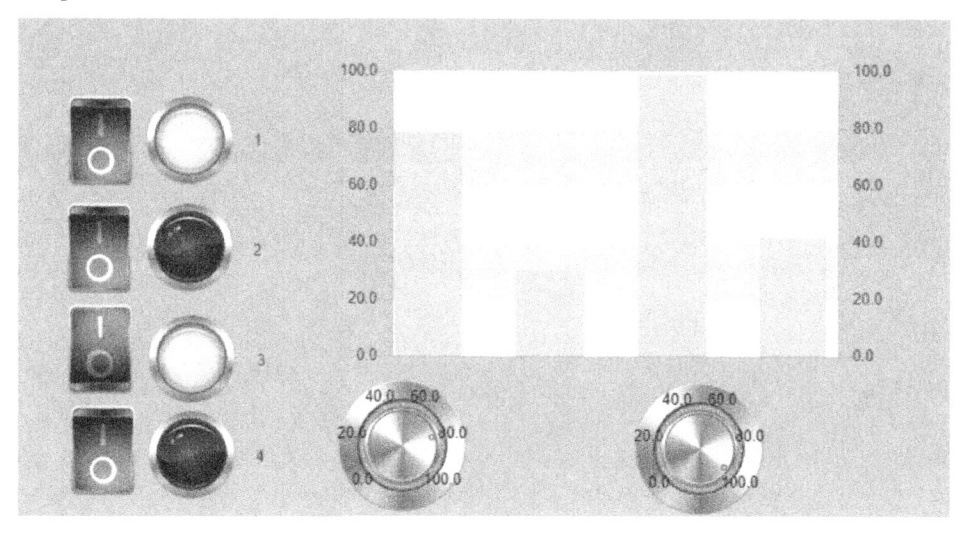

Figure 12.26 – Working HMI

In this case, we have two pots turned off and two turned on. We can move the pots and watch the chart change.

Now, what we have works, but it is not the best solution. For this HMI to work, we have to have code that inverts the switch's state. To accomplish this, we created PLC logic. This is a bit unnecessary and somewhat bad as we now have the PLC doing a menial task. For our purposes, a better solution would be to get rid of the pot variables altogether and simply put the NOT keyword in front of the switch variable, similar to what is in *Figure 12.27*.

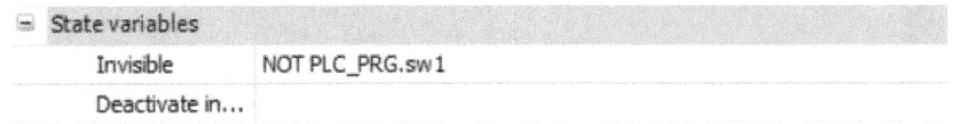

| State variables | |
|---|---|
| Invisible | NOT PLC_PRG.sw1 |
| Deactivate in... | |

Figure 12.27 – Switch variable with the NOT operator

When the variables are set with the NOT operator and run, we can observe the same behavior.

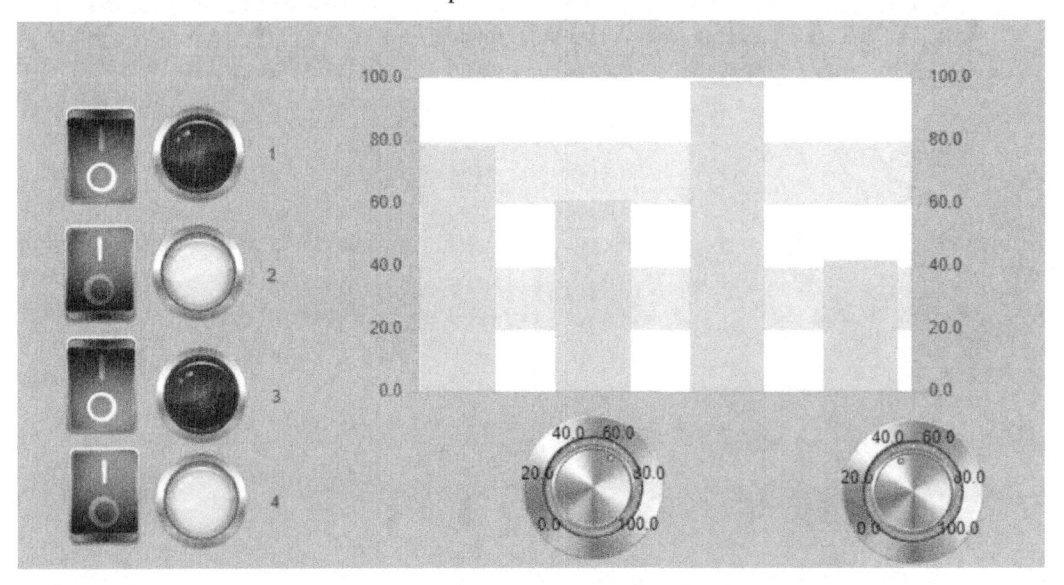

Figure 12.28 – HMI output with the NOT operator

In short, what we have learned is we can manipulate an HMI via the PLC code or through the HMI properties. For our purposes, adding in the PLC code was not the optimal solution. However, there will be times when manipulation of the HMI from the PLC code will be the optimal solution. A general rule I like to go by is to try to keep the HMI control manipulation on the HMI side. This isn't always possible but it should be strived for as it'll cause less code bloat and frees up your PLC to do more important tasks.

So, we have now created a functioning HMI. At this point, I would recommend modifying the current HMI to get experience using the low-code tooling that is provided in CODESYS. Most of the major HMI and SCADA development packages will use a similar method for creating HMIs, so you're going to want to get used to using the tools as they all work similarly.

## Summary

In conclusion, we have explored common HMI components such as switches, buttons, LEDs, pots, sliders, spinners, and more. We have also learned how to hook up HMI components and how PLC code can manipulate those components. We also explored how simply using commands in the **Properties** fields can allow us to manipulate the controls without the need for the PLC.

In all, this chapter has served as the basis for future HMI development exploration. This chapter demonstrated how to use basic controls and how to string them together to form a simple, yet functional, HMI.

As I have stated before, HMI development is as much an art as it is a science. The next chapter will be dedicated to the best practice of laying out an HMI so that your operators can effectively use them. For now, I strongly recommend getting used to the controls and the layout of the **Properties** menu.

## Questions

Answer the following questions based on what you've learned in this chapter. Cross-check your answers with those provided at the end of the book, under *Assessments*.

1. What is a button?
2. What kind of data fields do histograms take?
3. Can we add keywords to a property field?
4. Can we manipulate an HMI via PLC code? If so, when should we?

## Further reading

Have a look at the following resources to further your knowledge:

- CODESYS HMI overview: `https://www.codesys.com/products/codesys-visualization/hmi.html`

# 13
# Layouts — Making HMIs User-Friendly

HMI development has a lot in common with graphic design. This means, there are a few rules that should be followed as closely as possible to ensure that the HMI is user-friendly. There is a difference between laying out an HMI and something akin to a website. I usually like to consider HMIs as the cousins to traditional user interfaces. Both types of interface have certain things in common, such as logical layout, efficient coloring, and so on.

Though these types of user interfaces are cousins to one another, an HMI will have a person staring at it much more often. As such, certain factors must be considered that would normally be ignored when developing something like a website. As such, certain rules should be followed when developing an HMI.

Due to operators using the HMI more frequently and in a much more high-paced and mission-critical environment, HMIs need to be easy to use, easy to look at, well organized, and provide just enough information for the operator to do their job without overloading them with too much information. This means that things as simple as color selection are vital to the success of the HMI. To create a successful, functional HMI, we are going to explore the following concepts:

- Colors
- Size of controls
- Grouping/positions
- Blinking
- Organizing the HMI into multiple screens

To round out the chapter, we are going to enhance the HMI from the last chapter to give it more functionality and make it more user-friendly.

## Technical requirements

Similar to the previous chapters, the only technical requirement you will need to follow along with is a working copy of CODESYS. This chapter will expand and modify the HMI from the last chapter; as such, you will want to pull down that HMI or review *Chapter 11*, or build one from scratch. You can pull down the code for this chapter at the following URL: `https://github.com/PacktPublishing/Mastering-PLC-programming/tree/master/Chapter%2013`.

## The importance of colors

Believe it or not, colors can utterly sink an HMI. Colors are one of the most important aspects of an HMI in my opinion. Choosing the wrong colors for your HMI will literally hurt your operator's eyes. A general, but not normally followed, rule is that you want to use dark, pastel colors for your HMI. This will reduce the contrast of the HMI screen and make it easier to operate. Generally, you want to avoid bright colors. Normally, HMI developers will opt for colors such as black or gray for backgrounds and different shades of gray for control colors. To start the color discussion, let's look at backgrounds.

### Backgrounds

In terms of backgrounds, I like to stick with shades of gray, depending on what I'm doing or if specified otherwise. However, some organizations I have worked for have primarily used black or shades of dark blue backgrounds and have used them to great success.

Black backgrounds are excellent; however, they do require a bit more work when there is heavy use of labels. To put that in perspective, you'll probably have to adjust label colors for any background but, in my opinion, black requires a bit more drastic change. Consider *Figure 13.1*; upon studying the figure, the first thing you may notice is that there is a heavy contrast between the components, and the labels (especially the labels on the pots) are hard to see. However, once you adjust the coloring on the labels, this HMI will be easy on the eyes due to the dark background.

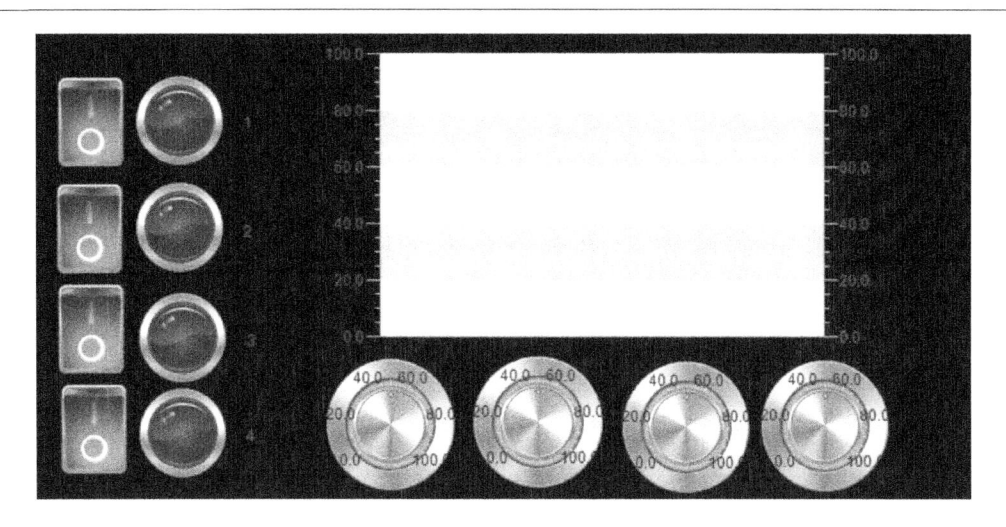

Figure 13.1 – Black background HMI

Comparing *Figure 13.1* to *Figure 13.2*, the dark gray creates less contrast and the labels are easier to see but the lighter background will be a little harder on the eyes. Overall, you'll have a bit of a balancing act between contrast and possible eye strain.

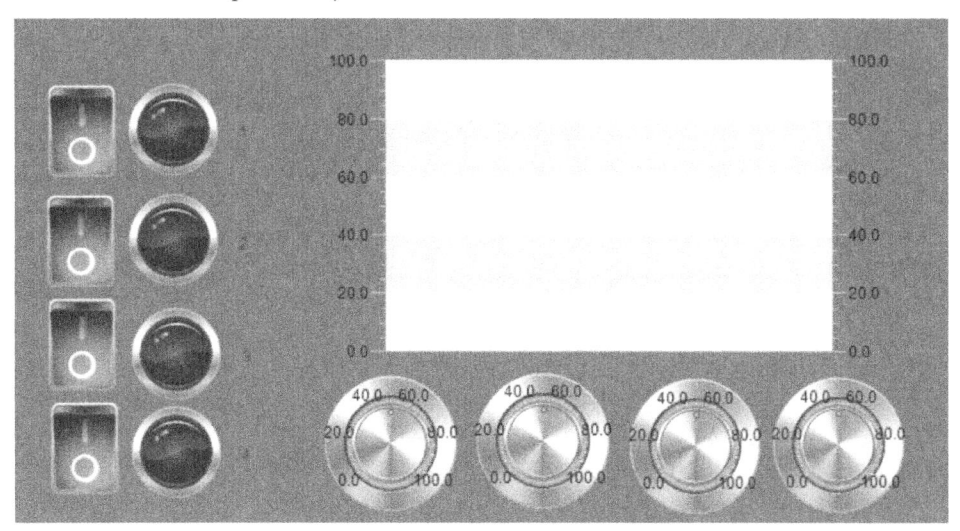

Figure 13.2 – Dark gray background HMI

In short, the black background is a bit easier to look at but the gray background is easier to work with in terms of contrast. In all, you're going to have to do some color matching with both but, in my opinion, the gray background is easier to work with. Now, we need to switch over to the colors red, green, and yellow.

### Red, yellow, and green

The colors red, yellow, and green are everywhere, from street lights to industrial machinery. They are so common that their meaning is almost ingrained in us. However, the following list is going to explore the three colors and how to use them to great effect in an HMI:

- **Red**: The color red is usually an indicator that something is wrong or something has stopped, and is usually associated with an alarm. The color red on any component should be used at a minimum unless the control is indicating something is stopped or in an erroneous state. As such, you usually want to reserve red for alarms, controls such as buttons that can go into an erroneous state, stop/off LEDs, and the like.

  Another use of the color red is when a part is reaching its upper operational limit. In this case, you will usually rig your HMI to have red text, an LED strip with green, yellow, or red, or do something of the like. In all, be careful with the color red in your HMI. By instinct, anytime the operator sees red, they will assume that there is something wrong.

- **Yellow**: Yellow is an odd color. Yellow is the in-between of green and red in terms of meaning. Yellow will usually signal that your machine is still in an operational state, and all systems are still functioning, but you need to be cautious because something could go wrong at any time. In other words, yellow is a warning color.

  Now, I have seen yellow used a bit more liberally in HMIs. When next to a switch as in the HMI we have built so far, or for button colors, you can generally get away with using yellow without confusing anyone. The only thing that you do need to be mindful of is color consistency. If you're using yellow to mean a warning, you don't want to color a control a similar shade of yellow, as that will cause confusion. Though you can be a little more liberal with yellow than you can with red, it is advisable to use this color sparingly as well.

- **Green**: Green means *go*. The short of it is that the color green means everything is running and working. Generally, green can be used liberally as well. However, consistency also matters with the color green. If you're using green in your HMI to signify a working state, you still want to be careful using it for non-indicator components. Seeing green will usually signal that all is working, but if the operator takes a glance at the screen and sees a yellow or red button in a sea of green, they may get confused. As such, though it is common to see green controls and LEDs, you generally want to reserve this color to indicate the following: on, working, ready, normal operation, and so on.

### Control colors

Choosing the correct coloring for controls, in general, is a bit of an art. Many of the best practice documents will tell you to pick shades of gray. However, this rule is rarely followed. Most of the time, people will select colors for controls based on aesthetics. I have personally seen controls colored in every color of the rainbow. Usually, this isn't a problem. The only time that color matters is when you are using red, green, or yellow due to it possibly being confusing to the operator. However, with that being said, it is usually a good idea to make things such as buttons and controls a shade of gray, with the colors green, red, or yellow acting as a status indicator. However, no matter what color you choose, you will want to gray out controls that are not active.

With all that being said, colors are never enough. Without proper labeling, no matter what color you choose, confusion can still set in, and the only way to remove the confusion is with proper labeling.

## Labeling colors

Regardless of the color that you choose, you always want to clearly state the status of the machine on the HMI. If you look at *Figure 13.3*, you will notice that there are no labels under the LEDs. In this case, an operator who is not familiar with the machine may not know off the top of their head what the LED is supposed to signify. They could assume any number of things, such as the machine is in shutdown mode or there is a broken part.

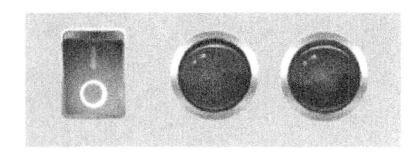

Figure 13.3 – Unlabeled LEDs

However, in *Figure 13.4*, the LEDs are labeled in such a way that the operator will know exactly what the LED means. In short, colors matter but without proper labels, they can mean anything.

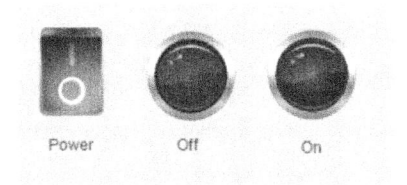

Figure 13.4 – Labeled LEDs

There is a lot to the art of color selection. This short tutorial is just to give you an idea of selecting the proper color and why it is important to label the controls, no matter what color they are.

## Understanding grouping/position

Another key aspect of HMI design is grouping. Controls and readouts need to be logically grouped so the operator can easily control the machine and take necessary readings. When it comes to grouping, I have heard two schools of thought. The first one is to stack the controls vertically, as in *Figure 13.5*.

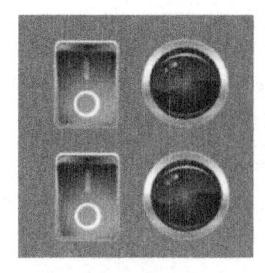

Figure 13.5 – Vertical stacking

With the controls laid out as they are in *Figure 13.5*, the operator scans the controls in a top-to-bottom motion. This configuration is known as **side navigation**. Normally the side navigation is on the left of the screen. Left navigation is considered more efficient and faster for the operator. The key to this layout is that each component gets equal weight. This means that, visually, the bottom switch is as important as the top switch.

Left-side layouts like these are common for things such as selecting submenus, homing different machine parts, and so on. This layout will free up a lot of space on the sides of the bar. This layout is also handy for configurations like the one seen in *Figure 13.5*. Since all the controls have the same visual weight, the operator is less likely to overlook a control. The other type of layout is where you place the controls on either the top or bottom of the screen. In terms of HMI development, it is usually more common to have fields for data input toward the top of the HMI or a data entry page. Typically, these fields will be laid in a horizontal pattern at the top of the screen, or in a stacked pattern. In terms of actual controls, such as buttons, I feel that it is best to put these toward the bottom of the screen, with the middle of the screen reserved for current data readouts. Consider the following figure:

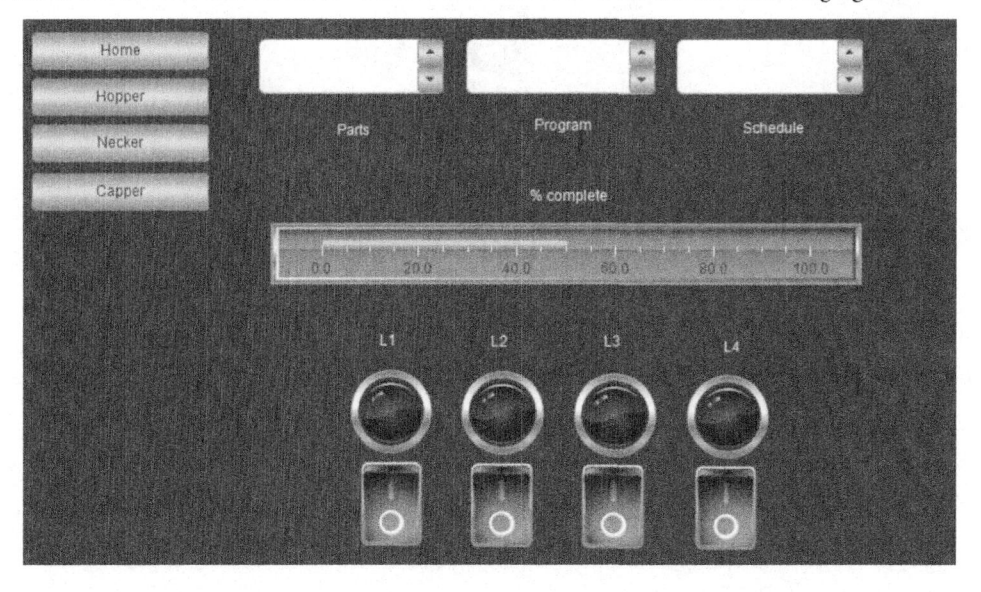

Figure 13.6 – Example HMI

In the example HMI, we have four buttons to the left of the screen. In a live HMI, these would allow us to navigate to those submenus. These are to the left so the operator can efficiently scan them to select the menu they want to navigate to without having to look around the screen. In short, it is the first thing they will see. The menu buttons are also stacked vertically. This means that they will be easy for the operator to scan through.

Moving to the right, we can see we have three input fields labeled **Parts**, **Program**, and **Schedule**. These fields are input fields. These fields are positioned so the operator can select their menu, then easily scan to the right, and start inputting the data for the run.

In the middle of the screen, we have a **% complete** bar. This bar provides pivotal information to the operator, mainly how far the job is along. This readout is placed squarely in the center of the screen. The reason for this is that it draws the operator's attention. In the case of the operator needing a readout, they simply have to stare at the middle of the screen to get their data. This reduces the extra scan time and, ultimately, makes the HMI easier to use.

At the bottom of the screen, we have four switches and LEDs. These controls will turn on **L1**, **L2**, **L3**, and **L4**. They have a LED placed on top so the operator can easily scan to see whether the light is on.

*Figure 13.6* is by no means a perfect HMI screen; it can be improved upon. However, it does demonstrate some basic layout principles. When developing an HMI, it is very important to make them as easy on the eye as possible. In other words, you don't want the operator to have to search for their controls or readouts. With that in mind, let's switch gears and talk about blinking.

# Best practices for blinking

Nothing says hi-tech and advanced like blinking lights. Everyone loves blinking lights. However, much like many other features that we have seen, blinking can be as much a curse as it can be a blessing. When used properly, blinking can be used to indicate an emergency (such as an issue that could cause harm to personnel or property) or it could mean that a job is loaded and ready to go. In either case, blinking is distracting.

If you blink a component such as an LED, button, popup, or whatever, you need to be aware that this action will take the operator's attention away from the controls and put their focus on the blinking component. For some things such as issues or emergencies, this is welcomed. However, blinking components for the sake of blinking components is bad. Generally, I will only blink a component under the following conditions:

- Machine malfunctions
- Safety-related issues (open door, safety sensor tripped, etc.)
- E-stop has been engaged

For the most part, you can blink any component you want as long as it has an `Invisible` field similar to *Figure 13.7*:

Figure 13.7 – Invisible field

However, just because you can blink a component does not mean you should. Generally, you should only blink components such as LEDs, banners with messages in them, and so on. On the contrary, you never want to blink one of the following:

- A switch
- A button
- An input field
- A popup
- Anything with text

Blinking something in the list can create not only an annoying situation for the operator but also a potentially dangerous one. For example, if you were to blink a popup with an error message, you could potentially make it to where the operator cannot tell whether there is a safety issue or a component issue. You could also make it difficult for the operator to acknowledge the popup. The same can be said for something akin to a blinking switch. If the switch has to be flipped, the blinking could interfere with the operator's ability to do so.

In cases of blinking popups, input fields, and so on, it is okay to have a blinking element to them. For example, perhaps the operator needs to press a button; it is okay to maybe have it alternate colors. In the case of blinking colors, you can still get the operator's attention without the need to make the control hard to use. Regardless of what you blink, you must remember to do it tastefully and ensure that it is assisting as opposed to distracting the operator.

## Blinking a component

To blink something, you can use a series of timers or you can use the **Util library**. The Util library is a library packed with a lot of different function blocks. One such function block is `Blink`. As the name suggests, the `Blink` function is an abstraction layer that can be used to easily blink a component.

To demonstrate blinking a component, the first thing we will need to do is to import the Util library. As such, double-click **Library Manager** in the tree and click the **Add Library** button. Once you do this, you will be met with the following screen:

Figure 13.8 – Library search screen

From here, you will want to click on the *expand* button next to **Application**. From there, you will need to expand **Common**, similar to what is in *Figure 13.9*.

Figure 13.9 – Util library

Double-click on **Util** and the library should be imported. From there, add an HMI screen to the project and drop in an LED and a switch, as in *Figure 13.10*.

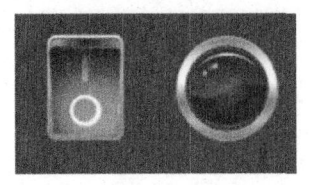

Figure 13.10 – Blinking LED HMI setup

Next, navigate to the PLC_PRG file and add the following variables:

```
PROGRAM PLC_PRG
VAR
        blink  : Blink;
        led    : BOOL;
        enable : BOOL;
END_VAR
```

These three variables are all that are needed to blink the LED. The blink variable references the Blink function, the led variable will be assigned to the LED in the HMI, while enable will be tied to the switch and will dictate whether the LED is blinking or not.

The logic to flash the LED is as follows:

```
blink(ENABLE:=enable,TIMELOW:=T#500MS, TIMEHIGH:=T#500MS, OUT
=> led);
```

This function has multiple inputs and a single output. The last argument is the output (OUT) that will dictate whether or not the LED is on. TIMEHIGH will determine how long the LED is on, while TIMELOW will determine how long the LED is off. Finally, enable will determine whether the blink function is active or not.

To see the blink function in action, start the code and flip the switch. Observe the LED is blinking at a steady rhythm. The LED will be on for 500 ms and off for 500 ms and will repeat.

As stated before, you should never blink a component such as a button. However, it is common to do something along the lines of blinking the color of a button. You'll do something like this when whatever the button controls goes into an error state. To demonstrate this, we are going to add a simple button to the HMI so that the HMI looks like *Figure 13.11*:

Figure 13.11 – Button

To alternate the button color, we need to set three settings as in the following:

Figure 13.12 – Button color and alarm color

In the **Colors** dropdown, you will see the **Color** and **Alarm color** fields. The **Color** field will set the default color, similar to *Figure 13.11*. The **Alarm color** field will set the secondary color, or in other words, the color that the button will be toggled to.

The third field that we need to set is the **Toggle color** field, as in *Figure 13.13*:

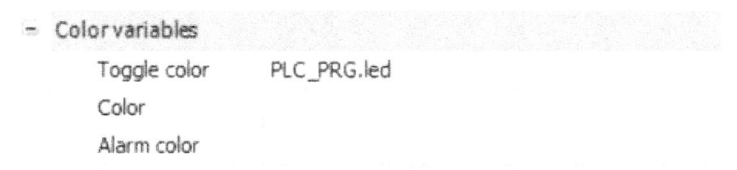

Figure 13.13 – Toggle Setting

For this example, we are going to use the same code that we used to blink the LED. As such, to alternate the color, all you have to do is set the `led` variable in the **Toggle color** field.

When you run the example, observe that the button will change between red and green. You can also use this feature to simply change the color of the button. Either way, you will set the **Color** and **Alarm color** fields and the **Toggle color** field of the control the same way. What is causing the blinking is the blink code we are using. As such, if you simply set the color in the PLC logic or the like, you can change the color without blinking.

## Animation

A cousin to blinking is animation. It is common to have animation on HMI screens. However, much like blinking, this must be used wisely. Animation is used quite often to simulate a process in as close to real time as possible. This can be handy as operators can easily track the process. Much like blinking, animation can be very distracting.

In all, blinking and animation can be used to great effect; however, both animation and blinking must be used wisely.

Now that we have explored blinking and have touched on animation, we need to switch over to one of the most important concepts in HMI development: screen organization.

## Organizing the screen into multiple layouts

One very common, but very poor, design decision in HMI development is to group multiple different screen responsibilities or way too much information on a single screen. There are many reasons why this is bad. Some reasons are as follows:

- Screen disorganization
- Cluttered appearance
- Poor usability
- Overloading the operator with irrelevant information

These are just a few reasons why screen organization is very important. However, one of the most important is overloading the operator with information. Generally, you only want to display the information that is relevant to the operator. If you include too much information, the operator can easily become confused or they can tune the information out and, ultimately, ignore important developments. One common way to combat this is to split an HMI application into multiple different screens.

Generally, screen organization can be determined with the one-sentence rule. You usually want to be able to describe the layout's responsibility in one sentence without the word *and*. Much like with the functions or methods, if you have to use the word *and*, you will want to split everything after the *and* into a layout of its own. HMIs that follow the one-sentence rule will generally produce cleaner HMIs that are easier for the operator to use.

In CODESYS, HMIs are broken out into what is called visualizations. Essentially, visualizations are individual screens packaged into a single HMI. It is common to have a home screen that is the main entry point for the HMI and the user can navigate to other screens from there. As such, the first step to this is creating the screens.

## Creating visualizations screens

Adding a new screen is quite simple in CODESYS:

1. You simply right-click on the **Application** manager, then **Add Object**, and finally, select **Visualization**, as shown in *Figure 13.14*.

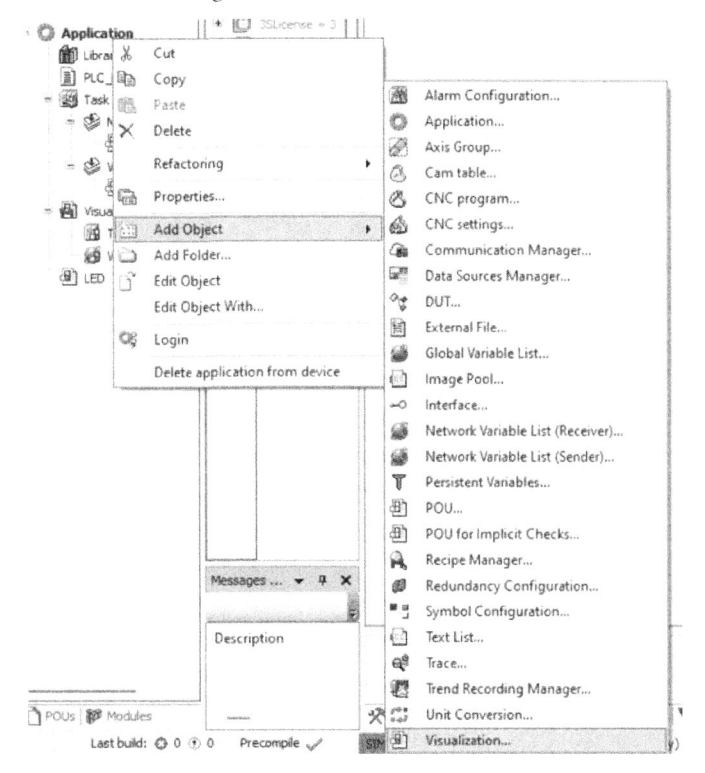

Figure 13.14 – Add visualization

Once you complete this step, you should be met with a screen similar to what is in the following screenshot:

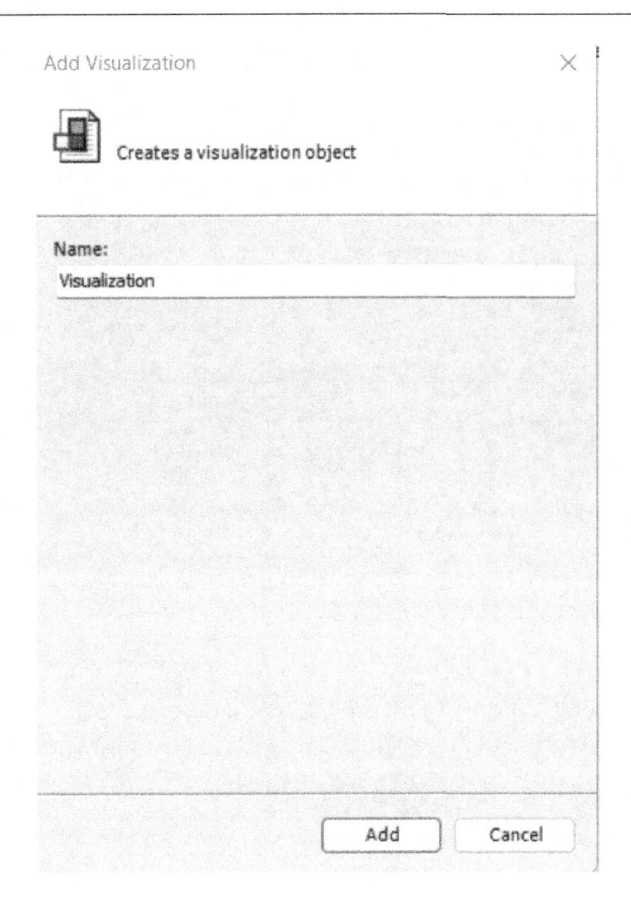

Figure 13.15 – Add Visualization screen

2. In the case of this example, we are going to name the screen pumps. Hence, change the name to pumps and click **Add**, following which the new screen will be added.

Once you follow these steps, the new HMI screen will be added to the project. An example of this can be seen in *Figure 13.16*. By default, the first screen that is added to the project, which is the one that is generated with **Visualization Manager**, will be the first screen that appears when the HMI is run. As such, it is common to make this default screen the landing screen, or as most often called in automation, the home screen.

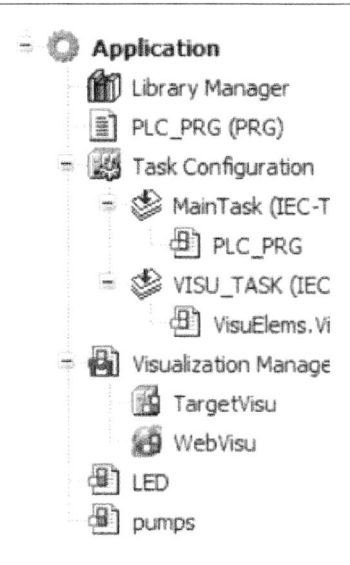

Organizing the screen into multiple layouts    271

Figure 13.16 – The pumps screen added to the HMI

*Figure 13.16* shows that a new screen has been added to the tree. As it stands right now, the screen is blank and even if we did add something to it, it would never load. As stated before, by default the screen that is generated with **Visualization Manager** is the default screen that will be loaded when the program loads. Usually, this is fine, as the default screen will simply serve as the program's home screen; however, there are times when we need to change the default screen. As such, the following section will be dedicated to setting the default screen when the program loads.

## Changing the default screen

Luckily, changing the default HMI view in CODESYS is simple. To demonstrate this concept, follow these steps:

1. Ensure that there is a **Visualization Manager** attached to your project. If you have an HMI already setup, you should have a Visualization Manager add the following controls to it:

Figure 13.17 – Visualization Screen

2. Now, add another screen called **V2** and add the following controls to it:

272    Layouts — Making HMIs User-Friendly

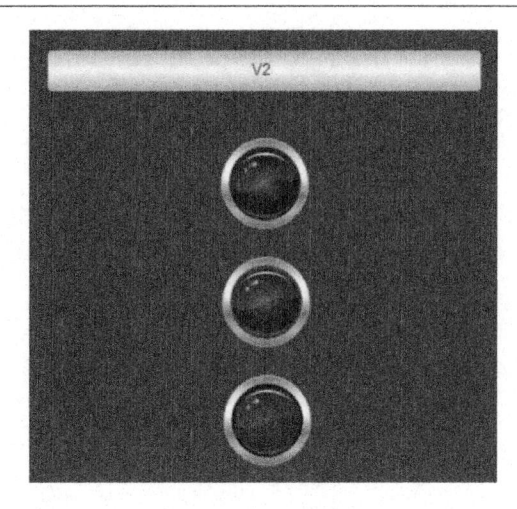

Figure 13.18 – V2 screen

When you run the application, you'll see **Visualization Screen** by default.

3. Now, if you want the **V2** screen to load, all you need to do is click on **Visualization Manager** and navigate to the **TargetVisu** tab. The key to setting the proper screen as the default resides in that tab.

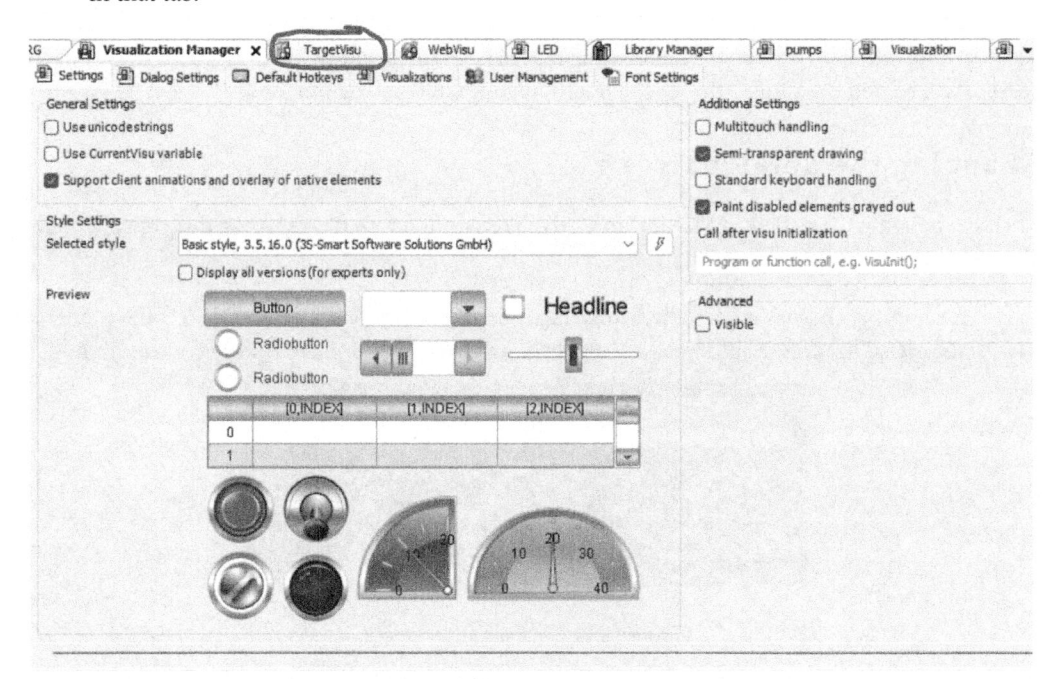

Figure 13.19 – Visualization Manager

4. Once you click on the tab, you should see a screen similar to what is in *Figure 13.20*. If you study that figure, you will see a button with three dots in the **Start visualization** row. To select the proper screen, click on that button and you will see a selection menu. Select the screen you need to run at startup.

Figure 13.20 – Visualization selection

5. For this example, change the screen to **V2** and run the program. You should see that you are met with *Figure 13.18* as opposed to *Figure 13.17*.

In normal HMI development, you will need to navigate between screens quite often. As it stands right now, we cannot do that and we are only able to choose the startup screen. As such, after you run this example, we are going to switch gears and start the process of screen navigation.

## Navigating between screens

Usually, you will navigate between screens using buttons. I have found it best to stack the screen navigation buttons on the left side of the screen. As stated before, this type of layout is very efficient for the operator, and since the navigation menu will be on the left of the screen, they will see it first:

1. As such, add the following layout to your default visualization screen:

274　Layouts — Making HMIs User-Friendly

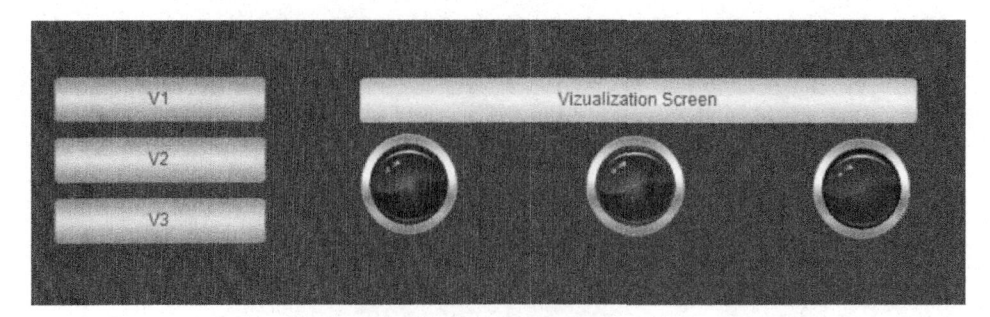

Figure 13.21 – Default layout

2. Next, you're going to want to add three visualizations called **V1_1**, **V2_1**, and **V3_1** to the project. For demonstration purposes, simply add a button called **Home** on each of the screens, as well as a label to indicate the screen, as shown in *Figure 13.22*:

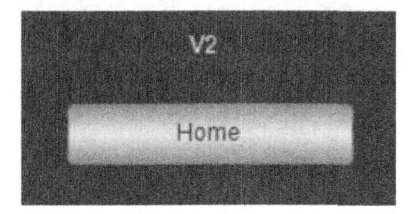

Figure 13.22 – V2 HMI

3. With that set up, we now need to configure the button to navigate to the proper screen. There are many ways to do this, but the easiest, in my opinion, is to set up the navigation as a button click. The first set of buttons we're going to configure will be the button on the home screen. To do this, double-click the button, scroll down to **Input configuration**, and click the **OnMouseClick** field, as shown in *Figure 13.23*:

| Input configuration | |
| --- | --- |
| OnDialogClosed | Configure… |
| OnMouseClick | Configure… |
| OnMouseDown | Configure… |
| OnMouseEnter | Configure… |
| OnMouseLeave | Configure… |
| OnMouseMove | Configure… |
| OnMouseUp | Configure… |
| Toggle | |

Figure 13.23 – Button menu

4. Once you click on that, you should be met with *Figure 13.24*. Click on **Change Shown Visualization** and double-click the button with three dots next to the **Assign** field. You will be met with another window; select the corresponding visualization for the button you are working on.

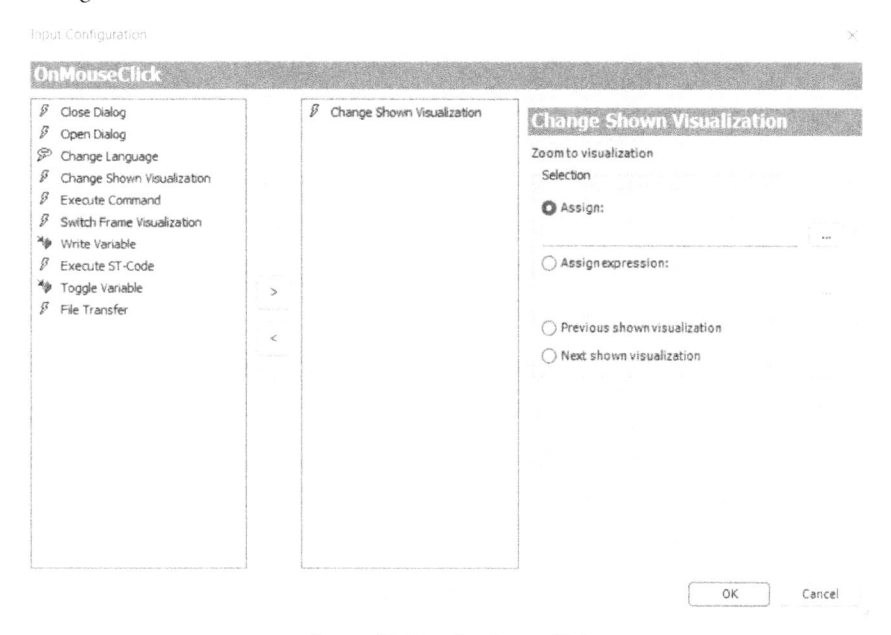

Figure 13.24 – OnMouseClick

5. Once these steps are complete, you can set the button in the **V1**, **V2**, and **V3** screens to the default screen. After you run the program, press the button on the default screen and notice how it will switch between the different screens.

In short, this is one way of navigating between screens. Of all the methods, this is, in my opinion, the easiest.

Thus far, we have explored enough to make a quality HMI. As such, we are going to take the HMI from the last chapter and modify it to make it more user-friendly and add some functionality to it.

## Final project – creating a user-friendly HMI

For our final project, we are going to expand and modify our HMI from the previous chapter to accommodate a home screen, a calibration screen, and a general health screen. As such, the first thing we are going to do is add the screens to the project. Therefore, we're going to add calibration, home, health, and operator visualization to the project. All of the screens are going to be newly created, except for the operator screen, which is the HMI from the last chapter. With all that said, the first screen we are going to make is the home screen.

276　Layouts — Making HMIs User-Friendly

The home screen should be very simple and only serve as a navigation menu. With that in mind, the home screen is usually akin to a welcome page, with simple navigation and maybe pictures, such as company logos and so on. For this project, we are going to lay out the screen like the following:

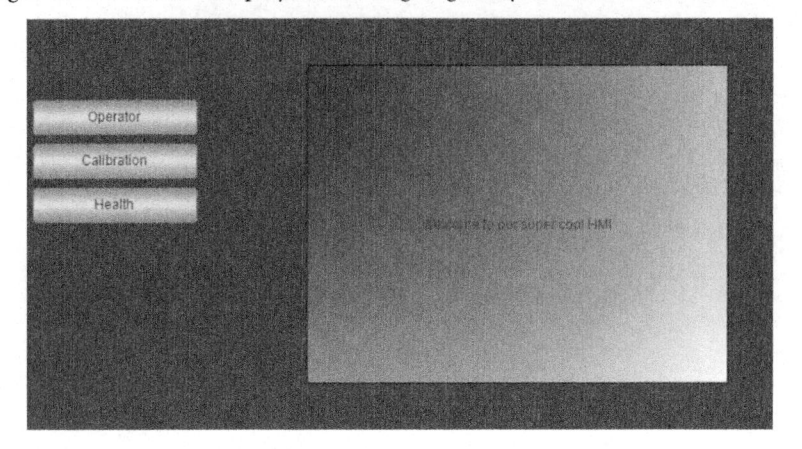

Figure 13.25 – Home screen

This is a very minimal home screen. It is simply a screen that is used to navigate to the subscreens. This means that all you need to do is set the button to navigate to the proper screen, as we have done previously in this chapter. If you notice, the buttons are stacked in a specific order. Since the operator will spend most of their time on the **Operator** screen, that button is on top, followed by the **Calibration** screen, and finally, the **Health** screen for troubleshooting.

As such, the next thing we going to tackle is the **Operator** screen, which is simply the screen from the last chapter, with the addition of a **Home** button to get back to the home screen.

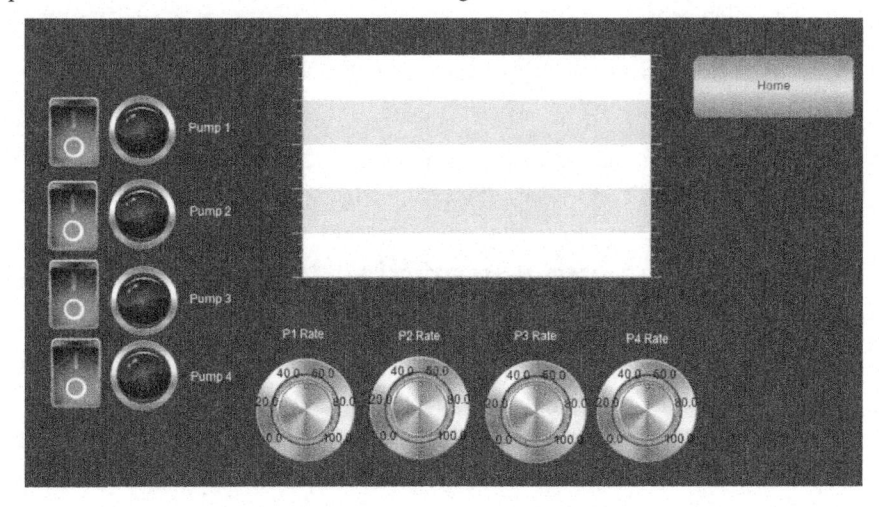

Figure 13.26 – Operator screen

The code for the HMI will be as follows:

```
PROGRAM PLC_PRG
VAR
    hist : ARRAY [1..4] OF INT;
    sw1  : BOOL;
    sw2  : BOOL;
    sw3  : BOOL;
    sw4  : BOOL;
    pot1 : BOOL;
    pot2 : BOOL;
    pot3 : BOOL;
    pot4 : BOOL;
END_VAR
```

If you copy and paste the HMI from the last chapter over to this project, you will not need to alter the settings, as they will be imported as well. The only thing you should have to do is set the **Home** button to navigate to the home screen.

The next screen we will implement is the **Calibration** screen. Calibration screens are very normal in HMIs, as parts often have to be calibrated to properly work. Generally, you want to abstract this into its very own screen for easy use. To do this, we're going to add some pots and spinners to the screen. When complete, your screen should look like the following:

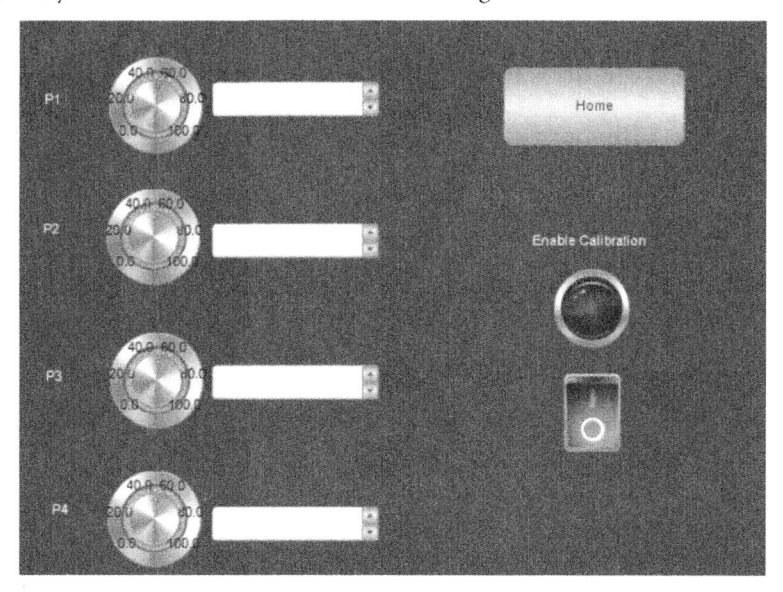

Figure 13.27 – Calibration screen

For this screen to work, you will need to add the following variables:

```
calibration : BOOL;
cal_pot1    : INT;
cal_pot2    : INT;
cal_pot3    : INT;
cal_pot4    : INT;
```

The `cal_pot` variable will correspond to the proper pot and spinner, while the calibration will be assigned to the LED and switch. Since we only want to calibrate when the switch is up, we are going to hide the pots by using the following:

Figure 13.28 – Pot visibility

The final screen we need to build is the **Health** screen. A real health screen will vary in complexity. Sometimes, they are simply a series of LEDs that indicate what is malfunctioning and what is not, while others are complex and give complex diagnostics.

For our diagnostics screen, we are going to keep it simple. We are going to simulate a broken pump. Our **Health** screen will consist of two LEDs for each pump: a green one that indicates that the pump is healthy and a red LED that will indicate the pump is malfunctioning.

Figure 13.29 – Health screen

In terms of logic, we are going to add the following variables to the `PLC_PRG` file:

```
h_led1 : BOOL;
h_led2 : BOOL;
```

```
h_led3 : BOOL;
h_led4 : BOOL;
```

We're going to hook up the variables to their corresponding LEDs. For the green LEDs, we will simply assign the variable to them; however, for the red LEDs, we will invert the variables with the NOT command, as we did with the pots on the **Calibration** screen.

To simulate the health of the pump, we will use the following logic:

```
h_led1 := TRUE;
h_led2 := TRUE;
h_led3 := TRUE;
h_led4 := FALSE;
```

When the code is run, you should see the following on the **Health** screen:

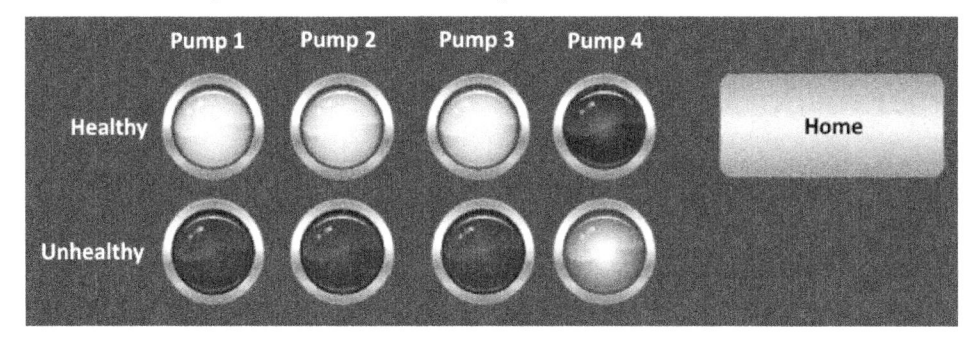

Figure 13.30 – Running the Health screen

As can be seen in this simulated situation, all the pumps are working except **Pump 4**, which is unhealthy.

In all, this is a stripped-down but very typical layout for a real-world HMI. At this point, you should have a very good idea of proper design practices that you can use to build your own HMIs in the future.

## Summary

This chapter was a crash course on HMI design practices. HMI design and development is a very important concept, as a bad HMI will sink a project very easily. Overall, by mastering the concepts such as screen navigation, blinking, colors, grouping, and so on, your HMIs will be easy to use and will allow your project to survive the test of time.

In this chapter, we have also explored error indications. Changing control colors and blinking is an excellent way to handle some issues. However, in automation engineering, we want to throw a true error in the event of an issue. Alarms are very closely tied in with HMI design; as such, our next chapter will be dedicated to exploring proper alarm usage and layouts.

## Questions

Answer the following questions based on what you've learned in this chapter. Cross-check your answers with those provided at the end of the book, under *Assessments*.

1. What do the colors red, green, and yellow mean?
2. What color backgrounds should you primarily use?
3. How many responsibilities should each HMI screen have?
4. How do you set a default screen?
5. What should your default screen be?
6. Add a navigation bar on each screen so you can navigate to each screen from any of the screens.

## Further reading

Have a look at the following resources to further your knowledge:

- *Standard Colors on HMI*: `https://www.mesta-automation.com/standard-colors-on-hmi/`
- *Design Tips to Create a More effective HMI*: `https://blog.isa.org/design-tips-effective-industrial-machine-process-automation-hmi`

# 14
# Alarms — Avoiding Catastrophic Issues with Alarms

Thus far in this book, we have covered the basics of catching errors, mainly by doing something such as blinking an LED or changing the color of a control. For many things, simply changing a control's color or blinking an LED is fine. However, there are times when a more dedicated HMI element is needed. With all that said, enter the world of alarms.

In many SCADA and HMI systems, alarms are dedicated controls that are specifically designed to warn operators about the status of the machine. Normally, alarms will allow you to change colors, display text, log issues, and more. Each HMI or SCADA package that offers alarms will offer different alarms, styles, functionality, and more. However, the core principles that govern most alarms are universal.

Much like HMIs, developing and properly implementing an alarm is as much a science as it is an art. This chapter is dedicated to implementing alarms logically and effectively. To do so, we are going to explore the following concepts:

- What are alarms?
- Where to use alarms
- Alarm configuration: Info, Warning, and Error setup
- Alarm HMI components
- Alarm PLC code
- How to acknowledge alarms

Alarms for motors are very common as motors can easily overheat, draw too much current, or have any other number of issues that need to be logged in an alarm or the program may need to be stopped until the problem is fixed. After exploring these concepts, we are going to round out the chapter by exploring an alarm for a motor.

## Technical requirements

As per every other chapter in the book, a full version of CODESYS will need to be installed. Also, as usual, the examples can be downloaded from the following URL:

`https://github.com/PacktPublishing/Mastering-PLC-programming/tree/master/Chapter%2014`

Unlike most of the other chapters, this one will require you to know about HMIs. Hence, it is highly recommended that you read *Chapters 11*, *12*, and *13* before exploring the concepts in this chapter.

## What are alarms?

The name alarm will normally conjure up images of flashing lights and critical errors on your machine. In many cases, this is exactly what alarms are used for – their main purpose is to relay information to the operator. As we will see later on in the chapter, in the section *Alarm configuration: Info, Warning, and Error setup*, alarms relay way more information than just catastrophic errors.

For most PLC applications, alarms have two sections. The first component of an alarm is the PLC logic itself. The PLC logic is the code that will trigger the alarm. The other section is the HMI component that will display the alarm in a logical, user-friendly manner. Therefore, before you can start using alarms, you're going to want to ensure that you have a decent understanding of PLC programming and HMI layouts for your custom alarms to work.

Alarms are multifaceted. Not only can they display information to the operator, but with the right PLC logic, they can be used to stop the machine in dangerous conditions. It is quite common for certain triggers to halt the operations of a machine. For example, assume you're programming an industrial furnace. If the furnace gets too hot, you may want to throw an alarm to alert the operator and shut the machine down. In short, alarms are there so the operator can avert a disaster and be informed of the current status of the machine. For this book, the term alarm will be used to denote any message that will appear in an alarm window or screen. With that in mind, a logical question would be when should we use an alarm?

## When should you use an alarm?

Much like with anything else, picking when and where to use an alarm is often up to the discretion of the developer. However, a few good rules of thumb are as follows:

- E-stop is engaged
- A machine or product is overheating or underheating
- The voltage/current is out of the operational range

- Machine components are not responding
- Broken machine parts
- Safety issues such as a person in a danger zone

Generally, anything that may cause an issue or cause unsafe working conditions should have an alarm associated with it. So, with that in mind, what should an alarm reflect?

## What should an alarm say?

For an alarm to be of any use, you will need the alarm to logically reflect the issue/warning at hand. An alarm should, at the minimum, include attributes such as the following:

- The issue that was detected
- Whether the alarm is an info, warning, or error alarm
- Timestamp (if possible)
- Whether the alarm was acknowledged (if possible)

Most HMI systems will give you options for these four attributes. However, some alarm systems do not give these options, especially if they are built using a traditional programming language. Regardless of the options available, you want to give the operator as much information as you can to pinpoint the cause of the message, especially if the alarm relates to a warning or an error message. It is also a common practice to log the alarms in a separate file so they can be analyzed later. Depending on the system that you are using, this can be easier said than done. This is usually easiest when you are developing an HMI that has the ability to write to files or if you're using a general-purpose programming language.

Ultimately, what your alarms need to be attached to and what the alarms should say is going to be up to the end user, your organization, and you. Much like HMIs, creating a decent alarm is a bit of an art. The next step in our journey is to learn how to implement an alarm.

## Alarm configuration – I, Warning, and Error setup

To do anything with alarms, the first thing we need to do is set up an alarm configuration. An alarm configuration is the configuration setup, such as the colors, the fonts, and so on that will govern the info, warning, or error alarms. In CODESYS, this is a relatively easy task. To add an alarm configuration, you will simply need to right-click **Applications** and follow the path in the following screenshot:

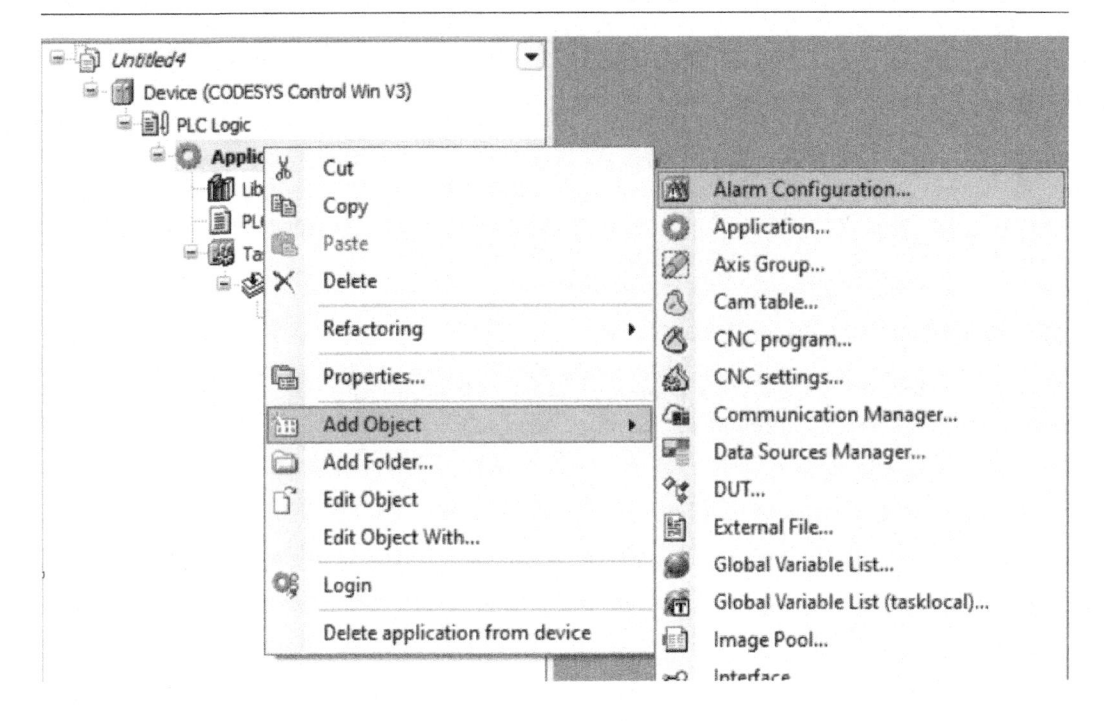

284　Alarms — Avoiding Catastrophic Issues with Alarms

Figure 14.1 – Path to add an alarm

This will bring you to a wizard like the one in the following screenshot. In the wizard, give the alarm configuration a name and click **Add**. For our alarm configuration, we will simply keep the default name. Once you click **Add**, you will see the **Error**, **Info**, and, **Warning** attributes under the **Alarm Configuration** attribute in the project tree.

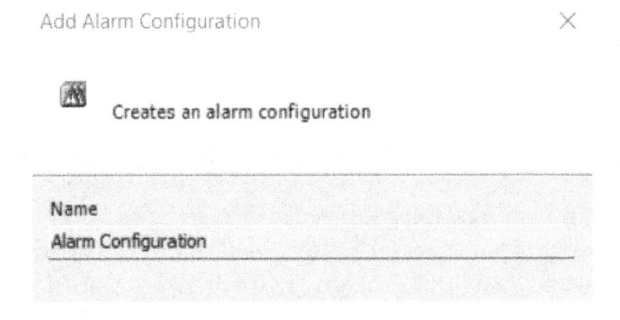

Figure 14.2 – Alarm configuration wizard

The **Info**, **Error**, and **Warning** attributes that are generated can be seen in the following screenshot:

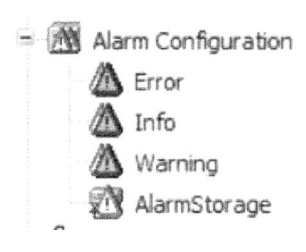

Figure 14.3 – Alarm Configuration tree

For alarms that utilize **Alarm Configuration**, each **Error**, **Info**, or **Warning** alarm will be identical in setup. In other words, they will have the same fonts, colors, and so on. You can technically choose any color you want to represent each of the states. However, in keeping with tradition, we will use the following color schemes:

- **Error**: Red
- **Warning**: Yellow
- **Info**: Green

In short, this is the color scheme that we established in *Chapter 13*. To set the configuration attributes, you will need to click on the **Error**, **Info**, and **Warning** tabs individually.

The first step is setting the configuration for each of the alarm types. You will first double-click on them. Once you double-click any of the alarm types, you should see a configuration screen similar to the following:

Figure 14.4 – Error configuration menu

286　Alarms — Avoiding Catastrophic Issues with Alarms

For our examples, you will want to click the **Archiving** checkbox and the **Acknowledge separately** checkbox. Next, you will want to navigate to the bottom and select your font and background color. *Figure 14.4* shows the configuration that will be used for errors. It is advisable to keep the same font but change the background color to the appropriate color for the appropriate alarms, that is, green or yellow. After this is complete, you will need to set up your alarm groups.

## Alarm groups

Alarm groups consume alarm configurations like the one that was just set up. To generate an alarm group, right-click on the **Alarm Configuration** button, select **Add Object**, then **Alarm Group…**, similar to what can be seen in the following screenshot:

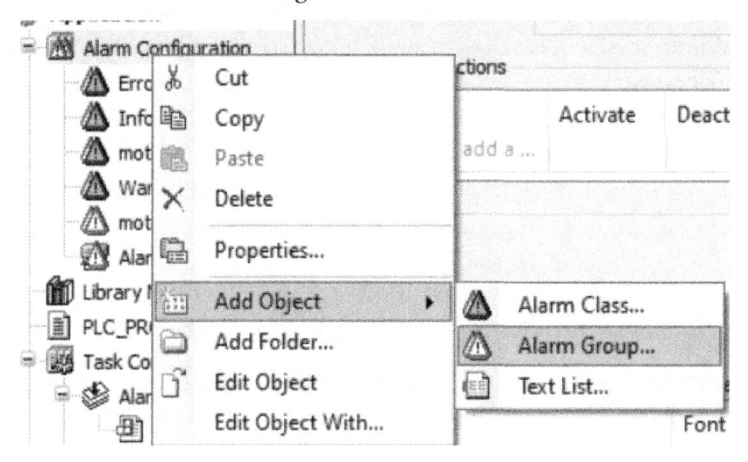

Figure 14.5 – Alarm group generation

After you click **Alarm Group…**, you will be met with a wizard similar to the one in the following screenshot:

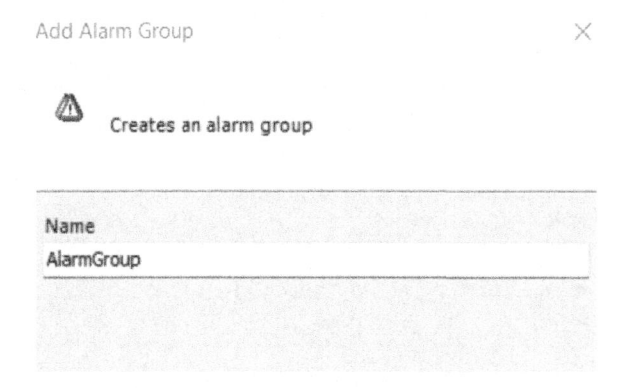

Figure 14.6 – Add alarm group wizard

For our example, we are going to give the alarm group the name `motor` and click the **Add** button. Once you do that, you should see the group appear in the tree, as in the following screenshot:

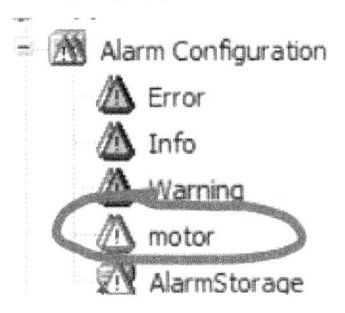

Figure 14.7 – Alarm Configuration group

Before you can fully set up the **motor** group, you will need to implement the following variables in the `PLC_PRG` file:

```
PROGRAM PLC_PRG
VAR
      info  : BOOL := TRUE;
      warn  : BOOL := FALSE;
      error : BOOL := FALSE;
      ack   : BOOL;
END_VAR
```

In this case, we have four variables that will be responsible for showing the alarm message. Now that those variables are set, double-click the **motor** alarm group. Once you click the group icon, configure the screen similar to the following screenshot:

| ID | Observation Type | Details | Deactiva... | Class | Message |
|---|---|---|---|---|---|
| 0 | Digital | (PLC_PRG.error) = (TRUE) | | Error | There is a motor error |
| 1 | Digital | (PLC_PRG.info) = (TRUE) | | Info | All Good! |
| 2 | Digital | (PLC_PRG.warn) = (TRUE) | | Warning | All Not so Good! |

Figure 14.8 – Motor group configuration

For our example, the **Observation Type** will be set to **Digital**. Next is the **Details** column. The **Details** section is the logic that will fire the alarm class, and by extension, the message attached to it. This is where the variables that we created come into play. As can be seen in *Figure 14.8*, the variables that we created are used here, minus the `ack` variable, which will be used for the alarm acknowledgment.

The next column we need to set is **Class**. The **Class** column is essentially the alarm type that will fire when the logic in the **Details** column is satisfied.

With all that, the alarm's configuration should be set up. However, as it stands, this is just the core logic and configuration. For this to be useful, we need to attach it to an HMI component so we can display our alarm on the HMI.

## Alarm HMI components

After we set up the alarm's configuration, we can drop in an HMI component. In terms of CODESYS, there are two types of HMI *controls*. One is the **Alarm Banner** and the other is the **Alarm Table**. Consider the following to understand the difference between a banner and a table:

- **Banner**: Shows one message at a time. It will prioritize alarms and only show the **Error**, **Warning**, or **Info** alarm in that order unless configured otherwise. In other words, a banner will show the most important alarm. No matter the type of alarm, the alarm can be toggled by toggling the variable in the alarm group.
- **Table**: An alarm table will show active alarms for an alarm group. New alarms will show at the top of the table. Alarms can be toggled simply by toggling the variable the alarm is tied to. Where banners are meant to be on every HMI screen, tables can be set on one diagnostic screen. Compared to banners, a table can give more information as it will show more alarms.

In short, a banner will allow you to see the most recent alarm while a table is similar to a log of current alarms. Whereas the banner will show only one message, the table will show several past messages.

Depending on the design, HMI tables will vary. However, at the very minimum, you're going to add a banner. The banner should be at the very top of each HMI screen and it should be in a place where the operator can easily see it. Usually, you'll want to place the banner in the center at the top, similar to the following:

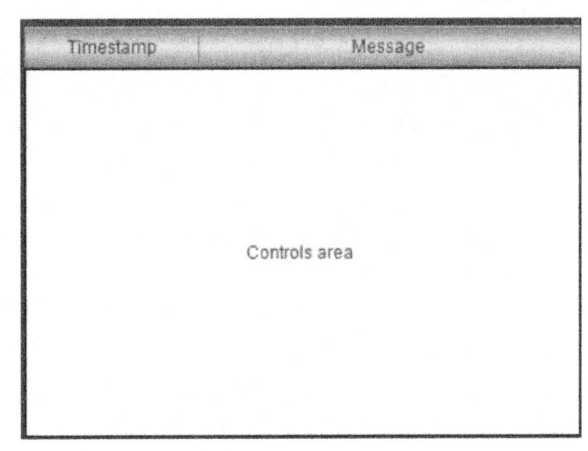

Figure 14.9 – Mock layout

The preceding layout shows the alert bar at the top of the screen. Under the bar is the controls area.

## Setting up an alarm banner

To set up the bar, firstly, we will need to set up the alarm configuration. For this example, we will use the same value that we set up in the **Alarm Configuration** section. After that step is complete, you will want to set up the following variables in the PLC_PRG file:

```
PROGRAM PLC_PRG
VAR
     info  : BOOL := TRUE;
     warn  : BOOL := FALSE;
     error : BOOL := FALSE;
     ack   : BOOL;
END_VAR
```

This example will use four variables. The info, warn, and error variables will display the associated alarm when they are set to TRUE. The other variable, ack, will serve as the acknowledgment variable.

If you were to force the info variable to TRUE, you would be met with the following:

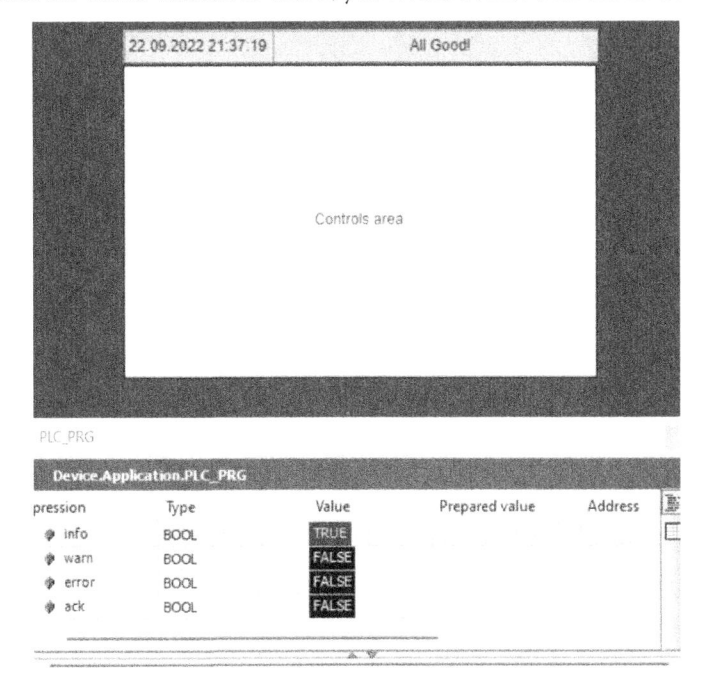

Figure 14.10 – Info alert with a variable list

To explore the other alarms, set the `info` variable to `FALSE` and then set one of the other variables to `TRUE` and observe the banner. If all goes well, you should see the banner change color and the message on the banner change. With that complete, we can now turn our attention to setting up alarm tables.

## Setting up an alarm table

The other type of alarm display is an alarm table. An alarm table is a layout that will show multiple alarms. This is good since the banner will only show one alarm at a time and there could be multiple issues at the same time. If you have to monitor multiple alarms, you'll usually want to opt for an alarm table over a banner. On the flip side of that, it is common to use both in an HMI.

An alarm table will resemble what is shown in the following screenshot:

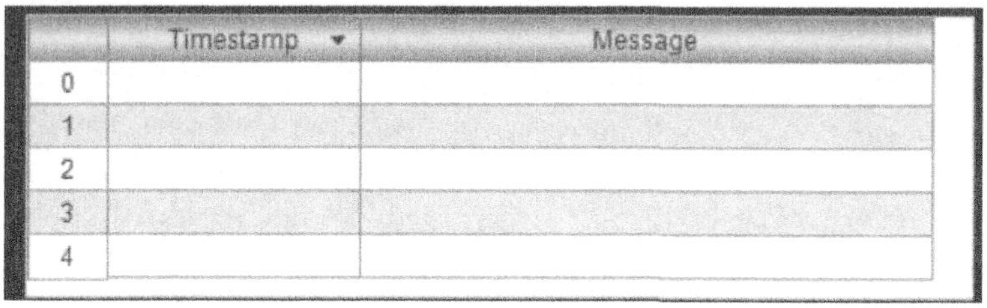

Figure 14.11 – Alarm table

As can be seen, the alarm table is split into rows and columns. The table will autogenerate a **Timestamp** column and a **Message** column; however, you will have the option of adding extra columns as you see fit. For this example, we're going to keep it simple and only use the **Timestamp** and **Message** columns.

Where and how you use the alarm table is ultimately up to you; however, I like to create a specialized HMI screen for alarms and diagnostic purposes. In terms of immediate alarms, I like to add a banner to each HMI screen. Due to the nature of the table, I feel that these types of controls are best used on these types of screens to make it easier to locate issues that may arise in the system. However, that is a personal preference and you may find it more suitable to go for another route. To emphasize, that is a personal preference and not a best practice or anything of the sort.

In terms of placement, since an alarm table is larger than a banner, I usually like to place these toward the middle of the HMI screen or off to one of the sides. Again, this placement is a personal preference. However, due to the size of the table, it can easily look out of place depending on where you place it.

Setting up an alarm table is very similar to setting up an alarm banner. To use an alarm table, you'll have to create an alarm configuration and add it to the alarm table setup. To properly set up the alarm table, your settings should match the following:

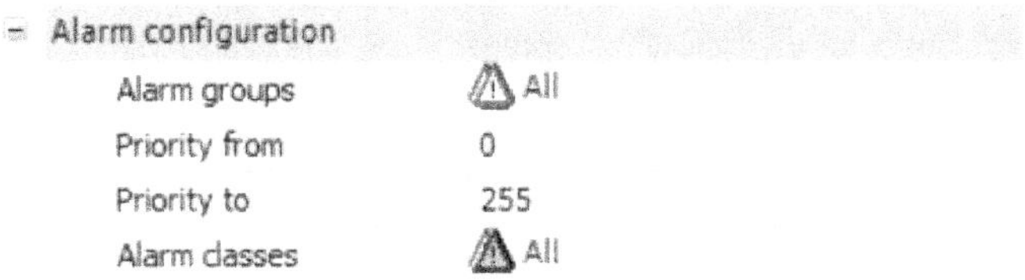

## Alarm HMI components

*Figure 14.12 – Alarm table alarm configuration settings*

In terms of the alarm configuration, we can simply reuse the one that we used for the banner for this example. We are also going to reuse the PLC code that we used for the banner in this example. We are going to remove the banner and simply place an alarm table on the screen. Once the program is running, set all the variables to TRUE, as in the following screenshot:

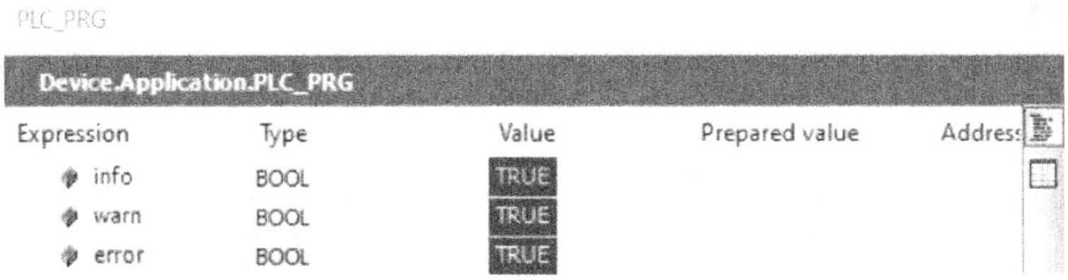

*Figure 14.13 – PLC variables*

When these variables are set, you should see what is in the following screenshot. As you can see, all the alarms are being shown at once. As we have stated before, this is a great tool to gain a deeper insight into the current state of the machine. The order that your alarms appear in may vary. In short, the newest alarm will appear in the **0** row.

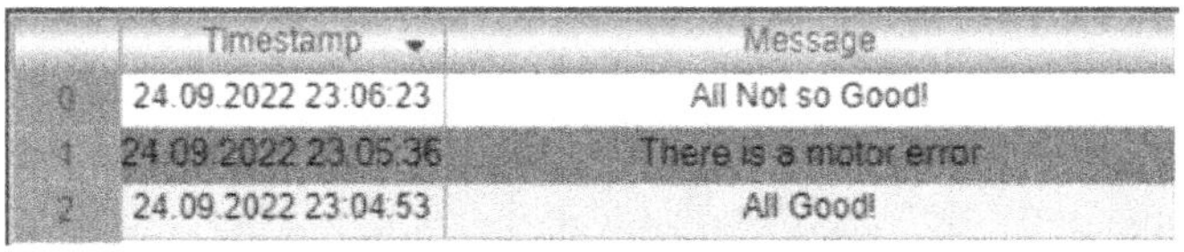

*Figure 14.14 – Alarm table output*

Now, alarms can be removed from the table by simply turning them off. In other words, setting the TRUE variable to FALSE will remove the alarm from the table.

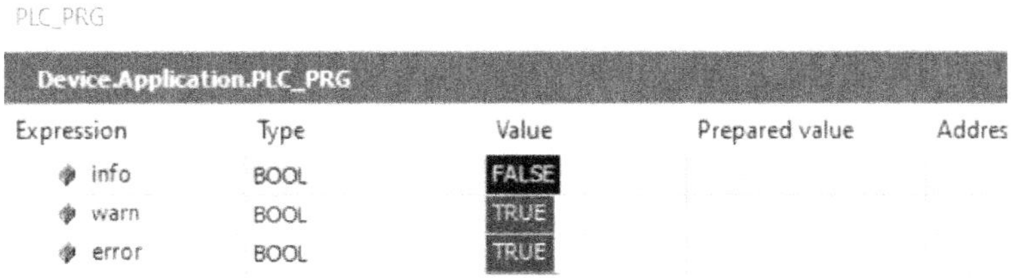

*Figure 14.15 – info variable set to FALSE*

Once you've done this, consider the following screenshot. As you can see, the **Info** alarm is now gone.

| Timestamp | Message |
|---|---|
| 24.09.2022 23:06:23 | All Not so Good! |
| 24.09.2022 23:05:36 | There is a motor error. |

Figure 14.16 – Info alarm removed

The preceding figure shows what happens when the variable that is tied to the alarm is turned off. In short, the alarm is removed from the table. Now, with that in mind, you can turn the alarm back on by simply toggling the variable again.

The chapter thus far has been dedicated to setting up the controls on the HMI and displaying messages on the HMI. However, we have only touched on the logic side. As we saw, triggering an alarm is usually as simple as setting a Boolean variable to `true` or `false`. Next, we will cover the PLC logic.

## PLC alarm logic

Now that we have set up an alarm, we need to get into the guts of the alarm, which is what I like to call the alarm logic. There is nothing fancy or complex about triggering an alarm. As we have seen, all we have to do is set a variable to `true` or `false`. However, understanding when to set the variable is the trick. For most things in automation, we use bounds or operating ranges to determine whether the part is in a healthy state or not. In other words, many things, such as heaters, motors, and so on, have an optimal operating range that they should always be in. Straying from the optimal range can easily affect the performance of the machine. Some of the most common situations that need immediate alarms are situations that can result in the injury of a person or the surrounding environment. In these cases, you will not want to set up a tolerance range; you will simply want to throw an alarm.

Typically, when you're working with a range, you will use the HMI to set up a lower bound and an upper bound. Though it is common to hardcode in the range value, it is usually better to set these bounds on something akin to a calibration screen. If a part is swapped out for something with a different operational range, the range will need to change based on the job run, or if another operational variable changes. Hence, our examples are going to have an accompanying HMI.

For this example, we are going to simulate a series of pumps. We are going to set up an operating range that looks like the following:

- **Normal**: 0 – 50 PSI
- **Approaching limit**: 50-75 PSI
- **Over limit**: >75 PSI

If the PSI is in the normal range, we will have a green banner. If the PSI is approaching the limit, it will be yellow. And anything over 75 PSI will trigger a red alarm. The PLC program will mostly consist of a series of control statements. To implement this program, we are going to use the following variables in the PLC_PRG file:

```
PROGRAM PLC_PRG
VAR
     info    : BOOL := FALSE;
     warn    : BOOL := FALSE;
     error   : BOOL := FALSE;

     info_pump    : BOOL := FALSE;
     warn_pump    : BOOL := FALSE;
     error_pump   : BOOL := FALSE;

     good_range  : INT;
     warn_range  : INT;
     error_range : INT;

     psi : INT;
END_VAR
```

In this example, we are setting alarms off by default. We do this so we can get an appropriate alarm based on the data we received from the sensor as opposed to assuming that everything is correct off the bat.

Below the alarm variables are the range variables. These are the variables that will be tied to an HMI control. In short, as the name suggests, these are the variables that will be used to set the good, warn, and error range, which will be tested against the psi variable.

After those are set, we will need to set up the logic to determine which alarm needs to be triggered. As can be seen in the code, this is mostly just a series of control statements. The PLC_PRG file should look like the following code snippet:

```
IF (psi < good_range) THEN
     info_pump  := TRUE;
     warn_pump  := FALSE;
     error_pump := FALSE;
ELSIF (psi > good_range AND psi <= warn_range) THEN
     info_pump  := FALSE;
     warn_pump  := TRUE;
     error_pump := FALSE;
ELSIF (psi > warn_range AND psi >= error_range) THEN
```

# Alarms — Avoiding Catastrophic Issues with Alarms

```
        info_pump   := FALSE;
        warn_pump   := FALSE;
        error_pump  := TRUE;
END_IF
```

Essentially, this is just a state machine. When the `psi` value falls into a certain range, the state, or alarm for that state, is triggered. For the sake of learning, this example was designed to be a little more complex than what you would actually do in the real world. Technically, you don't need the range. The range is only included to help visualize when the alarm will trigger. If you want to experiment, you can use the following code snippet and test the results:

```
info_pump   := (psi < warn_range);
warn_pump   := (psi >= warn_range AND psi <= error_range);
error_pump  := (psi > error_range);
```

For this example, we are going to add a new alarm configuration to give the appropriate message. In this case, we are going to add a new alarm group and configure it to match the following:

| ID | Observation Type | Details | Deactiva... | Class | Message |
|---|---|---|---|---|---|
| 0 | Digital | (PLC_PRG.info_pump) = (TRUE) | | Info | PSI Optimal |
| 1 | Digital | (PLC_PRG.warn_pump) = (TRUE) | | Warning | PSI Approching Upperlimit |
| 2 | Digital | (PLC_PRG.error_pump) = (TRUE) | | Error | PSI Over Range |

Figure 14.17 – Pump alarm group

When you have added the new alarm group, the **Alarm Configuration** tree should look like the following figure:

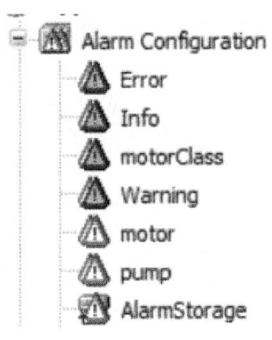

- Alarm Configuration
  - Error
  - Info
  - motorClass
  - Warning
  - motor
  - pump
  - AlarmStorage

Figure 14.18 – Alarm Configuration tree

In terms of the HMI, we will use the layout shown in the following screenshot:

PLC alarm logic 295

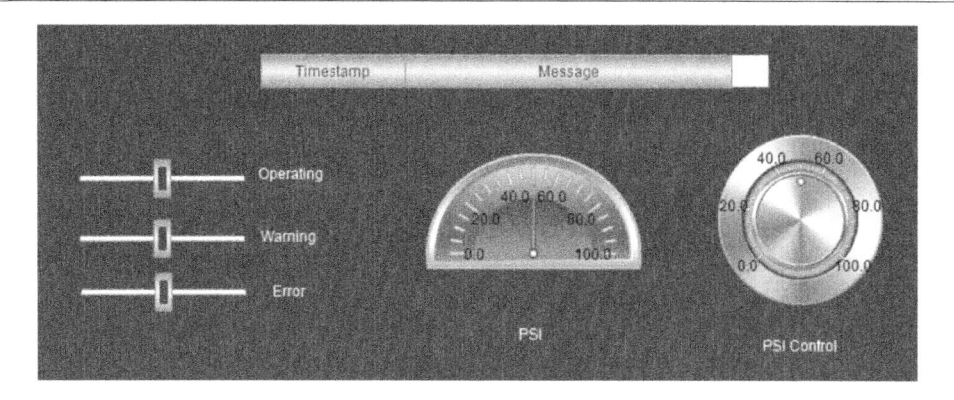

Figure 14.19 – HMI layout

The HMI will allow us to set our limits via the sliders on the left of the screen. The variables will be assigned with the following pattern:

- **Operating**: `info_pump`
- **Warning**: `warn_pump`
- **Error**: `error_pump`

The **PSI Control** pot will be attached to the PSI variable, as will the gauge. The gauge will be used to view the current PSI setting that is set with the pot. Finally, on top, we are going to use an alarm banner to display the current alarm.

The next thing we need to do is set the scales on the sliders. We are going to set the **Operating** slider with what is shown in the following screenshot:

```
- Scale
    Show scale
    Scale start      0
    Scale end       50
```

Figure 14.20 – Operating slider scale

Next, we're going to set the **Warning** slider. To do this, we're going to set the scale to what is shown in the following screenshot:

```
- Scale
    Show scale
    Scale start     51
    Scale end       75
```

Figure 14.21 – Warning slider scale

Lastly, we're going to set the **Error** slider's scale. We're going to set it to match the values shown in the following screenshot:

| Scale | |
|---|---|
| Show scale | |
| Scale start | 76 |
| Scale end | 100 |

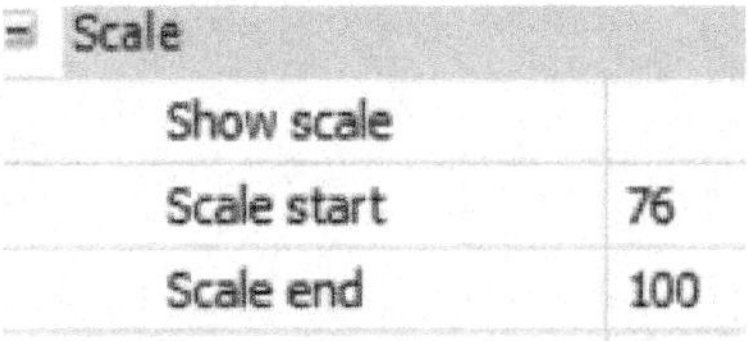

Figure 14.22 – Error slider scale

Next, we're going to set the pot. Since our scales are going to max out at 100 PSI, we're going to set the scale end to 120. We're going to do this just to give it some extra space for the error alarm. You will need to match the scale on the pot to match the values shown in the following screenshot:

| Scale | |
|---|---|
| Subscale position | Inside |
| Scale type | Lines |
| Scale start | 0 |
| Scale end | 120 |

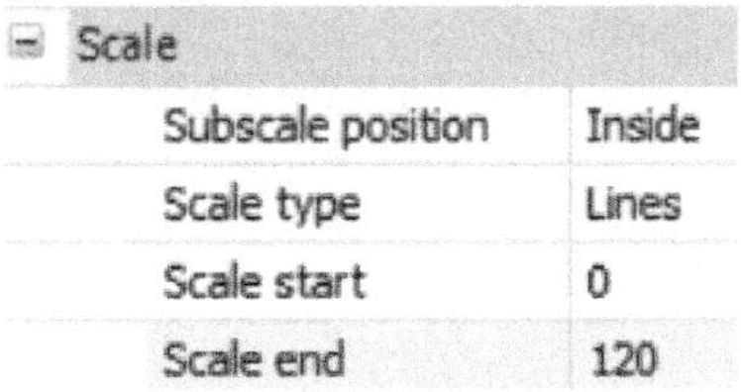

Figure 14.23 – Pot scale

We're going to want to set the scale on the gauge to the pot as well. We'll set the scale the same way we have set the scale on every other HMI component we have explored thus far in the book. Lastly, we need to set up the alarm banner. If you have not set up the alarm configuration for this example, you will need to do that now. After you ensure that the alarm configuration is set, you will want to set the banner to read the alarms from that group only. To do this, you will need to follow the steps outlined:

1. Expand the **Alarm configuration** tab and double-click **Alarm groups**. By default, it will be set to **All**, but we want to only target the **pumps** group.

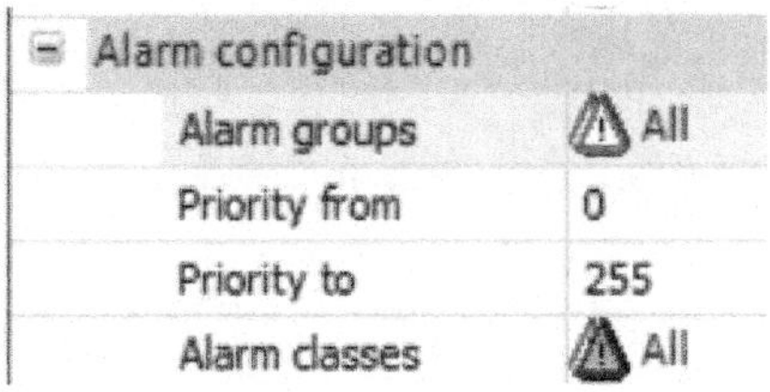

| Alarm configuration | |
|---|---|
| Alarm groups | All |
| Priority from | 0 |
| Priority to | 255 |
| Alarm classes | All |

Figure 14.24 – Alarm configuration

PLC alarm logic 297

2. Once you click the **Alarm groups** tab, you should be met with the screen that is shown in the following screenshot. By default, all the alarm groups are going to be selected. You will need to uncheck the **All** box (circled in the following screenshot), select the **pumps** group, and click the arrow button that is also circled. When the operation is complete, you should see the **pumps** group move from the left of the screen to the right. Once the operation is complete, you will see the alarm groups table reflect the change.

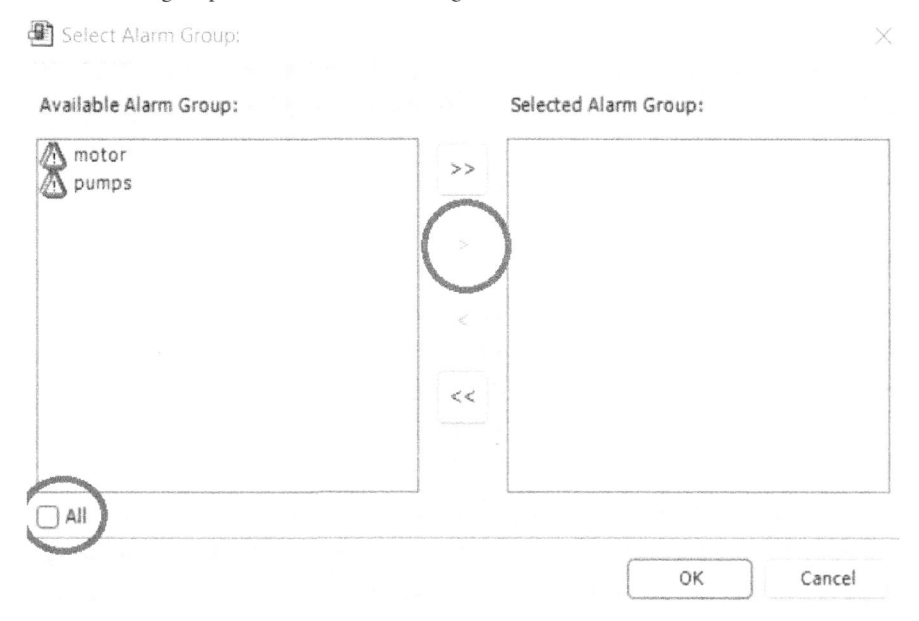

Figure 14.25 – Select Alarm Group wizard

The following is the result of setting the alarm banner to only register the **pumps** alarm group:

| Alarm configuration | |
|---|---|
| Alarm groups | pumps |
| Priority from | 0 |
| Priority to | 255 |
| Alarm classes | All |

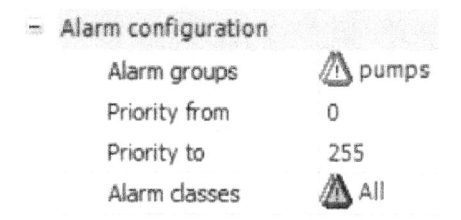

Figure 14.26 – Final alarm banner configuration

3. After all these components are configured, you can run the program. Once you run the program, you'll want to set all the slides to the right of the screen, as shown in the following screenshot. Once you do that, you will see the banner turn green.

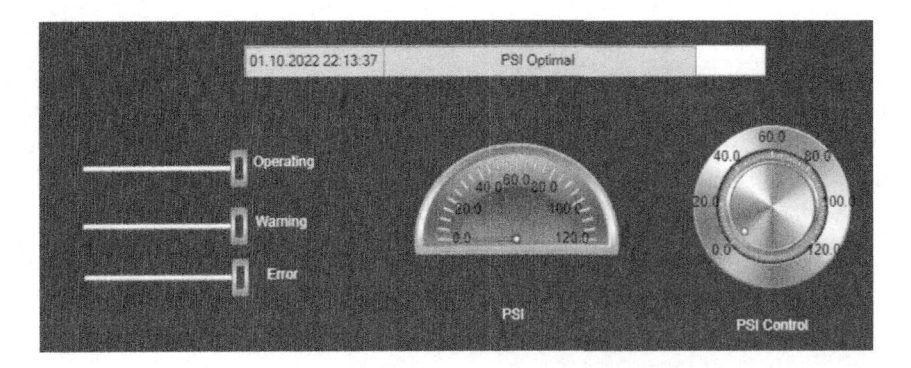

298   Alarms — Avoiding Catastrophic Issues with Alarms

Figure 14.27 – Green banner

4. Once the program is running, slowly rotate the pot to right. Notice that once you get past 50 PSI, the banner will turn yellow:

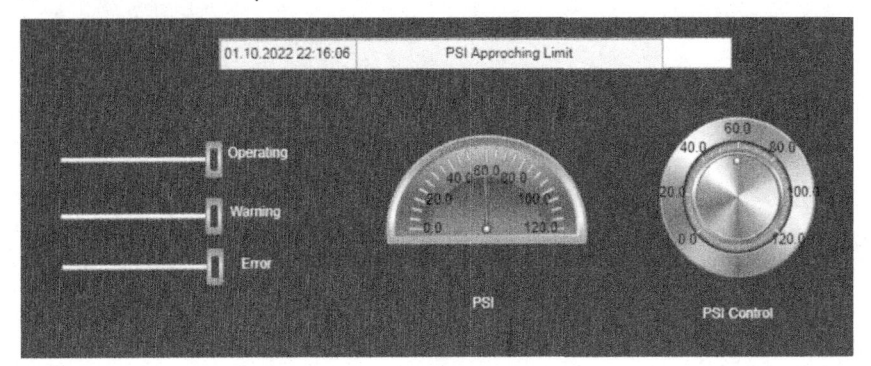

Figure 14.28 – Yellow banner

5. Finally, if you max out the pot, you will see the banner turn red. As stated before, the red banner will be the error banner.

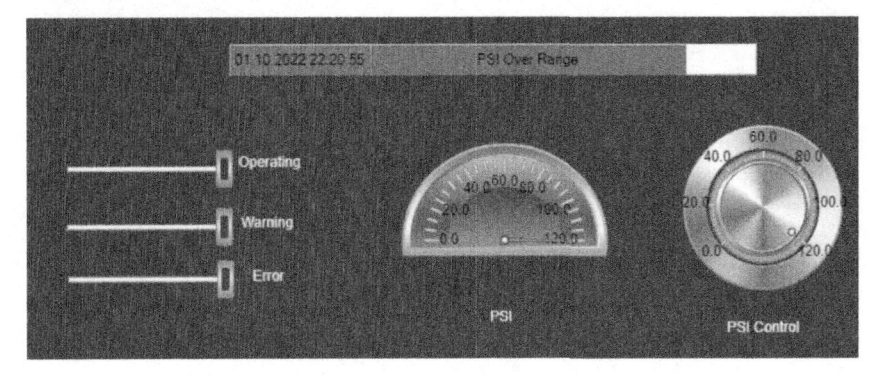

Figure 14.29 – Red banner

In a real-world application, the data that dictates which banner to show will most likely be fed in by a sensor of some type. Generally, alarms need to be dynamic and read data. However, you will most likely always use some type of input, such as sliders, to set the limits. Depending on what you're working on, it may be easier to trigger the alarms via the configuration in the **Alarm configuration** menu. For our example, we triggered the alarm programmatically, which is acceptable in many cases; however, it is important to explore using the GUI as well. Consider the following screenshot:

| ID | Observation Type | Details | Deactivation | Class |
|----|------------------|---------|--------------|-------|
| 0 | Digital | (PLC_PRG.info_pump) = (TRUE) | | Info |
| 1 | Digital | (PLC_PRG.warn_pump) = (TRUE) | | Warning |
| 2 | Upper limit | PLC_PRG.psi > PLC_PRG.error_range | | Error |

Upper limit

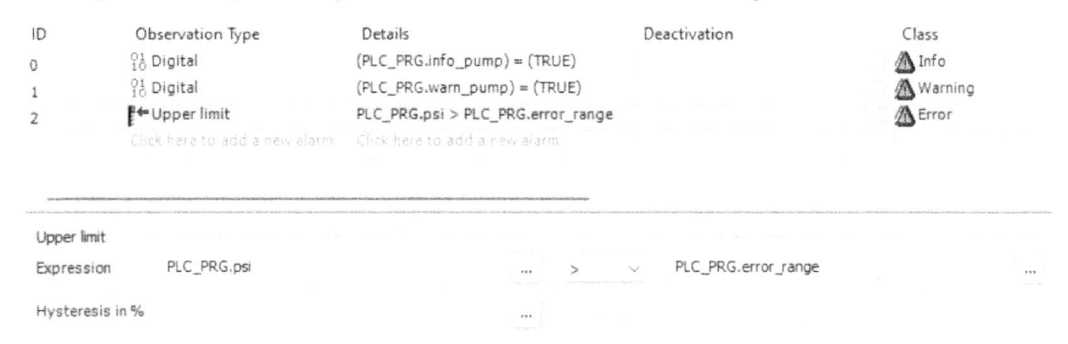

Expression        PLC_PRG.psi                    >      PLC_PRG.error_range

Hysteresis in %

Figure 14.30 – Set error limit via alarm configuration

Essentially, this will set the upper-limit logic statement for the alarm to trigger; however, using this methodology can be somewhat restrictive if you need to perform machine operations, such as a machine shutdown. You will need to set up the logic for that in the PLC program as well. With all that said, the next thing we need to look at is alarm acknowledgment.

## Alarm acknowledgment

If you've noticed, thus far, when you throw an error alarm, the text doesn't go away, whether it be an alarm in the banner or chart. This is because error alarms must be acknowledged. Essentially, if you throw an error, an alarm will be present, at least in text, if you do not acknowledge the alarm. An acknowledgment is basically a confirmation that an operator has seen the alarm and has decided to clear it. No matter whether you're using a table or a banner, you will clear alarms in the same way.

In short, there is an acknowledgment field that holds a variable. When the variable is `true`, the text in the alarm display will clear out. For our example, we are going to add a button to the HMI. In short, your HMI should be modified to look like the following:

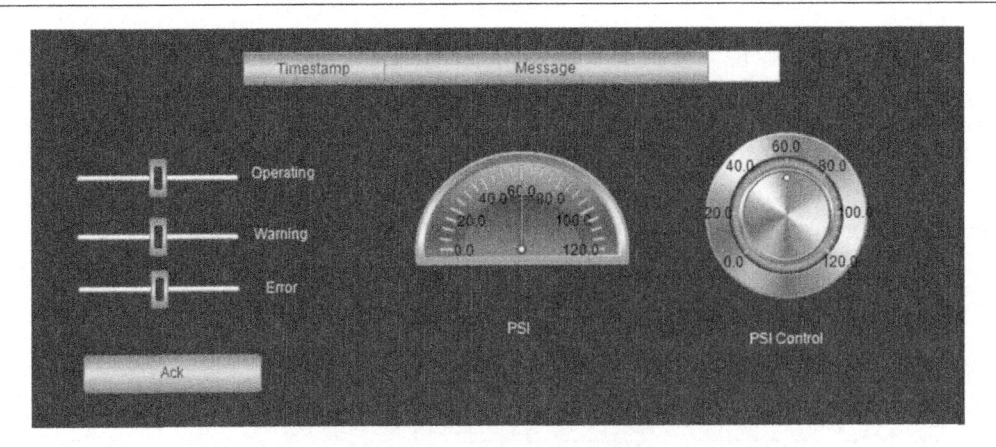

**300**  Alarms — Avoiding Catastrophic Issues with Alarms

Figure 14.31 – HMI with Ack button

Once you add the button, add a variable called `ack` of type `bool` to the `PLC_PRG` file. We're going to want to assign the button to the banner's acknowledgment field, as in the following screenshot.

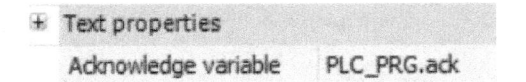

Figure 14.32 – Banner's Acknolwdege variable field

After you do that, you will need to set up the button to toggle the variable. To do this, click on the button, expand the **Input configuration** field, and select **OnMouseClick**.

Figure 14.33 – Input configuration fields

When you click on this field, you will be met with the following:

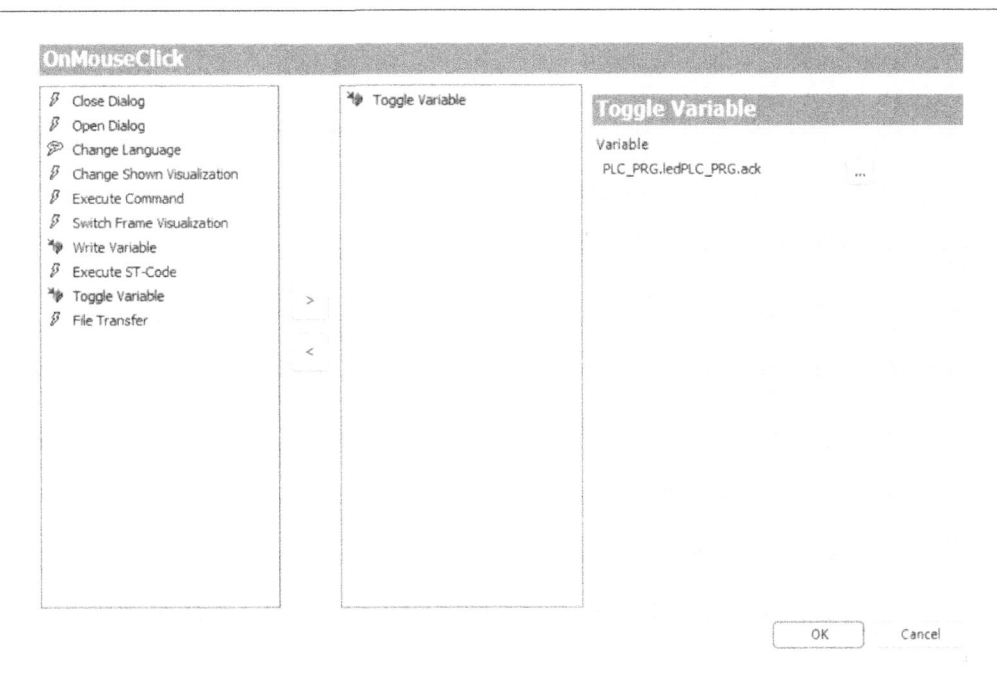

Figure 14.34 – OnMouseClick

Select **Toggle Variable** and click the right arrow button. Once you do that, you will need to click the button with three dots and select the `ack` variable that you added in the `PLC_PRG` file. You should then be all set to use the modified HMI. You will want to run the program and throw the error alarm. Once the alarm appears, you will want to dial back on the pot, preferably to zero, and click the **Ack** button. You should notice that the text is cleared and that the appropriate alarm is displayed. In a nutshell, the clicking of the **Ack** button can be thought of as an event. When the event happens, you are making an agreement with the HMI that you saw the alarm. If the issue that caused the alarm to trigger is clear, it will clear; however, if the problem has not been fixed, the alarm will appear again. The **Ack** button will not directly interact with the PLC code; it is purely an HMI control that is designed to interact with alarms.

To recap, an alarm acknowledgment is a way for the operator to confirm either that they have seen there is an issue and they are choosing to ignore it or are confirming that the problem is fixed. Acknowledgments are a vital part of HMI development and automation programming in general. At this point, it is recommended that you swap out the banner for a table and experiment. Notice that with the table, you will have to select the alarm that you want to acknowledge. With all that being said, we now know enough to move on to our final project, building an alarm system for our motors.

# Final project – motor alarm system

For the final project, we are going to create a motor alarm system. In the real world, motors are a pivotal part of automation. However, if a motor starts drawing too much or too little current, there could be a problem. Also, if the operating temperature is over or under range for the motor, there could be a problem. Therefore, we need alarms to indicate when these events occur and what they are. To round out the chapter, we are going to create an HMI similar to the one in the last section; however, we are going to add more alarms. So, the first thing we are going to do is lay out some requirements.

## Requirements

Motors have an optimal operating range for temperature, drawn voltage, and communication between the drive and PLC. We need to monitor these, and if there is any abnormal behavior, we need to trigger an alarm. Also, since there can be multiple issues all at once, we need to log all the issues so the technician can search the log and ensure they fix each issue. Our software needs to trigger the following:

- A warning if the voltage is less than 10 volts or greater than 20 volts
- An error alarm if the voltage is less than 4 volts or greater than 25 volts
- A warning alarm if the temperature is less than 65°F or greater than 100°F
- An error alarm if the temperature is less than 60°F or greater than 110°F
- An error alarm if there is no communication from the drive

To do this, we are going to need to build an HMI that can simulate temperature, communication, and voltage.

## Design/implementation of the HMI

This project is similar to the PSI alarm system in the last section. Hence, we can tweak that HMI design for this project, for which we are going to create an HMI to match the following:

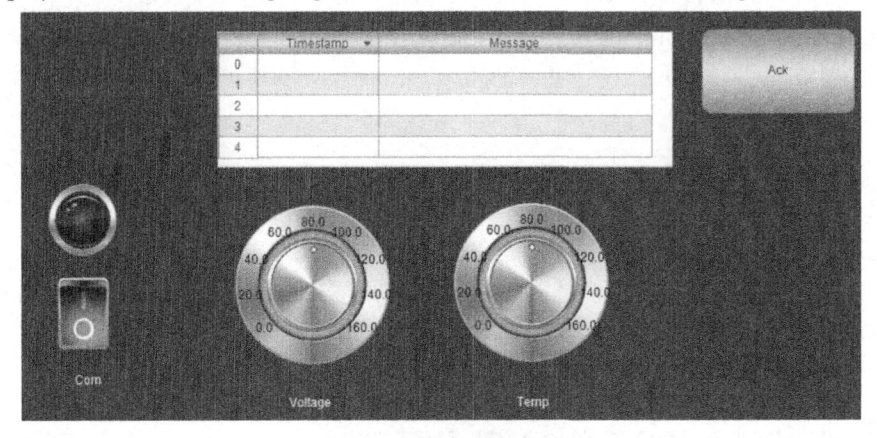

Figure 14.35 – Motor HMI

To get the scale of the controls, pull down the project and examine it. The variables we are going to use for this project are as follows:

```
PROGRAM PLC_PRG
VAR
    overTempErr   : INT := 110;
    overTempWar   : INT := 100;
    underTempWar  : INT := 65;
    underTempErr  : INT := 60;

    overVoltWar   : INT := 20;
    overVoltErr   : INT := 25;
    underVoltWar  : INT := 10;
    underVoltErr  : INT := 4;

    com : BOOL;
    ack : BOOL;
END_VAR
```

For motor alarms, you can sometimes get away with hardcoding many of the values, as long as you know in advance that the motor type won't change. For this example, we are going to assume that the motor type will be static, or at least static for a long while, and hardcode the values. After you implement the variables, assign the variables to their corresponding controls.

Next, we're going to create a new alarm group called `final_example`. Since we are hardcoding, we can set the alarm thresholds with the GUI. Your logic should look like the following screenshot:

| ID | Observatio... | Details | D... | Class | Message |
|---|---|---|---|---|---|
| 0 | Upper limit | PLC_PRG.pot_temp >= PLC_PRG.overTempWar | | Warning | Motor nearing over temp |
| 1 | Lower limit | PLC_PRG.pot_temp < PLC_PRG.underTempWar | | Warning | Motor nearing under temp |
| 3 | Lower limit | PLC_PRG.pot_temp < PLC_PRG.underTempErr | | Error | Motor under temp |
| 4 | Digital | (PLC_PRG.com) = (FALSE) | | Error | No com with motor |
| 5 | Upper limit | PLC_PRG.pot_voltage > PLC_PRG.overVoltWar | | Warning | Motor approaching volt limit |
| 6 | Lower limit | PLC_PRG.pot_voltage < PLC_PRG.underVoltWar | | Warning | Motor approching under v.. |
| 7 | Upper limit | PLC_PRG.pot_voltage > PLC_PRG.overVoltErr | | Error | Motor over voltage |
| 8 | Lower limit | PLC_PRG.pot_voltage < PLC_PRG.underVoltErr | | Error | Motor under voltage |
| 9 | Upper limit | PLC_PRG.pot_temp > PLC_PRG.overVoltErr | | Error | Motor over temp |

Figure 14.36 – Alarm thresholds

Next, set the `ack` variable for the alarm table the same way that it is set in the following screenshot:

- Control variables
    Acknowledge selected     PLC_PRG.ack

Figure 14.37 – Acknowledgment

You also want to only set the `final_example` alarm group and tie the `ack` variable to the button as we did in the PSI example. Once you do this, you are ready to run and play with the HMI. Turn the pots and watch which alarms are shown. When an alarm goes white, select it and click the **Ack** button on the HMI and watch how the text is cleared.

## Summary

In summary, this chapter has been a crash course on HMI alarms. We have covered the HMI and PLC side, as well as the setup of the alarms. We have also learned how to acknowledge alarms and more. By this point, you should know the basics of alarm systems. Overall, you will need to understand this chapter to be an automation programmer, so please ensure that you understand the material. In all, you should have a good grasp of advanced PLC programming in general at this point. This means we can move on to our final project of the book, creating an industrial oven.

## Questions

Answer the following questions based on what you've learned in this chapter. Cross-check your answers with those provided at the end of the book, under *Assessments*.

1. What is an alarm?
2. Name the three types of alarms?
3. What does a red alarm usually mean?
4. What is an alarm group?
5. What is an alarm acknowledgment?

## Further reading

Have a look at the following resources to further your knowledge:

- Configuring Alarm Management: `https://help.codesys.com/api-content/2/codesys/3.5.15.0/en/_cds_setting_up_an_alarm_configuration/`
- Visualizing Alarm Management: `https://help.codesys.com/api-content/2/core_visualization/3.5.16.0/en/_visu_struct_alarm_management/`

# Part 5 – Final Project and Thoughts

The final chapters of the book will end with building a simulated industrial oven and exploring the theory of communication systems. The project will include topics from each chapter of the book. The goal of the section is to combine topics from each chapter to explore how everything fits together, while the final chapter will explore at the conceptual level what communication systems are, how they work, and more.

This final section includes the following chapters:

- *Chapter 15, Putting It All Together — The Final Project*
- *Chapter 16, Distributed Control Systems, PLCs, and Networking*

# 15
# Putting It All Together — The Final Project

Congratulations on making it this far in the book. Hopefully, by this point, you have a good grasp of the more advanced concepts of PLC programming and software engineering in general. By this point, you should not only have become a better PLC programmer but also a better software developer in general. Thus far, we have explored OOP, advanced structured text, alarms, HMIs, the SDLC, and much, much more. In all, at this point, if you understand most of the material covered, you're probably light years ahead of the average automation programmer.

As far as programming and HMI development are concerned, we have reached a point where we can combine all these concepts into a fully working project. This chapter will be unlike the other chapters in this book as we will not be exploring new concepts. Instead, we are going to explore putting the concepts we have learned throughout the book together to make a simulated industrial oven. In a nutshell, we are going to cover the following:

- Project overview
- Getting the requirements
- HMI design
- HMI implementation
- PLC code design
- Implementing the PLC code
- Testing the application

The goal of this project is to integrate many of the concepts that we have learned in the previous chapters to form an industrial oven. Ovens are very common PLC-driven devices as they are used in many different manufacturing processes.

However, unlike most of the other projects we have built thus far, the code will be written in such a manner that it should be improved upon. In other words, the code in this chapter will be the first draft of a program. This twist stems from the fact that most software will usually need to be cleaned and refactored before release. Though we will apply skills we learned throughout the book, you as the reader should be constantly on the lookout for ways to improve and, if necessary, debug the software as you would do for a real-world application.

This chapter will attempt to follow the full SDLC in a Waterfall-like manner. However, since this is a learning example, the exact process that one would use for a real-world example will probably differ. However, we're going to keep the workflow as real-world as possible.

## Technical requirements

This chapter will require a comprehensive knowledge of all the topics covered in the previous chapters. If you have been skipping around the book, it is best to go back and read the chapters that you did skip. If you feel comfortable with the material already covered, you are free to proceed.

The source code for this chapter can be found at the following URL:

`https://github.com/PacktPublishing/Mastering-PLC-programming/tree/master/Chapter%2015`

As with all previous chapters, you can pull the source code down for the aforementioned URL. It is recommended that you pull down the code and attempt to modify it once you have a thorough understanding of the material presented in this chapter.

## Project overview

Before we dive into building the project, it is important to understand what we are developing and why. For our final project, we are going to build an industrial oven. Industrial ovens are often used in the manufacturing process for various applications such as curing paint, baking in chemicals, drying parts, or any number of other applications.

Our simulated customer is requesting an oven system for drying metal fixtures after they come from being washed. The way the manufacturing process works is that once a part is washed, it is placed in the oven for a variable amount of time depending on the fixture so that all excess moisture can be burned off. We have to be careful because there are rubber O-rings in the fixtures that will melt if the O-rings experience temperatures above their rated limit. The customer will want to be able to dry different parts that will require different dry times, and each fixture will have an O-ring with a different temperature limit. With all that in mind, we can now move on to gather our requirements.

# Getting the requirements

Now, that we have a general overview of the goals of the project, we can work on figuring out the requirements. From the overview, we can get the following general project user stories:

- As an operator, I want to be able to manually set the optimal temperature of the oven so that I can use the oven for different fixtures
- As an operator, I want to know when the oven is too hot to enter so that I know not to enter the heated area
- As an operator, I want the door to automatically lock and unlock
- As an operator, I want to know when the oven's temperature is at room temperature so that I can safely enter the oven to remove the dry fixtures
- As an operator, I want to view an alarm when the temperature is over the O-rings' rated temperature so that I know when the O-ring has been compromised
- As an operator, I want the PLC to automatically shut down when the oven's temperature reaches 10°F over the O-rings' rated temperature so that I can retrieve the parts as quickly as possible

These are the basic requirements for the project. Since these are high-level requirements, we will probably run into more questions as we start developing the project. However, these requirements are adequate for us to start hammering out the PLC and HMI side of the system.

Chances are, if you were developing this application for an organization, the workload would probably be split between a PLC programmer or a set of programmers and an HMI developer or a set of HMI developers; however, in cases such as this one, you will be responsible for both the HMI and PLC side of the project. Depending on your thought process, you may want to either start working on the HMI or PLC side of the application first. For me, it has always been easier to work from the HMI backward to the PLC code. This is mainly because once you have a decent outline of what the HMI is responsible for, it is easier to hammer out the PLC code. However, this is a personal preference, and you may find it easier to do the work in the opposite manner. In real life, you can plan out your workflow any way you want. With that, the first thing we are going to do is lay out the design of our HMI.

# HMI design

The first thing we should do is lay out our HMI. Based on the requirements, we are going to need the following at the minimum:

- An alarm table
- A series of inputs to allow the user to input the temperature of the oven
- A gauge to show the current temperature of the oven
- A power switch and LED for the oven

- An LED for the following:
  - Oven ramping up to temperature
  - Oven at temperature
- An alarm acknowledgment button

With the requirements, we can lay out our HMI to look like the following screenshot:

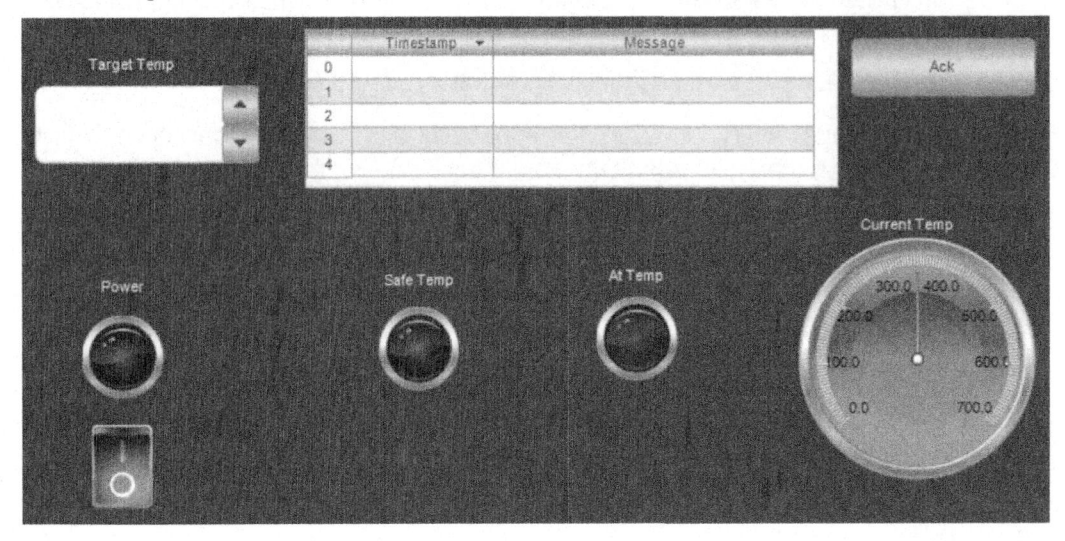

Figure 15.1 – Oven HMI

This is a simple HMI layout for our project. We have a simple **Power** button in the lower left-hand corner with a ramp-up and target temperature spinner above it. In the center of the screen, we have an alarm table for our alarm readout as well as three LEDs to indicate that the oven is on, another LED to indicate that the temperature of the oven is human-safe, and finally, an LED to indicate that the temperature of the oven is at the set temperature. We also have a temperature gauge to read the exact temperature of the oven and an acknowledgment button to acknowledge the alarms. Now that we have a rough layout for the HMI, we can go on and start implementing the logic for the HMI.

## HMI implementation

The first thing we need to do is start declaring variables. For this example, we are going to put all the variables that control the HMI in a **global variable list** (**GVL**) called `vars` for ease of use. For this project, we are going to declare the variables in groups such as LEDs and so on to make it easier for you, the reader, to follow along with the code. The first set of variables we are going to work on are the LED variables.

## LED variables

We have three LEDs that are used as temperature indicators and one LED that is used as a power indicator. We are going to create four Boolean variables, as follows:

```
PROGRAM PLC_PRG
VAR
    //LEDs
    power        : BOOL;
    safe_temp    : BOOL;
    target_temp  : BOOL;
END_VAR
```

The following will show you which variables to map to which LEDs:

- The `power` variable will be assigned to the switch and the power LED
- The `safe_temp` variable will be assigned to the `safe_temp` LED
- The `target_temp` variable will be assigned to the `target_temp` LED

Once you are complete with hooking up those variables, you can move on to the declaration and assignments of the acknowledgment variable.

## Acknowledgment variable

The next variable that we need to set up is the acknowledgment variable. As in the past chapters, we will create a Boolean variable called `ack`, as follows:

```
ack : BOOL;
```

This variable will need to be assigned to the following:

- The **Ack** button
- The **Acknowledgement** field in the alarm table

The button configuration should look like this:

Figure 15.2 – Button setup

As the preceding screenshot depicts, you will want to toggle the `ack` variable when the button is clicked. As for the alarm table, you will need to set up a field similar to what is shown in the following screenshot:

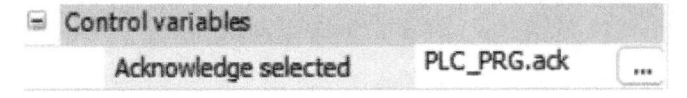

Figure 15.3 – Alarm table acknowledgment configuration

If you followed the steps correctly, you should now have both the button and part of the alarm table set up and ready to go. Once you feel you have everything set up, you can now move on to setting up the spinners.

## Spinner variables/setup

The spinner variable is going to be an integer. In short, the variable will be responsible for providing a target temperature for the oven. The variable will look like this:

```
target_temp_value  : INT;
```

The variable assignment will be as follows. The `target_temp_value` variable will be assigned to the target temperature spinner.

We also need to set the range on the spinners as well. For the sake of simplicity, we're going to set the range on the target temperature spinner to 500, as in the following screenshot:

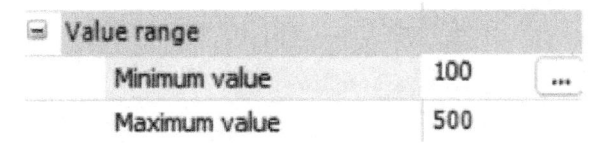

Figure 15.4 – Target temperature value range

For this example, we're going to set a minimum value of 100°F and a maximum temperature of 500°F. Once you complete these operations, you can move on to creating the variable for the gauge.

## Gauge variable/setup

Much as with the spinner, the gauge is going to be attached to an integer as well. We're going to use the following variable for the gauge:

```
oven_temp : INT;
```

# HMI implementation 313

In a real-world application, all the values would be floating points such as a `REAL` data type. However, for this project, we are going to use `INT` data types to avoid using decimals for the sake of simplicity. Much as with the spinner, we will also need to set the range on the gauge as well. We're going to set the range to 700°F to indicate overheating. The extra 200° is an arbitrary number; however, when you're working with things such as gauges, you will usually want to set the range over the maximum value just in case the part experiences values over the expected maximum value. You will want to set the values as shown in the following screenshot:

| Scale | |
|---|---|
| Subscale position | Outside |
| Scale type | Lines |
| Scale start | 0 |
| Scale end | 700 |
| Main scale | 100 |

Figure 15.5 – Gauge configuration

In this case, we set the maximum value on the gauge to 700; however, we also adjusted the main scale to 100. This is so the gauge lines are not bunched up and the gauge is not cluttered. When you're complete with these operations, your gauge should look like the one shown in *Figure 15.6*:

Figure 15.6 – Configured gauge

The final component that we need to set up is the alarm table. Once you are sure you are done with setting up the gauge, you can move on to the alarm table.

## Alarm table variables/configuration

To configure the alarm table, the first thing we're going to do is create an `Alarm_configuration` object and add an alarm group called **Temperature**. When you're done, your alarm configuration tree should look like this:

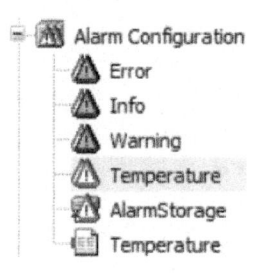

Figure 15.7 – Alarm configuration tree

In the case of this example, we're going to trigger the alarm with a set of Boolean variables that will be set in the PLC code; therefore, we're going to need to declare three more variables, as follows:

```
oven_overTemp  : BOOL;
oven_atTemp    : BOOL;
oven_safeTemp  : BOOL;
```

From the variables, we can see that there will be an `info` alarm that will tell the operator that the oven is safe to enter, a `warning` variable that will tell the operator when the oven is at the set temperature, and an `error` alarm that will tell the operator that the oven is overheating.

After you declare these variables, you will need to set up the alarm configuration. As such, you will want to double-click **Temperature** and match the setup to the following:

| ID | Observation Type | Details | D... | Class | Message |
|---|---|---|---|---|---|
| 0 | Digital | (vars.oven_overTemp) = (TRUE) |  | Error | Oven is overheating |
| 1 | Digital | (vars.oven_atTemp) = (TRUE) |  | Warning | Oven is at temp |
| 2 | Digital | (vars.oven_safeTemp) = (TRUE) |  | Info | Oven is safe to enter |

Figure 15.8 – Alarm configuration setting

Once that is done, we will need to configure our `error`, `warning`, and `info` classes.

## Error class setup

The error class will consist of the following configuration:

| State | Font | Background Color |
|---|---|---|
| Normal | | |
| Active | ■ Microsoft Sans Serif, 9.75pt, style=Bold | ▇ Red |
| Waiting for confirmation | | |

Figure 15.9 – Error class configuration

Once you have completed the error setup, double-click on the warning class.

## Warning class configuration

The warning class will consist of the following configuration:

| State | Font | Background Color |
|---|---|---|
| Normal | | |
| Active | ■ Microsoft Sans Serif, 9.75pt, style=Bold | 255, 255, 0 |

Figure 15.10 – Warning class configuration

Once you finish the configuration for this class, you will need to set up the info class.

## Info class configuration

The final class that we will need to set up is the info class. This class will consist of the following settings:

| State | Font | Background Color |
|---|---|---|
| Normal | | |
| Active | ■ Microsoft Sans Serif, 9.75pt, style=Bold | 0, 255, 0 |

Figure 15.11 – Info class configuration

After you complete the configuration for this class, you can move on to assigning the alarm group to the alarm table.

### Alarm table configuration

The steps to assign the alarm group to the table will be the same as the ones outlined in *Chapter 14*. Your table configuration should match the following screenshot:

| Alarm configuration | |
|---|---|
| Alarm groups | Temperature |
| Priority from | 0 |
| Priority to | 255 |
| Alarm classes | All |

Figure 15.12 – Alarm table configuration

At this point, the control HMI should be complete. All of the controls should be hooked up and configured. Therefore, the next phase in the development cycle is to implement the PLC code.

## PLC code design

Since we are now moving into the PLC code development, we need to start looking at a design. Normally, the design of the PLC code is done in conjunction with the HMI design; however, we designed and implemented the HMI first because we are completing all the steps ourselves and this is an education project. Therefore, as stated before, it is common and best practice to perform the whole design phase for both the HMI and the PLC code at the same time if possible. Again, we only implemented the project the way we did so that we can have an HMI ready to start testing our PLC code.

To keep the design simple, let's break the project down into the following function blocks:

1. `Oven`: This function block will handle turning the oven on and off, as well as ramping the oven up to temperature.
2. `Alarms`: This function block will trigger error, warning, and info alarms.
3. `Door`: This function block will be responsible for locking and unlocking the oven door.

You can see an illustration of the function blocks in the following diagram:

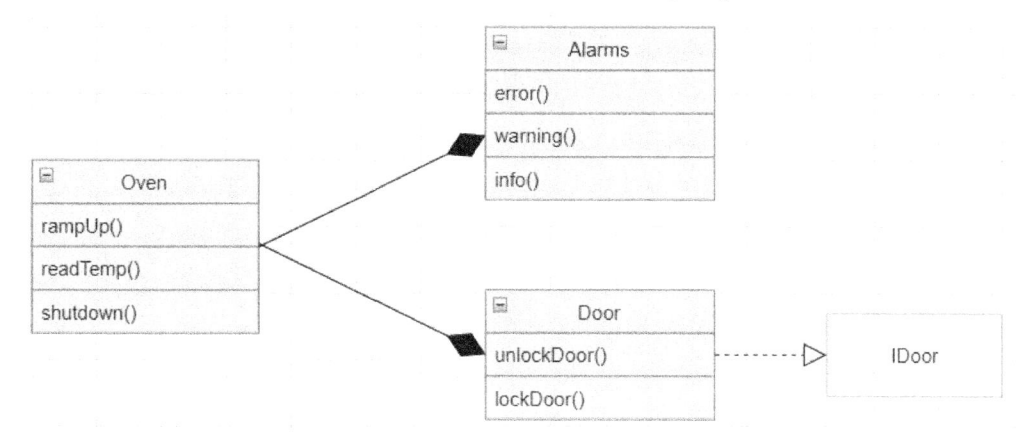

```
                          ┌─ Alarms ──────────┐
                          │ error()           │
                          │ warning()         │
┌─ Oven ──────────┐      │ info()            │
│ rampUp()        │──────└───────────────────┘
│ readTemp()      │
│ shutdown()      │──────┌─ Door ────────────┐
└─────────────────┘      │ unlockDoor()      │- - - - -▷  IDoor
                          │ lockDoor()        │
                          └───────────────────┘
```

Figure 15.13 – PLC code UML

As can be seen in the preceding diagram, the PLC side will consist of an Oven, Alarms, and Door function block as well as a Door interface. The Oven function block will be the workhorse of the PLC program. Essentially, the PLC side will be built around the composition principle. In short, we can justify this with the following statements:

- The oven has a series of alarms that need to be triggered
- The oven has a door that needs to be locked and unlocked

Finally, the purpose of the Door interface is so that we can model the door. There are many types of doors that we can use but they will all automatically lock and unlock. Therefore, to properly model the door, we will use an interface and simply implement the methods for the specific door that we use.

In short, this design is very simple and requires minimal code. Also, since we are using composition and all the function blocks/methods are following the **single-responsibility principle** (**SRP**), this design will allow for future expansion and easy maintenance.

Though simple, the PLC design will be quality enough to implement. As with the theme of this book, since we have a decent design, we should be able to easily implement the code. With all that being said, we can now implement the PLC code.

## Implementing the PLC code

Now that we have a design, we can implement the code. The code implementation should be relatively minimal. The first thing we are going to do is declare our function blocks. For this, we are going to create a folder named `FunctionBlocks` and use it to house the `Oven`, `Alarms`, and `Door` function blocks. When all the function blocks and methods are implemented, your tree should look like this:

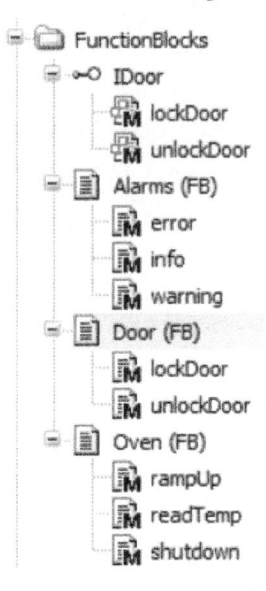

Figure 15.14 – Function blocks

Once you create the tree, you can start to implement the methods. The first set of methods you will want to implement are the methods of the `PLC_PRG` file.

## PLC_PRG file

The first place we're going to start implementing code is in the `PLC_PRG` file. Since it is our entry point, we're going to put our starting logic here. In short, you should have your variables all implemented in the `vars` GVL at this point, except a reference variable for the `Oven`, `Alarms`, and `Door` classes, which will look like the following:

```
PROGRAM PLC_PRG
VAR
    oven : Oven;
    alarms : Alarms;
    door : Door;
END_VAR
```

Once you add the oven variable, you should only need to add the following code to the file:

```
vars.safe_temp := TRUE;
alarms.info();
IF vars.power = TRUE THEN
     oven.readTemp();
     door.lockDoor();
END_IF
//shutdown
IF vars.power = FALSE THEN
     oven.shutdown();
END_IF
```

For this program, we are going to make an assumption for the sake of the simulated project. When the program starts, we are going to assume it is safe to enter, hence setting the safe_temp variable to TRUE. In short, if the power is on, a warning message will be displayed on the alarm table. The oven at temp LED will be displayed, and the *door locked* message will trigger as well. If the power is off, the oven.shutdown method will be called, and if the temperature is below 85°F, the door will unlock and the safe LED will turn on. An alarm will also be displayed saying the oven is safe to enter. With this complete, we can now move on to implementing the Alarms function block. Before you implement the code, there is a common naming issue in the variables. Can you find it and improve the code to better reflect what it is supposed to do? *Hint*—it is one of the LEDs that were mentioned.

## Alarms function block

The code for the Method blocks should be relatively simple. Essentially, whichever method is called will set the appropriate alarm. The error method's code should look like the following:

```
vars.oven_overTemp := TRUE;
vars.oven_safeTemp := FALSE;
vars.oven_atTemp   := FALSE;
```

With that, we can move on to implementing the info method with the following code:

```
vars.oven_overTemp := FALSE;
vars.oven_safeTemp := TRUE;
vars.oven_atTemp   := FALSE;
```

As can be deduced from the `error` and `info` methods, all we are doing is setting the correct variable to TRUE. Finally, the `warning` method should look like the following code snippet:

```
vars.oven_overTemp  := FALSE;
vars.oven_safeTemp  := FALSE;
vars.oven_atTemp    := TRUE;
```

Now, the implementation of the methods in the `Alarms` function block is probably not the most effective. The code can be simplified to one method. The methods were designed like that on purpose so that you, the reader, can modify and improve upon the code. After you finish implementing the project, come back to this section and try to condense and improve upon the code.

After you fully implement this code, we are going to move on to implementing the other function blocks. Moving down the tree, we are going to implement the method in the `Door` function block.

## Door function block

Now that we have our `Alarms` function block fully implemented, we can start to implement the `Door` function block. For things such as doors, it is common to have a large light on the outside of the door as a safety feature. It is also common to put an LED on the HMI; however, since this is a simulation, for now, we are just going to have a variable to indicate with the LED that the door is locked. We are going to add the following variable to the `vars` GVL file:

```
door_status : WSTRING;
```

After you add that variable to the GVL file, you can start to implement the methods in the `Door` function block. For the current iteration of the project, we are only going to display the status of the door, so the code for these two methods will also be relatively simple. With that being said, the `unlockDoor` method should look like the following code snippet:

```
vars.door_status := "Door is unlocked";
```

As can be deduced by looking at the `unlockDoor` method implementation, the `doorLock` method will be equally simple, with the following implementation:

```
vars.door_status := "Door locked";
```

This should, for the most part, do it for the `Door` function block. However, much as with the `alarm` class, see if you can modify this. See if you can address the following:

- Can you condense these two methods into a single method using control statements? Does doing this make more or less sense?
- How can you modify the HMI and PLC code to support an LED on the HMI screen?

- Should you create a separate HMI visualization for the door?

Before you address these questions, let's move on and implement the Oven function block logic.

## Oven function block

The final function block that we have to implement is the Oven function block. This function block will be more complex than the other function blocks as this will be the workhorse of the program. Ensure that you are carefully following along.

The first method that we are going to implement is the rampUp method. In a real-world application, you assume that when the oven is on, it is dangerous. Similarly, you will want to turn on the red LED and turn off the green one. This will signal to the operator that the oven is potentially hot and that they must not touch or enter any dangerous areas. To accommodate this, we are going to implement the following two lines of code to simulate this:

```
vars.ovenOn := TRUE; //Sends hypothetical signal to oven heater
vars.safe_temp := FALSE; //at a safe to enter temperature
vars.target_temp := TRUE; //At target temperature
```

Once that logic is in place, we need to move on to our readTemp method. This method is essentially going to be the workhorse of the program. This method will be responsible for firing alarms to give the temperature status to the operator as well as triggering the rampUp phase when the oven is not already at temperature. The readTemp method will simply consist of a series of control statements, as follows:

```
METHOD PUBLIC readTemp : BOOL
VAR
    alarms : Alarms;
    door   : Door;
END_VAR
VAR_INPUT
END_VAR
```

Once you create the alarms variable, you can move on to implementing the logic for the rest of the method with the following code:

```
IF vars.oven_temp < vars.target_temp_value THEN
    rampUp();
    alarms.warning();
ELSIF vars.oven_temp = vars.target_temp_value THEN
    alarms.warning();
```

```
    ELSIF vars.oven_temp > vars.target_temp_value THEN
            alarms.error();
    END_IF
```

The final function will simply be responsible for putting the oven back into a shutdown mode. Depending on the type of oven and the shutdown sequence, this method will vary. However, for this project, we are going to keep it simple; we will reset the red LED to *off* and the green one to *on* when the temperature of the oven is less than 85°F.

To accomplish this, the variables should look like the ones in the following snippet:

```
METHOD PUBLIC shutdown : BOOL
VAR
        door : Door;
END_VAR
VAR_INPUT
END_VAR
```

The method will also unlock the door. In real life, there would be things such as motion detectors and so on in the oven to prevent the oven from heating up in case someone or something is inside. For this project, we are going to keep it simple and ignore that; however, it is recommended that you go back and add a similar feature to enhance the project. The code to do this will look like the following:

```
vars.ovenOn := FALSE;
IF vars.oven_temp < 85 THEN
        vars.safe_temp := TRUE;
        vars.target_temp := FALSE;
        door.unlockDoor();
END_IF
```

At this point, the code should be implemented well enough for us to start testing and debugging it. Hence, we can now move on to testing to ensure the application works.

## Testing the application

Now that we have the code implemented, we can run a few test cases to see if the code works as expected. If you look at the code, we have an `oven_temp` variable that in real life would be tied to some type of thermal sensor. For our purposes, we are going to control it manually to simulate the conditions inside the oven. In real-world automation programming, this is a common technique. We don't always want to heat the oven to the target temperature until we know for sure the software is

Testing the application | 323

working; simply writing values to that variable will suffice. With all that being said, we can now start testing the application by testing out the door.

## Testing the door lock

We are going to start with the most basic and safety-critical part: testing the door. Essentially, we want to ensure the door is locking and unlocking properly. For this, we are going to use the following test case:

| Functionality | Input | Expected Value | Actual Value | Date | Pass(y/n) |
|---|---|---|---|---|---|
| door lock | power on | door locked | | 10/21/2022 | |

Figure 15.15 – Test case

The test case in the preceding screenshot is relatively simple as we are testing a Boolean state. In other words, the door is either locked or unlocked.

When we run the program and switch the power on, we get the following output:

door_status        WSTRING        "Door locked"

Figure 15.16 – Actual door output

As can be seen, the door is locked. The full test case should look like the following:

| Functionality | Input | Expected Value | Actual Value | Date | Pass(y/n) |
|---|---|---|---|---|---|
| door lock | power on | door locked | door locked | 10/21/2022 | y |

Figure 15.17 – Completed test case

Now that we've established that the door automatically locks, we need to ensure that the door unlocks properly. Testing if the door unlocks will be a bit more in-depth as the temperature will be a factor as well. We will need to create a few test cases to ensure the door unlocks properly. To test the functionality, we can use the test cases in the following screenshot:

| Functionality | Input / temp | Expected Value | Actual Value | Date | Pass(y/n) |
|---|---|---|---|---|---|
| door unlock | power off / 100 | door locked | | 10/21/2022 | |
| door unlock | power off / 82 | door locked | | 10/21/2022 | |
| door unlock | power on / 80 | door locked | | | |

Figure 15.18 – Unlock test cases

To test this functionality, we're going to turn the `power` variable on, then set the `oven_temp` variable to `100`, and then finally write the `power` variable back to `false`. When you're done, you should see the door in a locked state, similar to what can be seen in the following screenshot:

| door_status | WSTRING | "Door locked" |

Figure 15.19 – Door state for the first test

As can be seen, the test passed in this case. Now, you can repeat the process with the other temperatures:

| Functionality | Input / temp | Expected Value | Actual Value | Date | Pass(y/n) |
|---|---|---|---|---|---|
| door unlock | power off / 100 | door locked | door locked | 10/21/2022 | y |
| door unlock | power off / 82 | door locked | door locked | 10/21/2022 | y |
| door unlock | power on / 80 | door locked | door locked | 10/21/2022 | y |

Figure 15.20 – Door test cases

As can be seen, each test case should pass. We can mark the `Door` function block as working, at least for the door locking. Now, we also need to test the door unlocking. We have the test cases established, so go ahead and test if the door unlocks properly on your own.

## Testing the gauge

Another vital safety component of the oven is the gauge. The gauge is of vital importance as it will keep the operator informed of the internal temperature of the oven. In theory, the gauge should show the temperature of the oven. In other words, the gauge should match what the `oven_temp` variable is set to. We can come up with a few test cases to verify the functionality of the gauge. Essentially, what we want to verify is that the value we set the `oven_temp` variable to is the same value that is displayed on the gauge.

With the test criteria established, we can use the test cases in the following screenshot:

| Functionality | temp | Expected Value | Actual Value | Date | Pass(y/n) |
|---|---|---|---|---|---|
| gauge | | 200 | 200 | 10/21/2022 | |
| gauge | | 500 | 500 | 10/21/2022 | |
| gauge | | 100 | 100 | 10/21/2022 | |

Figure 15.21 – Gauge test cases

To execute the test, we will set the `oven_temp` variable and observe the gauge. When you set the variable to `100`, your gauge should match the following reading:

Figure 15.22 – Gauge reading for 100°F test case

Repeat the process with the other values, and you should see that the gauge will reflect the proper value:

| Functionality | temp | Expected Value | Actual Value | Date | Pass(y/n) |
|---|---|---|---|---|---|
| gauge | | 200 | 200 | 200 | 10/21/2022 y |
| gauge | | 500 | 500 | 500 | 10/21/2022 y |
| gauge | | 100 | 100 | 100 | 10/21/2022 y |

Figure 15.23 – Completed gauge test cases

Next, we're going to want to test the alarm system. This is another safety-critical functionality as it will alert the operator to issues.

We should get the following messages in these situations:

- An info message when the oven is safe to enter
- A warning message when the oven is at the target temperature
- An error message when the oven is 10°F over temperature

With this, we are going to create three basic test cases to test this functionality. Now, in the real world, you would want at least a few cases for each alarm. This will be up to you as the reader to take what you have learned thus far and apply it to create more cases to test each message alarm. For this example, we are going to create three test cases, as follows:

| Method | target temp | Oven temp | Expected Value | Actual Value | Date | Pass(y/n) |
|---|---|---|---|---|---|---|
| error | | 90 | 95 No change | | 10/22/2022 | |
| info | | 70 | 80 Oven is safe to enter | | 10/22/2022 | |
| warn | | 90 | 90 Oven is at temp | | 10/22/2022 | |

Figure 15.24 – Alarm test cases

For the sake of practice, only the first test case will be run. You, as the reader, will be responsible for running the remainder of the test cases.

According to our requirements, the error alarm should only be on if the oven is 10° over the set value. When we run the values, we get the following output:

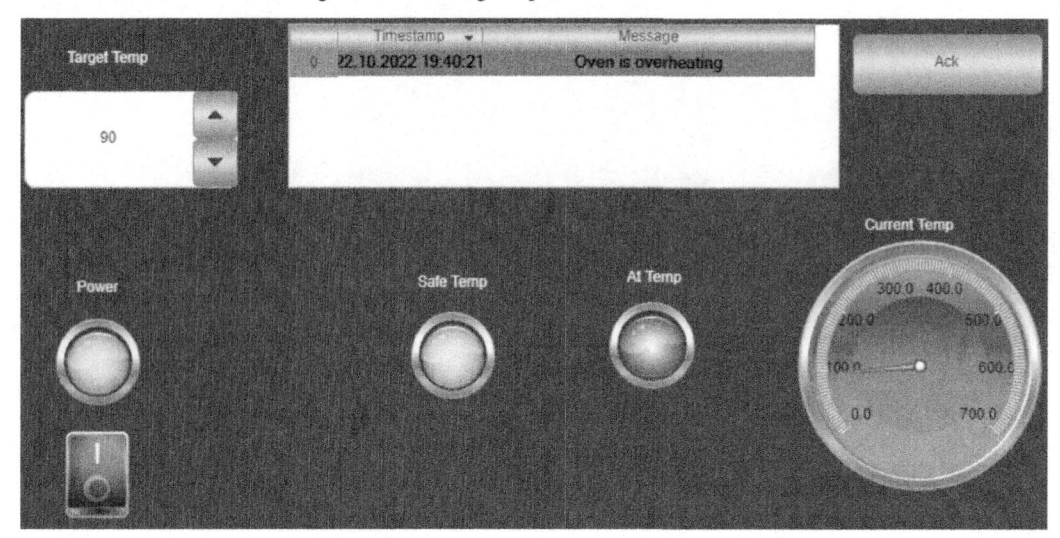

Figure 15.25 – HMI status for error test case

As we can see, these values cause a failure for the test case. The error alarm should only trigger when the `oven_temp` variable is at least 10° over the set value, not 5. Therefore, we have at least one bug in the program. Now that we found one bug, perform the rest of the test cases to see if there are bugs in the software. Moving forward, we have not tested the LED status. Observe *Figure 15.25*—do the LEDs seem to work as one would expect? If not, do you think there is a bug there? Now that we have tested a few test cases, try to think of some things that we have not covered and write a few test cases for those conditions to see if bugs exist. By this point, after you debug and add a few more test cases, you should be done with the project.

## Summary

Congratulations—you have now completed all the technical sections of this book! In this chapter, we have explored creating a sample oven. We have built this project using a Waterfall-like methodology, and we have gone through most of the SDLC sections. In this section, we have built the code and HMI for a simulated real-world project. Now, we did find bugs in the code, and you will be responsible for finding bugs and retesting them. There are no right or wrong ways to solve these bugs and test cases; you are free to use your intuition and what we have covered to fix them. If you are completely stuck, I would recommend looking at the questions for a punch list of things to fix and a few more test cases to create. Once you are done with all that, you can move on to the next chapter and explore distributed systems.

## Questions

Answer the following questions based on what you've learned in this chapter. The questions in this chapter are open-ended with no wrong or right answer. They are meant to be exploratory so readers can come to their own conclusions.

1. Write a test case to test the LEDs.

    A. Does the LEDs' behavior make sense for what they are meant for?

    B. If there is a bug, can you debug it?

2. Test the **Ack** button.

    A. Write a few test cases for the **Ack** button.

    B. Does the button work as intended?

3. Can you refactor the `Alarms` function block and condense everything into one method?
4. Can you add extra functionality to the program or HMI? This is an open-ended question, so use your imagination.
5. What should you do if there is no set value in the spinner? Should you set a default value, throw an alarm, or do something else? Use your imagination!
6. Can the variables be renamed to better reflect what they do?
7. Do the info, warning, and error alarm messages make sense? Can you improve them?

# 16
# Distributed Control Systems, PLCs, and Networking

Whether it be with a customer, a hiring manager, or even your pet, communication is key, and the automation realm is no different. Since the dawn of the computer age, the goal of all IT systems is to relay information from one electrical device to another. The early 2000s saw this concept explode with the widespread adoption of the internet. With the cost of computing drastically decreasing and automation controllers becoming significantly more powerful, point-to-point communication within an automation system has become paramount.

With the way most manufacturing environments now operate, it is not uncommon for many different types of automated controllers to be networked together for coordination. Even isolated systems still use networking technologies to communicate with different parts of the machine such as motor drives, power supplies, and so on. If you want to be an automation engineer, or at least grow as an automation developer, networking is a vital technology to understand. With that said, we are going to explore the following:

- What are networks?
- Common protocols
- PLC device communication
- Distributed control systems
- The difference between distributed control systems and PLCs

By the end of this chapter, you should have a basic understanding of the complexity that goes into networking devices.

This chapter is going to be a little different than other chapters. No project will be created and the topics explored here will be explored at a very high level. This chapter is merely meant to expose you to communication protocols and basic networking at a high level. There are many popular and widely used protocols out there, and each one is different. This chapter talks about just a few of the protocols and concepts that I have used most in my career. This chapter is only meant to expose you to networking and, hopefully, get you interested in diving deeper into the topic.

## Technical requirements

Unlike the other chapters in this book, this chapter will be only theoretical in nature. We will explore concepts but not develop code. There is no code that you need to worry about pulling down.

## What are computer networks?

If you're reading this book, chances are you know what a computer network is. This is because to function in the 21st century, you have to be able to use the internet, and the internet is nothing more than a global computer network. So, with that being said, what is a computer network? In short, a network is a way for computers and other electrical devices (such as printers, smartphones, tablets, or any other devices with either wired or wireless capabilities) to be able to communicate and share data. In other words, a computer network can best be thought of as a bunch of devices that are wired together so they can talk to one another.

### Network topology

To get into a more advanced networking concept, we need to explore topology. In short, a network topology is a way a network is built. So, you can think of a topology as a layout of all the devices on a network and the way they are interconnected. In other words, a topology is a network blueprint. Each topology has its own strengths and weaknesses, and the best one for your project will depend on what you're trying to accomplish. However, in terms of basic networks, the most common are as follows:

- Star topology
- Bus topology
- Ring topology
- Tree topology
- Mesh topology

To get an idea of what a topology would look like on paper, consider the ring topology that is represented in *Figure 16.1*. As can be seen, the devices form a ring-like structure. Many ring networks will only allow data to follow in one direction and are hence known as unidirectional networks. However, some will allow data to travel in both directions and are known as bidirectional networks. In short, you can get a good feel for the way a network will behave from the topology; however, it is always a good idea to do a little research on the networks as certain things such as ring's uni- and bi-direction attributes can often be hard to determine simply from looking at the layout.

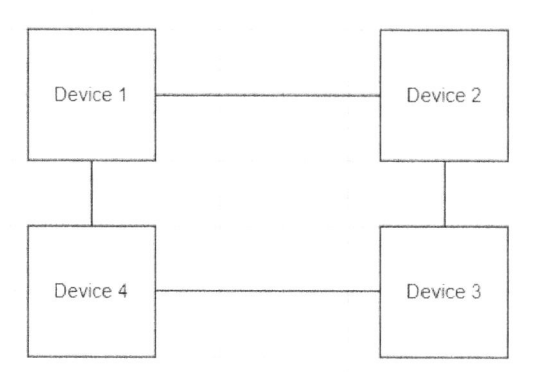

Figure 16.1 – Ring topology

Each one of these topologies has a different configuration and is meant for different tasks. However, just because you pick one type of topology does not mean your network is exclusively married to it, as you can combine multiple topologies to form what is known as a hybrid topology.

Speaking from personal experience, hybrid topologies are usually the most common, especially in the automation world. With the ever-increasing complexity of factories and other environments, it is very hard to pick a one-size-fits-all approach to networking. It is, therefore, very common to combine different network configurations to accomplish your overall goal.

An in-depth discussion of network topologies is well beyond the scope of this book. However, what you need to know is that devices can be networked in various configurations based on what you're doing. So, with that being said, we are going to switch gears and talk about the mechanisms that devices use to pass data from one device to another.

## Common IT protocols

Electrical devices inherently do not understand any form of language that we humans can speak. All a device such as a computer or PLC understands is whether a particular pin is energized at a given time or not. Specific languages known as protocols have to be used so that one device knows what the other device is saying. There are many different protocols out there. Some are very common and are used everywhere, while some are proprietary and are only used with specific devices. The following section is going to be dedicated to exploring two common, everyday protocols that can be used with PLCs and other automation devices. These protocols can be used directly or as an underlying system for more specific protocols. As such, the first protocol stack that we will explore is TCP/IP.

### TCP/IP

Arguably the most common form of computer communication is called the **Transmission Control Protocol** (**TCP**). TCP is one of, if not the most commonly used communication protocols around, mainly due to the internet. TCP is one of the main transmission protocols of the **Internet Protocol**

(**IP**) suite and is often referred to as TCP/IP. It should be noted that TCP and IP are two individual protocols; however, they are often used together. Compared to many other protocols, TCP is a very reliable communication protocol. However, although it is very reliable, it is often slower than many other protocols.

For TCP to work, it requires a three-way handshake between the two devices. The device that initiates the communication process is called the client and the other device is called the server. Essentially, when the two devices connect, the client will send a synchronization request to the server; in turn, the server will send a synchronization/acknowledgment signal back to the client, and finally, the client will send a final acknowledgment to the server. The process can be seen in the following figure:

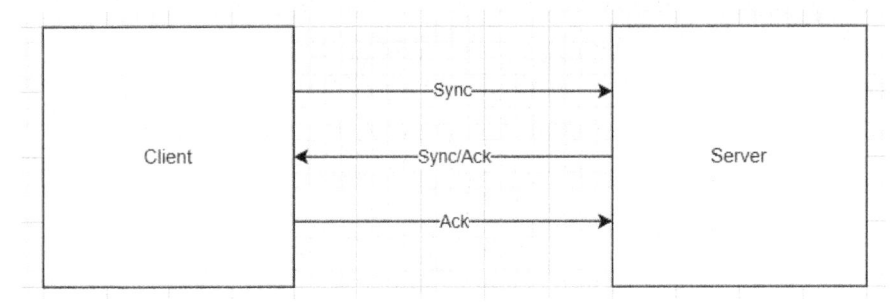

Figure 16.2 – Three-way handshake

Compared to other communications protocols, such as the **User Datagram Protocol** (**UDP**), which will be explored next, TCP is slower. The reason for the slow communication stems from the amount of information that is transmitted. There is more than just the requested data being transmitted going on with TCP.

To understand why TCP is so much slower than other communication protocols, you must first understand that TCP is much more reliable than many other communication protocols. In short, outside of the three-way handshake, when data is transmitted via TCP, it will sequence the data packets, perform acknowledgments, perform error detection, and, lastly, corrections. In all, this means that TCP will (more or less) ensure that the data is transmitted successfully and in the correct order.

Now, for many applications, TCP will be either too slow for the given application or unnecessary for some applications. Another alternative that can be used when TCP is either too slow or unnecessary is UDP.

## UDP

Compared to TCP, UDP is much, much faster but much less reliable. Much like TCP, UDP allows a client and a server to communicate with each other; however, unlike TCP, there is no handshake. One device will just send data across the line as soon as it is told to do so. Also unlike TCP, UDP will not perform any error checking, acknowledgments, sequencing, or so on. It will, however, conduct a checksum to ensure the integrity of the data, and if a data packet is damaged, the packet will be dropped. With UDP, you simply send and receive data; there is no guarantee that the data will arrive in the correct order, or whether the data packet will even arrive at all.

The process of sending and receiving data for a UDP system can be viewed in the following figure:

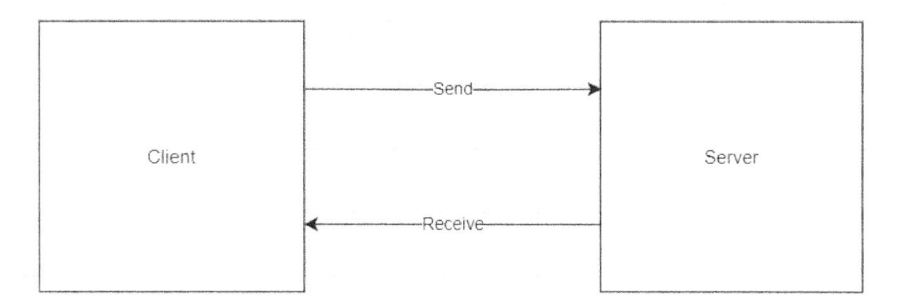

Figure 16.3 – UDP send/receive process

As can be seen, the UDP process is nothing more than sending and receiving data between the two devices. There are no intermittent steps; all the system is doing is sending and/or receiving data. The speed that UDP offers stems from the very simple transmission sequence and the fact that nothing is guaranteed.

When I was first starting out in the IT field and I learned about UDP and how unreliable it was, I couldn't fathom what it could be used for. For the life of me, I couldn't understand why anyone would want to use something as unreliable as UDP. However, I soon came to understand that there are many uses for UDP. In short, UDP is used for applications that do not depend on each data packet. This may seem a bit odd at first, as it may be hard to think of applications that do not depend on each data packet, but a few are digital streaming and digital communications.

To conceptualize this, consider streaming a movie. If a data packet is lost, the worst that will happen is you will experience a blip in the movie. In the case of streaming, it is more important to try to keep a smooth streaming experience. On the other hand, consider a video call. If you were to use TCP, the lag would make the call almost impossible. There would be a lag in the call and chances are that the call would be unintelligible. Again, with UDP, you may lose a few packets of data, which, at worst, would cause a blip or two, but you would still have a relatively smooth call.

As odd as it may sound, UDP is also used quite a bit in automation programming. Many devices use UDP as a communication method. I have seen UDP used for many different things. One area in which I have seen UDP used frequently is with PLC-to-device communication. By this, I mean the PLC talking to devices such as external power supplies and other devices that the PLC may need to control.

Now, TCP and UDP are used in many different IT applications, not just for automation applications. However, there are many other proprietary communication applications.

## PLC/automation device communication

UDP and TCP are general communication protocols. By this, I mean that they are used for many different types of IT applications, such as internet applications, common computer networks, and so on. However, many of the PLC manufacturers produce their own communication systems to be used with their PLCs and various types of industrial components. Some of these systems are very similar and use the same physical connectors as standard computers do – for example, Ethernet cables. However, some use exotic connectors and will be unique for certain devices. The first communication protocol we are going to discuss is one of the most popular, which is called Modbus.

### Modbus

**Modbus** is an industrial communication protocol. Modbus is a little different than the other protocols that we have discussed thus far. Where TCP and UDP are more agnostic in terms of IT applications, Modbus was developed in the late 70s for use in PLC communications by the company Modicon, which is now Schneider Electric. Modbus is what is known as an open protocol. This means, that even though it was developed by Modicon, the specs on how the protocol works are openly published and can be used in accordance with the license. For the most part, Modbus is the standard for industrial communications.

Modbus works off what is known as a master/slave configuration. Master/slave systems are very common for industrial communication. For these systems, the master will either query a slave or node device for information such as a sensor reading. The master can also request that the node device do something such as toggle a valve, turn on a motor, or the like. In short, with Modbus, only the master can initiate communication with the node devices. The node devices cannot initiate communication with the master device.

Modbus can be used for many different things. One thing that Modbus is used for is HMI communication. For example, there are third-party C# and Java libraries that can be used to orchestrate Modbus communication between devices. For a device such as the Velocio PLC, Modbus communication can be used for communication between the PLC and a C# HMI.

It is important to know that there are many different types of Modbus implementations. For example, there is Modbus ASCII and Modbus RTU. Both RTU and ASCII are serial connections. Though both will ultimately do the same job, they do differ in how they work. In terms of Modbus RTU, there is a 3.5-character space between the messages. In other words, the 3.5-character is used as a delimiter. On the other hand, ASCII uses two ASCII characters to distinguish messages. RTU uses a binary form to transmit data, whereas ASCII transmits data in ASCII form. This means that although ASCII Modbus is more readable, it is less efficient than RTU. It is also important to note that when setting up a Modbus network, each node will have a unique ID between 1 and 254 with the master node always being set to 255. Now, Modbus is Modbus; however, it is important to understand that there are different flavors of the protocol so you must ensure you are choosing the proper hardware and developing the correct software for compatibility.

Another common implementation of Modbus is Modbus TCP/IP, which is Modbus wrapped in Ethernet IP, AKA a TCP frame payload. In the case of Modbus TCP/IP, you can use standard switches and Ethernet cables for communication. Generally, Modbus TCP/IP is becoming more popular in newer systems as it is a newer technology.

Though Modbus is a very common protocol that is often touted as the industry's de facto protocol, as stated before, it is not the only one. As already stated, there are many other protocols, and the next one we are going to explore is called Profibus.

## Profibus

Another very common communication protocol for automation controllers is **Profibus**. Profibus was developed and promoted by Siemens to network things such as sensors to a controller. Profibus works off a master/slave network configuration. Usually, the master device will be some type of controller, such as a PLC. On the other hand, the slave nodes will be devices such as sensors, drives, and so on. Profibus network can experience speeds of up to 12 Mbps; however, most systems are set to a significantly lower speed, usually around the 1.5 Mbps range.

Unlike many other communication systems, Profibus requires the use of a specialized cable. Usually, the cable is a shielded purple single-pair RS-485 cable with a DB-9 connector at the end instead of something like a standard Ethernet cable. At first glance, the connector on the cable can seem odd as it has an on/off switch on it. This switch connects to a terminating resistor that, when placed in the on position, denotes the end of the device chain. This can be a cause for issues because if a switch is in the incorrect state, the chain can be prematurely cut short. If you do opt to use Profibus and you do encounter device communication issues, one of the first places you should look is at the terminating switches.

Another major difference between Profibus and Ethernet networks is that Profibus will usually support larger networks. However, great care must be taken when selecting the length of a Profibus cable. On the short end, it is recommended that there be a minimum cable length of about 3 feet (or 1 meter) between each of the nodes. A cable length of anything shorter can result in communication issues. It is common, even if the nodes are next to each other in the same cabinet, to have 3 feet of cable between each node. On the other hand, the length of the cable will dictate how fast you can transfer data. In terms of Profibus, the maximum length you will want to use is about 1,200 meters, which will allow up to about 9.6 kbps. On the other end of the spectrum, you can get up to 12,000 kbps with a length of 100 meters. The shorter the cable is, the faster the data transfer rate can be. This is a very important concept to remember when developing a Profibus network, as you will have to weigh up the pros and cons of having longer cables but slower transmission speeds, and vice versa.

There is also a limit to the number of devices that can be on a Profibus network. In short, each device on a Profibus network must have a unique device address. The drawback is that devices on a Profibus network can range from 1 to 127. At most, you can have 127 devices on the network. Depending on the type of device, the address will either be set with a physical dip switch on the device or via the configuration software.

Profibus is a very common communication system and it is still widely used. However, there is another type of Profi network called Profinet, which utilizes new Ethernet-based technologies. The next section will be dedicated to exploring Profinet.

## Profinet

Siemens also offers another major protocol called **Profinet**. Compared to Profibus, Profinet is based on newer, Ethernet-based technology. Profinet shares many similarities with Ethernet, even down to the cabling. Most that employ the communication system will use an industrial version of an Ethernet cable. Normally, you will be able to spot a Profinet cable due to its green color. However, it is common for some to use a standard Ethernet cable when in a pinch or for troubleshooting purposes.

Outside of being able to use off-the-shelf cables, Profinet is also faster than Profibus. The extra speed characteristic stems from its Ethernet roots. Similar to Profibus, Profinet also has length limitations. A Profinet cable can be up to 100 meters in length. However, Profinet is still on average faster than Profibus. Usually, the standard operating speed for a Profinet network is 100 Mbps. Generally, Profinet is favored in newer applications that require faster communication speeds and response times.

With all that being said, once you determine the length of the cable, it is important to understand the basics of the Profinet topology. By default, Profinet networks are often configured into a star topology similar to the following figure:

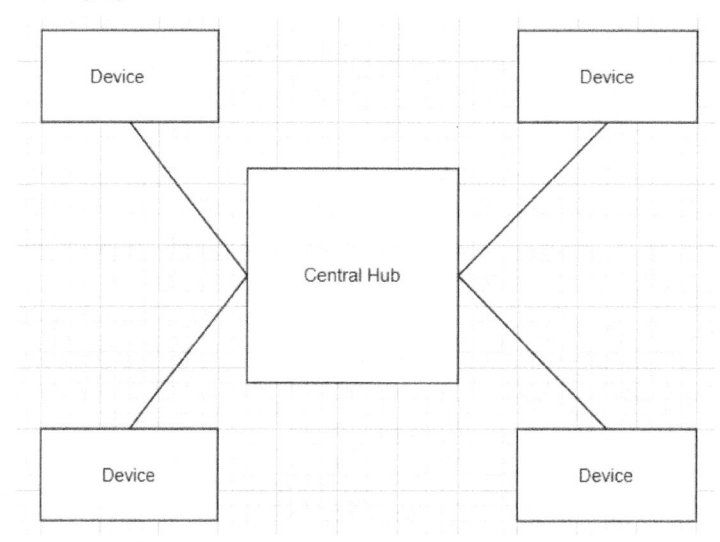

Figure 16.4 – Profinet star topology

The star topology is automatically created when multiple nodes are connected to a single switch. It is a very common topology when Profinet is employed, especially for smaller networks that only have a single switch. With that being said, Profinet can also be configured into a tree topology by networking multiple star networks together. Consider the diagram of a Profinet tree topology in *Figure 16.5* to see what a tree topology will look like.

The tree configuration is when you are trying to coordinate multiple sections of a plant or facility together. In short, each star network would be something like an individual part of a manufacturing process, and the master hub would be coordinating all the processes. Profinet can support the line topology as well. However, there is a lot to the line topology and, because of that, it will not be covered.

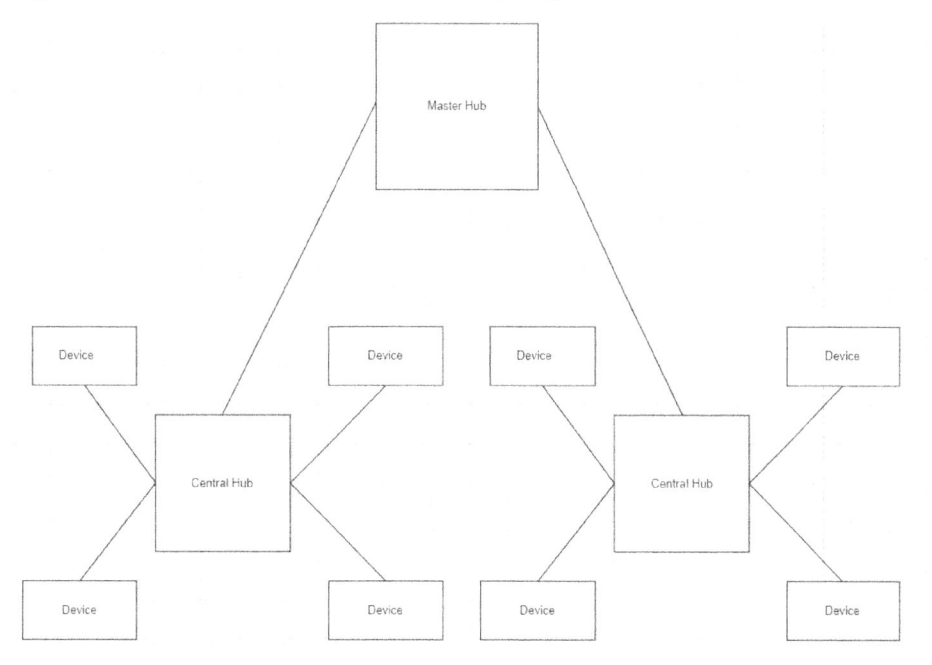

Figure 16.5 – Profinet tree

In all, the topology you choose will depend on what you're working on and trying to accomplish. As with many other things we have seen throughout this book, the route you chose to solve a problem will depend on the problem itself, as well as the desired solution.

A major difference between Profibus and Profinet stems from the addressing that the devices use. In short, where Profibus networks use a range of 1 to 127 for unique device identifiers, Profinet networks use the following types of addresses:

- IP address
- MAC address
- Device name

For the most part, you are going to use the device name the most to interact with the individual device.

Profibus and Profinet are both excellent communication systems. Both can be used to great success in the field, but most engineers are leaning towards using Profinet more due to its faster speeds, response time, and the fact that it is generally considered to be future-proof. However, many places still employ Profibus, and an understanding of how it works, at least at the practical level, is vital to the success of any automation engineer. With Profibus and Profinet having been explored, we can move on to another industrial communication system, which is known as EtherCAT.

## EtherCAT

Similar to Modbus, Profinet, and Profibus, **EtherCAT** is another proprietary communication protocol developed by Beckhoff. EtherCAT stands for **Ethernet for Control Automation Technology**. Similar to Profinet, it is an Ethernet-based communication protocol. EtherCAT is a communication system that is used for a wide range of applications, including industrial machinery, medical equipment, mobile machines, and a variety of other applications. Similar to Profinet, the physical connection, is standard Ethernet cabling. This means that much like Profinet, off-the-shelf Ethernet cables can be used when troubleshooting or in a pinch. Though the cabling is the same, the underlying communication system is different.

The way the EtherCAT system works is unique compared to the other communication protocols that we have explored thus far. Essentially, the EtherCAT master will send a data packet known as a frame to all the nodes on the network. The nodes will read the frame and will perform the instructions that were meant for it and ignore the instructions that were meant for the other devices on the network. The devices will also add their information to the frame. EtherCAT devices typically have two Ethernet ports on the device. One of the ports is for sending data and the other is used for receiving data.

Typically, the network is configured in a ring-like topology similar to what can be seen in *Figure 16.6*. With this configuration, as long as the communication hardware is intact and working, the frame will circulate throughout the network. Overall, EtherCAT will provide you with the following:

- Ability to use off-the-shelf Ethernet cables
- Allows for processing on the fly
- No need for hardware such as switches and so on as with Profinet
- A downed node will not necessarily kill the communication chain

PLC/automation device communication    339

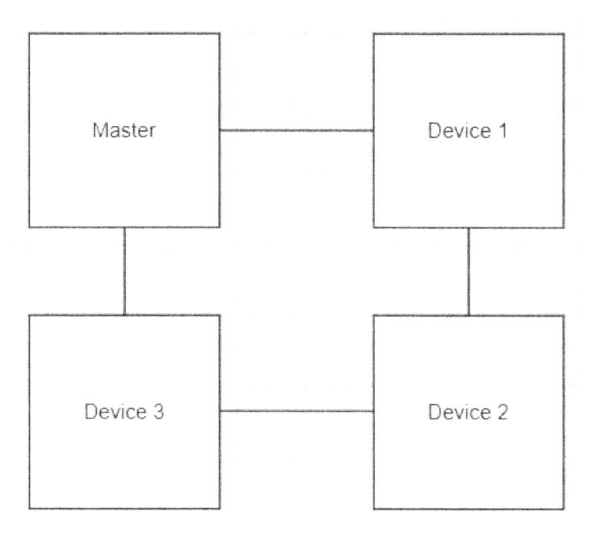

Figure 16.6 – Typical EtherCAT configuration

In all, EtherCAT is a very powerful and robust communication protocol that, due to using the lowest two layers of the Ethernet protocol, is significantly faster than Modbus or Profinet, which makes it more suitable for real-time applications. Now that we have a grasp on EtherCAT, the final industrial network that we are going to explore is known as DeviceNet.

## DeviceNet

Much like EtherCAT and the other industrial protocols that we have explored so far, **DeviceNet** is another common industrial communication system that is produced by Rockwell Automation, formerly Allen-Bradley. However, it is available to third-party vendors and can be used with non-Allen-Bradley devices.

Similar to systems such as Profibus, DeviceNet also has a maximum number of devices. The device IDs range from 0 to 63 for a total of 64 devices on the network. Also, like the Profi family of networks, the device IDs have to be unique to each device. DeviceNet will check for redundant IDs and throw an error if there are duplicate IDs on the network. In this regard, the system is foolproof as you'll know with little to no troubleshooting whether the device IDs are duplicated. Also similar to other communication systems, the whole network as well as the device IDs are set up via a software interface. With all that being said, let's move on to one of the major pinch points of DeviceNet.

The cabling for DeviceNet is also a unique cable. DeviceNet cables are usually four wires that supply the power and carry the signal. DeviceNet cables are broken down into the following wire components:

- +24 V
- -24 V

- White data wire
- Blue data wire

DeviceNet cables are generally broken down into the following categories:

- Round:
  - Thick round
  - Thin round
- Flat:
  - KwikLink Flat
  - KwikLink Lite Flat

Each one of these cables can be used for different purposes in a DeviceNet topology. A DeviceNet topology is somewhat unique, and to get a feel for how it works, the next section is going to be dedicated to understanding what each part of the topology does and what cables to use.

## DeviceNet topology

Though many of the other communication systems can have unique topologies and have their own rules, I've personally always found DeviceNet's to be some of the more interesting in their simplicity. In short, a DeviceNet network is composed of two parts: the trunkline and droplines.

To understand how these two parts fit together, let's start by exploring the trunkline. The easiest way to conceptualize the trunk is as the main data line. In short, this line is responsible for supplying the data throughout the network. You can have devices in the middle of the line or you can have droplines stemming off from the trunk that go to nodes. At each end of the trunk, you will have a 121-ohm resistor to mark the end of the trunk. The trunk is represented in the following figure:

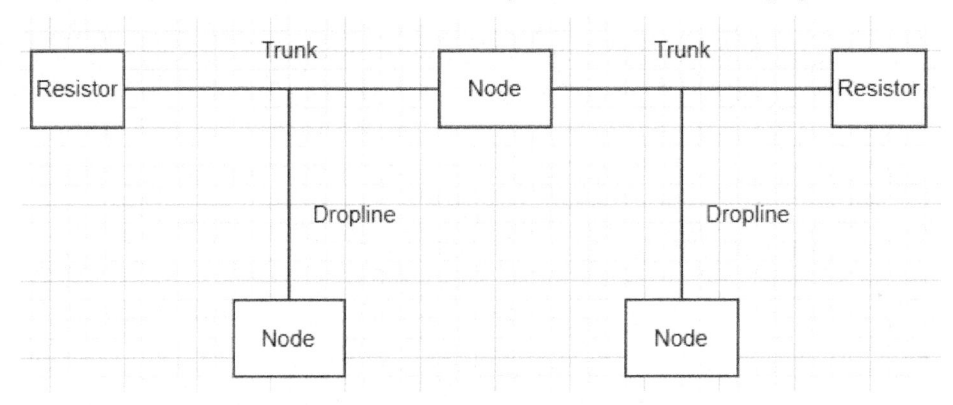

Figure 16.7 – DeviceNet topology

The resistors are very important for the operation of the network and if not properly installed, will lead to issues in the network. These resistors must be 121-ohm 1% resistors that are 0.25 watts or bigger. They must be connected directly to the signal wires, which, in the case of DeviceNet, are the blue and white wires. These resistors help reduce electrical noise in the network; hence, if your network appears to be noisy, it is worth looking at these resistors to see whether they are still operating as expected or are installed properly.

As can be seen in *Figure 16.7*, droplines can branch off the trunk. Typically, these lines will branch off to nodes (such as sensors) or components (such as motors, valves, pumps, and the like). Droplines can also branch off into trees as well, but if you do this, it is important to remember the maximum number of devices on your DeviceNet network.

In terms of which cable to use, you will usually use a thick wire cable. However, according to Mary Dixon in her article titled *What is DeviceNet?*, you can use the following cables from the trunk:

- Thick round
- Thin round
- KwikLink Flat
- KwikLink Lite Flat

Droplines can be composed of the following cables:

- Thick round
- Thin round

It is strongly recommended that you read Mary Dixon's article for further information about DeviceNet. The article will be linked in the *Further reading* section.

As can be seen in the bullet points, thick and thin cables are usually the most durable type of cables. Speaking from personal experience, when I worked on a DeviceNet network, I would usually run into round cables most often. Also, speaking from experience and agreeing with Mary Dixon, one of the most common issues that I would face were cabling issues. A pro tip from my personal experience would be to always check the resistors at the ends of the lines; many issues would arise from poorly installed resistors or resistors that would simply fail over a long time.

DeviceNet can support the following speeds:

- 125 Kbps
- 250 Kbps
- 500 Kbps

However, the data transmission speeds will vary based on the length of the cable and the type of cable used. For high-speed applications, it is important to factor in the cable length and type.

These are just a few of the many different types of communication protocols that exist for automation engineers. Many of these are built off common protocols such as TCP/IP and UDP. However, when developing applications it is common to have to convert between protocols, as many brands use a specific protocol by default. For example, Beckhoff uses EtherCAT, and Logix PLC uses Ethernet/IP. The next section will explore the basics of converting protocols.

## Protocol conversion

Consumers of automation systems (such as factories and plants in general) are usually slow to update their systems. It is common to have systems that are sometimes decades old already installed. This can become problematic as it is also common to have to install upgrades and add new hardware to interface with the old equipment, or even equipment of different brands that use a different communication protocol.

In situations where you have to interface with either older or different protocols, you will need to convert between protocols. In other words, if one part of a system uses EtherCAT and another part utilizes another protocol such as Profibus or DeviceNet, you will need to be able to convert between the two protocols to relay information to and from devices. Converting between two protocols can be accomplished in many different ways. There is no one-size-fits-all solution for converting between communication protocols. However, there are usually common elements that will be involved no matter what. Normally, converting between protocols will require some type of physical hardware and, in many cases, software to either configure or support the conversion. Normally, unique hardware must be used to convert between two specific communication protocols. As logic will dictate, the way the hardware works and is configured will vary from manufacturer to manufacturer and from device to device. Therefore, it is important to understand how the hardware works.

In all, this section has explored network communication protocols. For the most part, the protocols we explored are for communicating with devices such as sensors and parts to controls (such as valves, motors, and so on). However, communication can be used for much, much more.

## Other communication topics to explore

So far, we have touched on a few different communication protocols and networking principles. However, there are a lot more communication protocols to explore, and what we have explored so far has only scratched the surface of those specific topics. In fact, whole books have been dedicated to the subjects that we explored. For now, I would recommend exploring the following along with what we touched on in detail:

- The principle of serial communication
- RS-232
- RS-422
- RS-485

- CAN bus
- HART
- Ethernet/IP
- The OSI model
- Communication interfaces

The next section will be dedicated to understanding distributed control systems (DCSs).

## Understanding distributed control systems

For many industrial processes, individual processes can vary greatly across a geographical location. For example, consider a bottled water bottling center. Let's assume the treatment process involves the following steps:

1. Run a pump to collect water from the local lake.
2. Open the water intake valve and intake the water into a heating tank to boil the water.
3. Take the boiled water and add minerals to the treated water.
4. Bottle the water.

For a process like this, which has four intermittent steps, multiple PLCs would need to be used. However, there is a catch to this process. The four steps are going to take place over the bottling plant, which by definition is geographically dispersed. Since the bottling process can be thought of as a single process, we're going to need a way to control the whole process. Enter the world of DCSs.

A DCS is very similar to a SCADA system. A DCS is essentially a coordinating machine. The job of a DCS is to coordinate systems. To conceptualize this, consider the following figure:

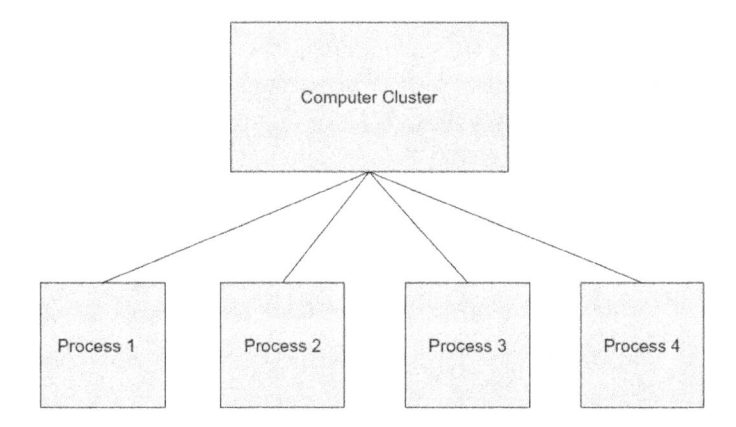

Figure 16.8 – DCS layout

As can be seen in *Figure 16.8*, the central cluster is coordinating four processes. At its heart, a DCS can simply be thought of as a central controller that supervises multiple processes. The cluster can be composed of multiple computing systems, such as operator terminals and so on. In all, a DCS is a supervisory system for a whole process or facility.

With an understanding of what DCS is, a logical question is where are DCSs utilized? The answer is, pretty much anywhere. Some common areas where DCSs are used are as follows:

- Chemical plants
- Manufacturing facilities
- Nuclear power plants
- Agricultural environments

DCSs can be used anywhere where whole processes need to be monitored or coordinated. With all that, what is the difference between a PLC and a DCS?

## The differences between DCSs and PLCs

The line between PLCs and DCSs is beginning to blur. However, DCSs and PLCs are two separate types of controllers. Due to the differences in the controllers, the overall applications for the controllers are different. To start the exploration between the two types of controllers, it is important to remember that a DCS controller is designed to supervise many PLCs, and as such, they are used to oversee entire processes.

Due to the supervisory nature of DCSs, they are usually not suited for controlling an individual process; they are much too slow. In terms of individual processes, a PLC is much more responsive and capable to make close to real-time adjustments. A DCS, on the other hand, is meant to (and usually does) supervise multiple systems, so it usually cannot handle the quick response time necessary to oversee a singular process.

Though response time is important, a DCS is much more scalable. In short, the number of I/O ports that a DCS can handle is much greater. If you need to drastically scale your application, especially over a large geographical area, and quick response time is not needed, a DCS might be a better solution. It should be noted that since a DCS is composed of many different controllers, communication between different brands, and communication protocols, you will need to be able to interface between them. This means you are almost always going to need to convert between protocols.

These aforementioned differences are just a couple of surface-level differences that show some differences between the two types of controllers. The best way to demonstrate the differences between the two types of controllers is to remember what they are used for. In short, if you need scalability and the ability to control multiple processes, a DCS is probably the best. You can use the following definitions to help select the controller:

- **PLC**: A PLC is used when there is a need for a fast response time and the process that it controls is singular. You may also use a PLC when the application is not geographically dispersed.
- **DCS**: A DCS is used when you need to supervise a whole process.

In all, a DCS is a system that will oversee many processes where a PLC will only oversee a single process. With that, we can wrap up our final technical overview.

## Summary

In conclusion, we have covered communication protocols, network topologies, DCSs, PLCs, and when to use each one. Networking is the backbone of many distributed systems. If you opt to use a DCS, networking will be used for many different distributed tasks. DCSs are also widely used in automation systems where distance matters. In all, to be an automation engineer, you must understand these concepts.

You have reached the end of this book; congratulations on that! This book was a crash course on the more advanced concepts that you will encounter in an advanced automation programming/engineering job. This book mostly focused on the software and IT side in general. With the current trend in automation, these are the concepts that you will want to focus on.

By this point, you should be exposed to many of the more advanced concepts of automation programming. Most of the concepts that were explored in this book are traditional software engineering concepts. Due to the nature of automation programming, these concepts are mostly unknown or ignored. However, as someone who has applied the concepts to automation projects in the past, I can testify that they will help improve the quality of your software and the speed at which you develop. Overall, this book covered a lot of material at a fast pace. I would certainly recommend exploring the concepts presented here in greater detail. Once you fully master these concepts and learn to apply them, your software will never be the same again.

## Questions

Answer the following questions based on what you've learned in this chapter. Cross-check your answers with those provided at the end of the book, under *Assessments*.

1. What is a DCS?
2. When should you use a PLC over a DCS?

3. What data transmission rates does DeviceNet transmit data at?
4. What is the difference between Profinet and Profibus?
5. What is the default topology for Profinet?
6. What is the difference between TCP and UDP?
7. Which is faster: TCP/IP or UDP?
8. Which communication protocols can use standard Ethernet cables?
9. What are the resistor sizes that are required at the end of a DeviceNet trunk?

# Further reading

Have a look at the following resources to further your knowledge:

- *What is network topology? Best guide to Type & Diagrams*: https://www.dnsstuff.com/what-is-network-topology
- *What is TCP?*: https://www.fortinet.com/resources/cyberglossary/tcp-ip
- *The User Datagram Protocol (UDP)*: https://erg.abdn.ac.uk/users/gorry/course/inet-pages/udp.html
- Mary Dixon, *What is DeviceNet?*: https://realpars.com/devicenet/
- *DeviceNet – designed for factory automation*: https://www.can-cia.org/can-knowledge/hlp/devicenet/
- Mondi Anderson, *What is EtherCAT?*: https://realpars.com/ethercat/
- Michael Bowne, *The difference between Profibus and Profinet*: https://us.profinet.com/the-difference-between-profibus-and-profinet/
- *Modbus ASCII vs Modbus RTU vs Modbus TCP/IP*: https://theautomization.com/modbus-ascii-vs-modbus-rtu-vs-modbus-tcpip/
- *Distributed control systems (DCS)*: https://www.techtarget.com/whatis/definition/distributed-control-system
- *What Is Distributed Control System (DCS)*: https://www.electricaltechnology.org/2016/08/distributed-control-system-dcs.html

# Assessments

This section contains answers to questions from all chapters. Go ahead and check if you've got them right.

## Chapter 1: Software Engineering for PLCs

1. C
2. A, B, E
3. B
4. D
5. A
6. C

## Chapter 2: Advanced Structured Text — Programming a PLC in Easy-to-Read English

1. A pointer points to a memory address where a reference is similar to a pointer with less syntax and references another variable.
2. The ^ symbol dereferences a pointer.
3. `TRY, CATCH, FINALLY`
4. Self-documenting code is logically named program attributes such as variable names.
5. A good comment is a comment that is short and adds context to the code without cluttering it. A bad comment will not add any context to the code, is long, or will unnecessarily clutter the code.
6. You should code to a variable so you only need to change values in one place, which will reduce the number of bugs in a program. Coding to a variable will also add context to what the value represents.

## Chapter 3: Debugging — Making Your Code Work

1. Print debugging is where messages are put in the program to help the developer see where they are at in the program's execution.
2. Using a tool such as the debugging tool.

3. The process of finding and eliminating a bug in a program.
4. Functional Error

## Chapter 4: Complex Variable Declaration — Using Variables to Their Fullest

1. C
2. A GVL is a global variable list where a struct is a data type.
3. A constant is a variable that does not change where an ENUM is a user-defined data type that is composed of constants.
4. There are many different errors you can get with an array; however, the most common one stems from trying to access an element that is not present.
5. `ArrayName[1..5, 1..6] OF <TYPE>;`

## Chapter 5: Functions — Making Code Modular and Maintainable

1. A function is a callable block of code that provides modularity to a program, can accept parameters/arguments, and will only run when invoked.
2. Arguments that are pre-assigned and do not need to be provided when the function is called.
3. A parameter that is assigned a value when the function is called based on its name.
4. The function's parameters, return type, and other attributes that distinguish it.
5. By default in a one-to-one fashion.
6. Any amount of code as long as the code's intended purpose can be described in one sentence without the word *and*.
7. The type of value the function will return.
8. Technically yes, depending on the system. However, certain return types cannot be assigned to certain variables.

## Chapter 6: OOP — Reducing, Reusing, and Recycling Code

1. Class
2. A method calls itself.
3. 3. A pointer of a function block to its own function block instance.
4. Getter and Setter

5. A getter will retrieve a value where setter will set a value.

## Chapter 7: OOP — The Power of Objects

1. Abstraction, Encapsulation, Inheritance, Polymorphism
2. No
3. No limit
4. Private hides attributes such as methods from outside files, while public allows any file to access the attributes.

## Chapter 8: Libraries — Write Once, Use Anywhere

1. A set of prebuilt attributes that can be imported and called by your current program.
2. It shows users how to implement and use the library.
3. Via the library manager.
4. Singleton, Factory, Façade, anything that will simplify the use of the library.
5. /// is Declaration header while (**) is a member header.
6. Varies

## Chapter 9: The SDLC — Navigating the SDLC to Create Great Code

1. The number of lines tested in a unit test.
2. 80%
3. 50%
4. The steps involved for developing and deploying a piece of software (steps in the software development process).
5. Roughly six; however, that number can vary from person-to-person.
6. A way of showing the relationship between function blocks.
7. Unit test is where a developer will test out code modules. Regression testing ensures the system works as intended after a code change.
8. A test case is a set of criteria that is used to ensure a code works under a given circumstance.
9. Validation is ensuring the program solves the original problem. Verification ensures the software works as it was designed to work.

## Chapter 10: Advanced Coding — Using SOLID to Make Solid Code

1. Function blocks, methods, functions
2. Break the module up so it is describable in a complete sentence without the word *and*.
3. *Principle that states that an object of a part class/function block should be replaceable with objects of a child class/function without affecting the behavior.*
4. *Principle that states that a function block should not have to implement an interface it does not use, nor should it depend on a method that it does not use.*
5. *See the following:*
   - S: Single-responsibility Principle
   - O: Open-closed Principle
   - L: Liskov Substitution Principle
   - I: Interface Segregation Principle
   - D: Dependency inversion Principle

## Chapter 11: HMIs — UIs for PLCs

1. An HMI is a User Interface for an industrial machine.
2. Industrial User Interface for automation projects.
3. PLC, HMIS, Sensors, Logging devices, and so on.
4. A rough outline of a UI/HMI screen.
5. Yes: with the proper plugins and libraries

## Chapter 12: Industrial Controls — User Inputs and Outputs

1. An HMI control that affects a Boolean variable when pressed to trigger an event.
2. Arrays
3. Yes
4. Yes: it will depend on the situation.

# Chapter 13: Layouts — Making HMIs User Friendly

1. Red: error, Yellow: warning, Green: normal operation.
2. Dark colors such as black and dark grey.
3. An HMI should be tasked with running one and only one operation such as programming, monitoring health, and so on.
4. Visualization manager
5. The home screen.
6. Varies.

# Chapter 14: Alarms — Avoiding Catastrophic Issues with Alarms

1. An alarm is a machine health/operation status.
2. Info, Warning, Error
3. Error
4. Objects that consume alarm configurations
5. A confirmation to the system that the alarm has been seen, and in some cases, addressed.

# Chapter 15: Putting It All Together — The Final Project

The questions in this chapter are open-ended with no wrong or right answer. They are meant to be exploratory so readers can come to their own conclusions.

# Chapter 16: Distributed Control Systems, PLCs, and Networking

1. A cluster that supervises multiple other processes.
2. When a single process needs to be overseen.
3. 125 Kbps, 250 Kbps, 500 Kbps
4. One is ethernet based, the other is serial based, among other things such as speed.
5. Star
6. TCP/IP is slower but more reliable than UDP. UDP is faster but data integrity is not guaranteed.
7. UDP
8. Profinet, EtherCat
9. 121 ohm

# Index

## A

abstraction  133
access specifiers  130
acknowledgment variable  311, 312
ADR operator  27
Agile methodology  176
alarm acknowledgment  299-301
alarm banner  288
  setting up  289
alarm groups  286
  setting  286-288
alarm HMI components  288
  alarm banner  288
  alarm table  288
alarms  282
  configuration  283-285
  issue/warning, reflecting  283
  using  282
alarm table  288
  setting up  290, 291
alarm table variables  314
arguments  99, 100
  default arguments  102, 103
  named parameters  100, 101
arrays  71
  declaring  71, 72
  generating  72

initialized arrays  73, 75
investigating  71
multidimensional arrays  71, 75
Artificial Intelligence (AI)  3
Auto Declare tool  68
  used, for declaring variables  68, 69

## B

bad comments  36
bidirectional networks  330
blinking
  animation  268
  best practices  263, 264
  component  264-267
breakpoints  53-56
bugs  44
  functional errors  44
  logic errors  44
  syntax errors  44
buttons  239, 240

## C

Calculation program  130-132
classes  110

## Index

**code commenting** 35, 36
  bad comments 36
  good comments 36
**CODESYS** 4, 9, 10, 110
  download link 4
  testing 10, 11
**CODESYS debugger tool** 52, 53
  breakpoints 53-56
  stepping 56, 57
**colors**
  significance 258
**composition** 139, 140
  demonstrating 140-142
**computer networks** 330
**constants** 69-71
  declaring 70
  examples 69
**control properties** 247, 248
**custom libraries**
  building 164
  implementation 165-169
  requirements 164

## D

**Data Unit Type (DUT) wizard** 79-81
**debugger** 43
**debugging** 44
  CODESYS debugger tool 52, 53
  forcing variables 57
  print debugging 47-51
  techniques 47
  tools 47
  versus testing 45
**debugging process** 45, 46
  issue, fixing 46
  problem, analyzing 46
  problem, isolating 46
  problem, reproducing 46
  solution, validating 47
**default arguments** 102, 103
**dependency inversion principle (DIP)** 215, 216
  implementing 216-219
**deployment phase, SDLC** 188
  delivery of system 188
  modifications 188
  training 188
  user acceptance testing 188
**dereferencing**
  pointers 28
**design patterns** 147, 148
**DeviceNet** 339, 340
  cables 339-341
  data transmission speeds 341
  topology 340, 341
**distributed control systems (DCS)** 343, 345
  layout 344
  uses 344
  versus PLCs 344
**distributed control systems (DCSs)** 342
**division by 0 program**
  checking, for 0 code 21
  custom exceptions, handling 25
  division by 0 error 19, 20
  error handling 23
  errors, identifying 23
  exception, catching 24, 25
  exception variables 23
  FINALLY statement 23
  main program 19
  TRY-CATCH block 21, 22
  variables 19
  variables, for unique exceptions 24
**documentation** 33
  code commenting 35, 36

code to variables  34, 35
self-documenting code  33, 34

# E

**encapsulation**  133
**enums**  81
  declaring  81, 82
**error handling**  18
**errors**  18
**Ethernet for Control Automation Technology (EtherCAT)**  338, 339
  benefits  338
**exceptions**  18
**extreme programming (XP)**  176

# F

**facade pattern**  148, 160
**factory pattern**  160
**fatal error**  18
**FINALLY statement**  23
**finite state machines (FSMs)**  37
**flip switches**  238, 239
**forcing variables**  57
**frameworks**
  versus library  154
**function**  90, 91
  creating  92-95
  contents  91, 92
  PLC_PRG file  96
**functional errors**  44
**functional testing**  186, 187
**Function Block Diagrams (FBDs)**  8
**function blocks**  108, 110
  example  110-113

# G

**gauge variable**  312, 313
**getter method**  120, 121
**global variable list (GVL)**  76, 310
  creating  76, 77
  demonstrating  77, 78
**good comments**  36
**grouping/position**  261-263

# H

**Hello, World! ladder logic program**
  completed Hello, World! project  13
  input, toggling to false  14
  input, toggling to true  14
  Login button  13
  PLC_PRG file  11
  running ladder logic program  14
  variable code  12, 13
**histogram**  245, 246
**HMI**  225, 226, 237
  building  250-253
  creating  233-235, 249
  design  249, 250
  development tools  228
  functionalities  230
  grouping/position  261, 263
  need for  226-228
  programming languages, to develop  229, 230
  requirements  249
  SDLC  231, 232
  working  253
**HMI, colors**
  backgrounds  258, 259
  green  260
  labeling  261
  red  260

selecting, for controls 260
significance 258
yellow 260
**HMI controls 238**
   buttons 239, 240
   flip switches 238, 239
   histogram 245, 246
   LEDs 240, 241
   measurement controls 243-245
   potentiometer 241, 242
   properties 247, 248
   push switches 239
   sliders 242, 243
   spinners 243
   text field 246, 247
**HMI implementation, simulated industrial oven**
   acknowledgment variable 311, 312
   alarm table configuration 316
   alarm table variables 314
   error class setup 315
   gauge variable 312, 313
   info class configuration 315
   LED variables 311
   spinner variables 312
   warning class configuration 315
**hybrid topology 331**

# I

**IEC 61131-3 compliant PLCs 6**
**IEC 61131-3 standard PLCs 3, 6, 7, 108, 110**
**IEC languages 7**
   Instruction List (IL) 8
   ladder logic 7
   Sequential Flow Charts (SFCs) 8
   Structured Text 9
**inheritance 134-140**

**inheritance chain 137**
**initialized arrays 73-75**
**Instruction List (IL) 8**
**Integrated Development Environment (IDE) 10, 44**
**integration testing 185, 186**
**interfaces 143**
   examining 143-147
**interface segregation principle (ISP) 213**
   implementing 214, 215
**Internet Protocol (IP) 332**
**invalid pointers**
   catching 29
   handling 29
**invalid pointer variables**
   TRY-CATCH for 30, 31
**invalid references**
   checking for 32, 33
**IT protocols 331**
   TCP/IP 332
   UDP 332, 333

# K

**Keep it simple, stupid (KISS) 158, 159**

# L

**ladder logic 7**
   ToolBox 12
**LEDs 240, 241**
**LED variables 311**
**library 154**
   distribution 155
   installing 156-158
   need for 154
   versus frameworks 154

**library development**
  guiding principles  158
**library development, rules**
  abstraction and encapsulation  159
  documentation  160-164
  Keep it simple, stupid (KISS)  158, 159
  patterns  160
**lightbulb state machine  37**
**Liskov substitution principle (LSP)  208**
  implementing  208-213
**logic errors  44**

# M

**measurement controls  243-245**
**methods  114, 115**
  adding  115-118
**Modbus  334**
**Modbus ASCII  334**
**Modbus RTU  334**
**Modbus TCP/IP  335**
**Model View Controller (MVC) pattern  37, 147**
**modular code  90**
  need for  90
**motor alarm system**
  creating  301
  HMI design implementation  302, 303
  requirements  302
**motor control program**
  building  83-85
**multidimensional arrays  71, 75**
  elements, accessing  75, 76
**MVVM pattern  147**

# N

**named parameters  100, 101**
**network topology  330**

# O

**object-oriented design (OOD)  198**
**object-oriented programming (OOP)  107, 108, 129**
  abstraction  133
  benefits  109
  encapsulation  133
  inheritance  134-137
  pillars  110
  polymorphism  138, 139
**objects  108, 109, 113**
**open-closed principle (OCP)  203**
  implementing  203-207

# P

**painting machine**
  building  219-221
**part computation library  169**
  implementation  169-171
  requirements  169
**persistent variable list  83**
**persistent variables  82**
**playtime  46**
**PLC alarm logic**
  implementing  292-299
**PLC/automation device communication  334**
  DeviceNet  339, 340
  EtherCAT  338, 339
  Modbus  334
  Profibus  335, 336
  Profinet  336, 337
  protocol conversion  342
  topics  342
**PLC code design, simulated industrial oven  316, 317**

**PLC code implementation, simulated industrial oven** 318
   Alarms function block 319
   Door function block 320
   Oven function block 321, 322
   PLC_PRG file 318, 319
**PLC memory** 26
**PLC_PRG file** 11, 96
**PLCs** 345
   versus distributed control systems 344
**pointers** 25, 31
   dereferencing 28
   syntax 26
**polymorphism** 138, 139
**potentiometer** 241, 242
**print debugging** 47-51
**private access specifier** 130
**Profibus** 335
**Profinet** 336, 337
   star topology 336, 337
   tree topology 337
   versus Profibus 337, 338
**Programmable Logic Controller (PLC)** 3
**Program Organizational Unit (POU)** 56
**Program Organization Unit (POU)** 92
**properties** 118
   adding 119
**public access specifier** 130
**push switches** 239

# R

**recursion** 122, 123
   demonstrating 123, 124
**references** 31
   example program 32
   invalid references, checking 32, 33

**reference variable**
   declaring 31
**regression testing** 187
**RETURN statement**
   using 97, 98
**return types**
   examining 96, 97
**ring topology** 330

# S

**screen, into multiple layouts** 268
   default screen, modifying 271-273
   navigating, between screens 273-275
   visualizations screens, creating 269-271
**self-documenting code** 33, 34
**Sequential Flow Charts (SFCs)** 8
**setter method** 121, 122
**side navigation** 262
**simulated assembly line**
   creating 148-150
**simulated industrial oven**
   door lock, testing 323, 324
   gauge, testing 324-326
   HMI design 309, 310
   HMI implementation 310
   PLC code design 316, 317
   PLC code implementation 318
   project overview 308
   project requisites obtaining 309
**single-responsibility principle (SRP)** 199, 200, 317
   implementing 200-203
**sliders** 242, 243
**Software Development Life Cycle (SDLC)** 174
   Agile methodology 176
   build 182

deployment  188, 189
design  178
implementing  175
maintenance  189
requirements collecting, tips  178
requirements/planning  177
significance  174
test  183
waterfall methodology  175, 176
**software engineering  3, 4**
for PLCs  4, 5
**SOLID programming  198**
benefits  198
dependency inversion principle (DIP)  215, 216
governing, principles  199
interface segregation principle (ISP)  213
Liskov substitution principle (LSP)  208
open-closed principle (OCP)  203
single-responsibility principle  199, 200
**spinners  243**
**spinner variables  312**
**Standard Template Library (STL)  154**
**state machine  37**
variables  38
**state machine logic  38, 39**
non-running state machine  39
running state machine  40
state machine exception thrown  40
**stepping  56**
Step Into command  56
Step Out command  57
Step Over command  56
**structs  78**
creating, with DUT wizard  79-81
declaring  78
**Structured Text  9**

**Supervisory Control And Data Acquisition (SCADA)  230, 231**
**switch  238**
**syntax errors  44**

# T

**temperature conversion library, SDLC project**
building  189-192
deploying  194
designing  190
maintaining  194
requirements gathering  189, 190
testing  192, 193
**temperature unit converter  104-106**
**testing**
validation testing  183, 186
verification testing  183, 184
versus debugging  45
**text field  246, 247**
**third-party library  155**
**THIS keyword  123**
**three-way handshake  332**
**Transmission Control Protocol (TCP)  331**
**troubleshooting**
example  59-64
forcing, versus writing  58, 59
while loop  64, 65
**TRY-CATCH  21-23**
for invalid pointer variables  30, 31

# U

**unidirectional networks  330**
**Unified Modeling Language (UML)  179, 205**
diagrams  179, 180
diagrams, reading  180-182

**unit converter**
   creating  125-127
**unit testing  184, 185**
**User Datagram Protocol (UDP)  332, 333**
   send/receive process  333
**user-friendly HMI**
   creating  275-279
**User Interface (UI)  225**
**Util library  264**

# V

**validation testing  183, 186**
   functional testing  186, 187
   regression testing  187, 188
**values**
   forcing, versus writing  57, 58
**variables  26**
   declaring, with Auto Declare tool  68
   forcing  57
   for state machine  38
**verification testing  183, 184**
   integration testing  185, 186
   unit testing  184, 185

# W

**waterfall methodology  175, 176**
**well-written program  90**
**while loop  64, 65**
**Windows Presentation**
     **Foundation (WPF)  229**
**wireframing  232, 233**
**working control panel**
   requisites  227

Packtpub.com

Subscribe to our online digital library for full access to over 7,000 books and videos, as well as industry leading tools to help you plan your personal development and advance your career. For more information, please visit our website.

## Why subscribe?

- Spend less time learning and more time coding with practical eBooks and Videos from over 4,000 industry professionals
- Improve your learning with Skill Plans built especially for you
- Get a free eBook or video every month
- Fully searchable for easy access to vital information
- Copy and paste, print, and bookmark content

Did you know that Packt offers eBook versions of every book published, with PDF and ePub files available? You can upgrade to the eBook version at `packtpub.com` and as a print book customer, you are entitled to a discount on the eBook copy. Get in touch with us at `customercare@packtpub.com` for more details.

At `www.packtpub.com`, you can also read a collection of free technical articles, sign up for a range of free newsletters, and receive exclusive discounts and offers on Packt books and eBooks.

# Other Books You May Enjoy

If you enjoyed this book, you may be interested in these other books by Packt:

**PLC and HMI Development with Siemens TIA Portal**

Liam Bee

ISBN: 9781801817226

- Set up a Siemens Environment with TIA Portal
- Find out how to structure a project
- Carry out the simulation of a project, enhancing this further with structure
- Develop HMI screens that interact with PLC data
- Make the best use of all available languages
- Leverage TIA Portal's tools to manage the deployment and modification of projects

**Learning RSLogix 5000 Programming - Second Edition**

Austin Scott

ISBN: 9781789532463

- Gain insights into Rockwell Automation and the evolution of the Logix platform
- Find out the key platform changes in Studio 5000 and Logix Designer
- Explore a variety of ControlLogix and CompactLogix controllers
- Understand the Rockwell Automation industrial networking fundamentals
- Implement cybersecurity best practices using Rockwell Automation technologies
- Discover the key considerations for engineering a Rockwell Automation solution

## Packt is searching for authors like you

If you're interested in becoming an author for Packt, please visit `authors.packtpub.com` and apply today. We have worked with thousands of developers and tech professionals, just like you, to help them share their insight with the global tech community. You can make a general application, apply for a specific hot topic that we are recruiting an author for, or submit your own idea.

## Share Your Thoughts

Now you've finished *Mastering PLC Programming*, we'd love to hear your thoughts! Scan the QR code below to go straight to the Amazon review page for this book and share your feedback or leave a review on the site that you purchased it from.

`https://packt.link/r/180461288X`

Your review is important to us and the tech community and will help us make sure we're delivering excellent quality content.

# Download a free PDF copy of this book

Thanks for purchasing this book!

Do you like to read on the go but are unable to carry your print books everywhere? Is your eBook purchase not compatible with the device of your choice?

Don't worry, now with every Packt book you get a DRM-free PDF version of that book at no cost.

Read anywhere, any place, on any device. Search, copy, and paste code from your favorite technical books directly into your application.

The perks don't stop there, you can get exclusive access to discounts, newsletters, and great free content in your inbox daily

Follow these simple steps to get the benefits:

1. Scan the QR code or visit the link below

https://packt.link/free-ebook/9781804612880

2. Submit your proof of purchase
3. That's it! We'll send your free PDF and other benefits to your email directly

Printed in Great Britain
by Amazon